PLANT SECONDARY METABOLITES FOR HUMAN HEALTH

Extraction of Bioactive Compounds

Innovations in Plant Science for Better Health: From Soil to Fork

PLANT SECONDARY METABOLITES FOR HUMAN HEALTH

Extraction of Bioactive Compounds

Edited by

Megh R. Goyal, PhD, PE

P. P. Joy, PhD

Hafiz Ansar Rasul Suleria, PhD

Apple Academic Press Inc. | Apple Academic Press Inc.
3333 Mistwell Crescent | 1265 Goldenrod Circle NE
Oakville, ON L6L 0A2 | Palm Bay, Florida 32905
Canada USA | USA

First issued in paperback 2021

Exclusive worldwide distribution by CRC Press, a member of Taylor & Francis Group

No claim to original U.S. Government works

ISBN 13: 978-1-77463-439-4 (pbk)
ISBN 13: 978-1-77188-766-3 (hbk)

Library and Archives Canada Cataloguing in Publication

Title: Plant secondary metabolites for human health : extraction of bioactive compounds / edited by Megh R. Goyal, P.P. Joy, Hafiz Ansar Rasul Suleria.

Names: Goyal, Megh Raj, editor. | Joy, P. P., editor. | Suleria, Hafiz, editor.

Series: Innovations in plant science for better health.

Description: Series statement: Innovations in plant science for better health : from soil to fork | Includes bibliographical references and index.

Identifiers: Canadiana (print) 20190143134 | Canadiana (ebook) 20190143169 | ISBN 9781771887663 (hardcover) | ISBN 9780429425325 (ebook)

Subjects: LCSH: Phytochemicals. | LCSH: Plant metabolites. | LCSH: Bioactive compounds. | LCSH: Functional foods.

Classification: LCC QK861 .P63 2019 | DDC 572/.2—dc23

CIP data on file with US Library of Congress

INNOVATIONS IN PLANT SCIENCE FOR BETTER HEALTH: FROM SOIL TO FORK BOOK SERIES

Series Editor-in-Chief:

Hafiz Ansar Rasul Suleria, PhD
Honorary Fellow at the Diamantina Institute,
Faculty of Medicine, The University of Queensland (UQ), Australia
email: hafiz.suleria@uqconnect.edu.au

The objective of this book series is to offer academia, engineers, technologists, and users from different disciplines information to gain knowledge on the breadth and depth of this multifaceted field. The volumes will explore the fields of phytochemistry, along with its potential and extraction techniques. The volumes will discuss the therapeutic perspectives of biochemical compounds in plants and animal and marine sources in an interdisciplinary manner because the field requires knowledge of many areas, including agricultural, food, and chemical engineering; manufacturing technology along with applications from diverse fields like chemistry; herbal drug technology; microbiology; animal husbandry; and food science; etc. There is an urgent need to explore and investigate the innovations, current shortcomings, and future challenges in this growing area of research.

We welcome chapters on the following specialty areas (but not limited to):

- Food and function
- Nutritional composition of different foods materials
- Functional and nutraceutical perspectives of foods
- Extraction of bioactive molecules
- Phytomedicinal properties of different plants
- Traditional plants used as functional foods
- Phytopharmacology of plants used in metabolic disorders

- Importance of spices and medicinal and functional foods
- Natural products chemistry
- Food processing waste/byproducts: management and utilization
- Herbals as potential bioavailability enhancer and herbal cosmetics
- Phytopharmaceuticals for the delivery of bioactives
- Alternative and complementary medicines
- Ethnopharmacology and ethnomedicine
- Marine phytochemistry
- Marine microbial chemistry
- Other related includes nuts, seed spices, wild flora, etc.

About the Book Series Editor-in-Chief

Dr. Hafiz Suleria is an eminent young researcher in the field of food science and nutrition. Currently, he is an Honorary Fellow at the Diamantina Institute, Faculty of Medicine, The University of Queensland (UQ), Australia. Before joining the UQ, he worked as a lecturer in the Department of Food Sciences, Government College University Faisalabad, Pakistan. He also worked as a Research Associate in a PAK-US Joint Project funded by the Higher Education Commission, Pakistan, and Department of State, USA, with the collaboration of the University of Massachusetts, USA, and the National Institute of Food Science and Technology, University of Agriculture, Faisalabad, Pakistan.

Dr. Suleria's major research focus is on food science and nutrition, particularly in screening of bioactive molecules from different plant, marine, and animal sources, using various cutting-edge techniques, such as isolation, purification, and characterization. He also did research work on functional foods, nutraceuticals, and alternative medicine. He has published more than 60 peer-reviewed scientific papers in different reputed/impacted journals. He is also in collaboration with more than five universities where he is working as a co-supervisor/special member for PhD and postgraduate students and also involved in joint publications, projects, and grants.

BOOKS IN THE SERIES

Bioactive Compounds of Medicinal Plants: Properties and Potential for Human Health

Editors: Megh R. Goyal, PhD, and Ademola O. Ayeleso, PhD

Plant- and Marine-Based Phytochemicals for Human Health: Attributes, Potential, and Use
Editors: Megh R. Goyal, PhD, and Durgesh Nandini Chauhan, MPharm

Human Health Benefits of Plant Bioactive Compounds: Potentials and Prospects
Editors: Megh R. Goyal, PhD, and Hafiz Ansar Rasul Suleria, PhD

Plant Secondary Metabolites for Human Health: Extraction of Bioactive Compounds
Editors: Megh R. Goyal, PhD, P. P. Joy, and Hafiz Ansar Rasul Suleria, PhD

Bioactive Compounds from Plant Origins: Extraction, Applications, and Potential Health Benefits
Editors: Hafiz Ansar Rasul Suleria, PhD, and Colin Barrow, PhD

Phytochemicals from Medicinal Plants: Scope, Applications, and Potential Health Claims
Editors: Hafiz Ansar Rasul Suleria, PhD, Megh R. Goyal, PhD, and Masood Sadiq Butt, PhD

The Therapeutic Properties of Medicinal Plants: Health-Rejuvenating Bioactive Compounds of Native Flora
Editors: Megh R. Goyal, PhD, PE, Hafiz Ansar Rasul Suleria, PhD, Ademola Olabode Ayeleso, PhD, T. Jesse Joel, and Sujogya Kumar Panda

The Role of Phytoconstitutents in Health Care: Biocompounds in Medicinal Plants
Editors: Megh R. Goyal, PhD, Hafiz Ansar Rasul Suleria, PhD, and Ramasamy Harikrishnan, PhD

Assessment of Medicinal Plants for Human Health: Phytochemistry, Disease Management, and Novel Applications
Editors: Megh R. Goyal, PhD and Durgesh Nandini Chauhan, MPharm

ABOUT THE SENIOR EDITOR-IN-CHIEF

Megh R. Goyal, PhD, PE

Retired Professor in Agricultural and Biomedical Engineering, University of Puerto Rico, Mayaguez Campus; Senior Acquisitions Editor, Biomedical Engineering and Agricultural Science, Apple Academic Press, Inc.

Megh R. Goyal, PhD, PE, is a Retired Professor in Agricultural and Biomedical Engineering from the General Engineering Department in the College of Engineering at the University of Puerto Rico–Mayaguez Campus; and Senior Acquisitions Editor and Senior Technical Editor-in-Chief in Agriculture and Biomedical Engineering for Apple Academic Press, Inc. He has worked as a Soil Conservation Inspector and as a Research Assistant at Haryana Agricultural University and Ohio State University.

During his professional career of 49 years, Dr. Goyal has received many prestigious awards and honors. He was the first agricultural engineer to receive the professional license in Agricultural Engineering in 1986 from the College of Engineers and Surveyors of Puerto Rico. In 2005, he was proclaimed as "Father of Irrigation Engineering in Puerto Rico for the Twentieth Century" by the American Society of Agricultural and Biological Engineers (ASABE), Puerto Rico Section, for his pioneering work on micro irrigation, evapotranspiration, agroclimatology, and soil and water engineering. The Water Technology Centre of Tamil Nadu Agricultural University in Coimbatore, India, recognized Dr. Goyal as one of the experts "who rendered meritorious service for the development of micro irrigation sector in India" by bestowing the Award of Outstanding Contribution in Micro Irrigation. This award was presented to Dr. Goyal during the inaugural session of the National Congress on "New Challenges and Advances in Sustainable Micro Irrigation" on March 1, 2017, held at Tamil Nadu Agricultural University. Dr. Goyal received the Netafim

Award for Advancements in Microirrigation: 2018 from the American Society of Agricultural Engineers at the ASABE International Meeting in August 2018.

A prolific author and editor, he has written more than 200 journal articles and textbooks and has edited over 63 books. He is the editor of three book series published by Apple Academic Press: Innovations in Agricultural & Biological Engineering, Innovations and Challenges in Micro Irrigation, and Research Advances in Sustainable Micro Irrigation. He is also instrumental in the development of the new book series Innovations in Plant Science for Better Health: From Soil to Fork.

Dr. Goyal received his BSc degree in engineering from Punjab Agricultural University, Ludhiana, India; his MSc and PhD degrees from Ohio State University, Columbus; and his Master of Divinity degree from Puerto Rico Evangelical Seminary, Hato Rey, Puerto Rico, USA.

ABOUT THE EDITORS

P. P. Joy, PhD

Retired Professor of Agronomy and Former Head of the Pineapple Research Station, Kerala Agricultural University, Vazhakulam, Kerala, India

P. P. Joy, PhD, is an eminent researcher and Professor of Agronomy & Head, Pineapple Research Station, Kerala Agricultural University, Vazhakulam, Kerala, India. He joined the University in 1984 as an Assistant Professor (Agronomy), and undertook research and development of pulses initially, followed by rice, medicinal and aromatic plants and subsequently fruit crops.

Presently, he is involved in developing agro-technologies, processing technologies, and IT-enabled services for pineapple, passion fruit, and other economically important fruit crops of Kerala. With 32 years of service in agricultural research and development, teaching and extension, he has expertise in agronomy, soil, and water conservation; water, nutrient, weed and cropping system management; and research projects, databases, web, and ICT management. He is well conversant with the latest research methodology, analyses, and interpretation of research results and has ample experience in both field and laboratory experimentation.

The team lead by him developed a variety "Sugandhini" (ODC–130) in cinnamon, 134P in passion fruit, and agrotechnologies for rice, pulses, medicinal, and aromatic plants, and fruit crops. He was the principal investigator of 25 university research projects, and he was associated with another 32 projects of All India Coordinated Research Project on Spices (AICRPS) and Kerala Agricultural University. He was the Principal Investigator of the ICAR ad-hoc research projects, including "Standardization of agro-techniques for lesser-known medicinal and aromatic plants of Zingiberaceae," "Development of lemongrass oleoresin for flavoring," a Kerala State Council for Science, Technology & Environment project "Evaluation of passion fruit types for commercial cultivation in Kerala," and Kerala Pineapple Mission (KPM) project "Organic versus inorganic nutrient management of pineapple varieties." He was also associated with 12 projects funded by the Indian Council of Agricultural Research, Ministry of Agriculture, Ministry of Health and Family Welfare, National Watershed Development Project for Rainfed Areas (NWDPRA), Kerala Horticulture Development Programme (KHDP), Kerala State Council for Science, Technology and Environment (KSCSTE), and Kerala Pineapple Mission (KPM).

Dr. Joy was instrumental in developing websites for the research at which centers he worked. He has good research, development and management experience, and received many awards and honors. He is a member of several national bodies/societies. He has authored over 200 publications, including 25 books/chapters, 70 research papers, 25 bulletins, 35 seminars/symposia/conference papers, and 40 popular articles. He received his BSc (Agri.) degree from Kerala Agricultural University, MSc (Agri.) in Agronomy from the University of Agricultural Sciences, Bangalore, and PhD in Agronomy from Kerala Agricultural University, all with the first rank.

Hafiz Ansar Rasul Suleria, PhD

Alfred Deakin Research Fellow,
Deakin University, Melbourne,
Australia; Honorary Fellow,
Diamantina Institute Faculty of Medicine,
The University of Queensland, Australia

Hafiz Anasr Rasul Suleria, PhD, is currently working as the Alfred Deakin Research Fellow at Deakin University, Melbourne, Australia. He is also an Honorary Fellow at the Diamantina Institute, Faculty of Medicine, The University of Queensland, Australia.

Recently he worked as a postdoc research fellow in the Department of Food, Nutrition, Dietetic and Health at Kansas State University, USA.

Previously, he has been awarded an International Postgraduate Research Scholarship (IPRS) and an Australian Postgraduate Award (APA) for his PhD research at the University of Queens School of Medicine, the Translational Research Institute (TRI) in collaboration with Commonwealth and Scientific and Industrial Research Organization (CSIRO, Australia).

Before joining the UQ, he worked as a lecturer in the Department of Food Sciences, Government College University Faisalabad, Pakistan. He also worked as a research associate in the PAK-US Joint Project funded by the Higher Education Commission, Pakistan, and Department of State, USA, with the collaboration of the University of Massachusetts, USA, and National Institute of Food Science and Technology, University of Agriculture Faisalabad, Pakistan.

He has a significant research focus on food nutrition, particularly in the screening of bioactive molecules—isolation, purification, and characterization using various cutting-edge techniques from different plant, marine, and animal source, and *in vitro*, *in vivo* bioactivities, cell culture, and animal modeling. He has also done a reasonable amount of work on functional foods and nutraceutical, food and function, and alternative medicine.

Dr. Suleria has published more than 50 peer-reviewed scientific papers in different reputed/impacted journals. He is also in collaboration with more than ten universities where he is working as a co-supervisor/special member for PhD and postgraduate students and is also involved in joint publications, projects, and grants. He is Editor-in-Chief for the book series on *Innovations in Plant Science for Better Health: From Soil to Fork*, published by AAP. Readers may contact him at: hafiz.suleria@uqconnect.edu.au.

CONTENTS

CONTRIBUTORS

Annie Abraham
PhD, Professor and Head, Department of Biochemistry, University of Kerala, Kariavattom, 695581, Thiruvananthapuram, Kerala, India, Tel.: +91-9447246692, E-mail: annieab2001@gmail.com

Y. Anie
PhD, Assistant Professor, School of Biosciences, Mahatma Gandhi University, Priyadarshini Hills P.O., Kottayam – 686560, Kerala, India, Tel.: +91-9947090370, E-mail: aniey@mgu.ac.in

V. S. Binchu
M. Phil. Research Scholar, School of Biosciences, Mahatma Gandhi University, Priyadarshini Hills P.O., Kottayam – 686560, Kerala, India, Tel.: +91-9946537042, E-mail: binchuvshaji@gmail.com

V. Chandrasekar
PhD, Scientist, Agricultural Process Engineering, ICAR-CIPHET, Ludhiana – 141004 , Punjab, India, Tel.: +91-9442740534, E-mail: chandrufpe@gmail.com

K. C. Dhanya
PhD, Assistant Professor, Department of Microbiology, St. Mary's College, Thrissur – 680020, Kerala, India; Mobile: +91-9947496077; E-mail: dhanuchandra@gmail.com

Megh R. Goyal
PhD, PE, Senior Editor-in-Chief, Apple Academic Press Inc., PO Box 86, Rincon – PR – 00677–0086, E-mail: goyalmegh@gmail.com

V. H. Haritha
M. Phil. Research Scholar, School of Biosciences, Mahatma Gandhi University, Priyadarshini Hills P.O., Kottayam – 686560, Kerala, India, Tel.: +91-9446314151, E-mail: vhhari@yahoo.co.in

V. N. Hazeena
MSc, Research Scholar, School of Biosciences, Mahatma Gandhi University, Priyadarshini Hills P.O., Kottayam – 686560, Kerala, India, Tel.: +91-8156855687, E-mail: aneshazi@gmail.com

L. R. Helen
MSc, Research Scholar, School of Biosciences, Mahatma Gandhi University, Kottayam, Kerala, India–686560. Tel.: 9497377149, E-mail: lalhelenraisa@gmail.com

Swamy Gabriela John
PhD Candidate, Department of Agriculture and Biosystems Engineering, South Dakota State University, Brookings 57007, SD USA. Tel.: +1-4087593123, E-mail: gabrielafoodtech@gmail.com

P. P. Joy
PhD, Professor of Agronomy & Head, Kerala Agricultural University, Pineapple Research Station, Vazhakulam PO, Muvattupuzha, Ernakulam – 686670, Kerala India, Tel.: +918848096306, +919446010905

M. S. Latha
PhD, Professor, School of Biosciences, Mahatma Gandhi University, Kottayam, Kerala, India, 686560. Tel.: 9446190331, E-mail: mslathasbs@yahoo.com

Aditya Menon
PhD, Scientist, Pushpagiri Research Centre, Pushpagiri Institute of Medical Sciences and Research Centre, Tiruvalla - 689101, Pathanamthitta Dist., Kerala, India; E-mail: prc@pushpagiri.in

M. K. Preetha
MSc, Research Scholar, School of Biosciences, Mahatma Gandhi University, Kottayam, Kerala, India–686560. Tel.: 9495512014, E-mail: preethakmattathil111@gmail.com

A. Sangamithra
PhD, Assistant Professor, Department of Food Technology, Kongu Engineering College, Perundurai, 638060, Erode, Tamil Nadu, India, Tel.: +91-8680909333, E-mail: asokmithra@gmail.com

V. Sreelakshmi
MPhil, PhD Research Scholar, Department of Biochemistry, University of Kerala, Kariavattom – 695581, Thiruvananthapuram, Kerala, India, Tel.: +91-9747878424, E-mail: sreelakshmi.v.88@gmail.com

Hafiz Ansar Rasul Suleria
PhD, McKenzie Fellow, Food Nutrition Department of Agriculture and Food Systems, The University of Melbourne, Level 1, 142 Royal Parade, Parkville Victoria, 3010 Australia; Mobile: +61- 470439670; E-mail: hafiz.suleria@unimelb.edu.au

S. Syama
MSc, Research Scholar, School of Biosciences, Mahatma Gandhi University, Kottayam, Kerala, India–686560. Tel.: 9605635134, E-mail: syamasavani@gmail.com

K. M. Thara
PhD, Scientific Officer, Department of Biotechnology, University of Calicut, P.O. Calicut University, 673635, Malappuram (District), Kerala, India, Tel.: +91-9446342696, E-mail: tara_menon2003@yahoo.co.in

A. Vysakh
MSc, Research Scholar, School of Biosciences, Mahatma Gandhi University, Kottayam, Kerala, India–686560. Tel.: 9497358718, E-mail: vysakh15@gmail.com

ABBREVIATIONS

$^1O_2$	singlet oxygen
5mC	5-methylcytosine
ABTS	2,2'-azino-bis(3-ethylbenzothiazoline–6-sulphonic acid
ADA	American Diabetic Association
ADP	adenosine diphosphate
AGE	advanced glycation end products
AIM2	absent in melanoma 2
AKR1B1	aldo-keto reductase family 1, member B1
ALA	alpha-linolenic acid
ALR2	aldose reductase enzyme
ALS	amyotrophic lateral sclerosis
AMPK	activated protein kinase
AOM	acute otitis media
AP–1	activator protein–1
AR	aldose reductase
ARI	aldose reductase inhibitor
ASVD	arteriosclerotic vascular disease
ATP	adenosine triphosphate
ATPase	adenosine triphosphatase
bFGF	basic fibroblast growth factor
BMI	body mass index
Bp	base pair
BPA	bisphenol A
BRCA1	breast cancer 1 (Oncogene)
CBP	cAMP-response element binding protein
CCAT	enhancer binding protein α-C/EBPF α
CCK	cholecystokinin
CD	cluster of differentiation
cDNA	complementary deoxyribonucleic acid
ChREBP	carbohydrate-responsive element-binding protein
CIMP	CpG island methylator phenotype
CKF	chronic kidney failure
CLA	conjugated linoleic acid

CNS	central nervous system
COX	cyclooxygenase
COX–2	cyclooxygenase–2
CRP	C-reactive protein
CSE	conventional solvent extraction
CSF	cerebrospinal fluid
CTGF	connective tissue growth factor
CYP2D1	cytochrome P450 2D1
CYP3A4	cytochrome P450 3A4
Cys	cysteine
DAG	diacylglycerol
DAPK	death-associated protein kinase
DES	diethylstilbestrol
DHA	docosahexaenoic acid
DIC	instant controlled pressure drop
DKA	diabetic ketoacidosis
DM	diabetes mellitus
DMSO	dimethylsulphoxide
DNA	deoxyribonucleic acid
DNMTs	DNA methyltransferases
DOHaD	developmental origin of health and disease
DPPH	2,2-diphenyl–1-picrylhydrazyl
dU	DuBois
DW	dry weight
EAAE	enzyme-assisted aqueous extraction
EACP	enzyme assisted cold pressing
ECM	extracellular matrix
EDR	endothelial-dependent relaxation
EDTA	ethylene diamine tetraacetic acid
EFSA	European Food Safety Authority
EGCG	epigallocatechin–3-gallate
EPA	eicosapentaenoic acid
FAO	Food and Agriculture Organization of the United Nations
FC	Folin's–Ciocalteu reagent
FDA	Food and Drug Administration
FMR1	fragile X mental retardation 1
FOSHU	foods for specified health use
FRAP	fluorescence recovery after photobleaching

FTIR	Fourier transform infrared spectroscopy
FUFOSE	functional food science in Europe
FXN	frataxin
GAD	glutamic acid decarboxylase
GC-MS	gas chromatography-mass spectrometry
GDM	gestational diabetes mellitus
GEN	genistein
GFAT	glutamine: fructose–6-phosphate amidotransferase
GFR	glomerular filtration rate
GIT	gastrointestinal tract
Gln	glutamine
GLP–1	glucagon-like peptide 1
GPx	glutathione peroxidase
GR	glutathione reductase
GRAS	generally recognized as safe
GS-DHN	glutathionyl–1,4-dihydroxynonene
GSH	reduced glutathione
H_2O_2	hydrogen peroxide
HATs	histone acetyltransferases
HBA1c	glycated hemoglobin
HDL	high density lipoprotein
HHS	hyperglycaemic hyperosmolar state
His	histidine
HMEECs	human middle ear epithelial cell lines
HMG	3-hydroxy 5-methyl glutarate-coA
HNE	hydroxynonemal
HNF1 α	hepatocyte nuclear factor 1-α
HNF1b	hepatocyte nuclear factor 1-β
HNF4α	hepatocyte nuclear factor 4-α
HNO_2	nitrous acid
HPLC	high pressure liquid chromatography
HPTLC	high performance thin layer chromatography
hRSV	human respiratory syncytial virus
HS-SPME	headspace solid-phase microextraction
HUVECs	human umbilical vein endothelial cells
IC_{50} value	concentration required to inhibit half of the initial growth rate
ICCA	islet cell cytoplasmic antibodies
ICSA	islet cell surface antibodies

IGF-1	insulin-like growth factor-1
IGT	impaired glucose tolerance test
IL	interleukin
IL-1	interleukin-1
IL-1β	interleukin-1-β
IL-6	interleukin-6
IL-8	interleukin-8
IPF1	insulin promoter factor-1
JECFA	FAO/WHO expert committee on food additives
L•	lipid radical
LADA	latent autoimmune diabetes of adult
LC-MS	liquid chromatography-mass spectroscopy
LDL	low density lipoprotein
Leu	leucine
LOO•	lipid peroxyl radical
LOOH	lipid hydroperoxide
LOX	lipooxygenase
LPS	lipopolysaccharide
MAE	microwave assisted extraction
MAPK	mitogen-activated protein kinase
MBD	methyl-CpG-binding domain
MCP-1	monocyte chemotactic protein-1
MDR	multidrug resistance
mecA	methicillin-resistant gene
MECP2	methyl CpG binding protein 2
MG	methylglyoxal
MIC	minimum inhibitory concentration
miRNAs	microRNAs
MMP-2	matrix metalloproteinase-2
MODY	maturity-onset diabetes of the young
MPO	myeloperoxidase
mRNA	messenger RNA
MSRR	methionine sulphoxide reductase regulator gene
MTT	3-(4, 5-dimethylthiazol-2-yl)-2, 5-diphenyltetrazolium bromide
N_2O_3	dinitrogen trioxide
NAD	nicotinamide adenine dinucleotide
NADH	reduced nicotinamide adenine dinucleotide
NADPH	reduced nicotinamide adenine dinucleotide phosphate

NE	neutrophil elastase
NET	neutrophil extracellular traps
NEUROD1	neurogenic differentiation factor 1
NF	necrosis factor
NF-κB	nuclear factor kappa B
nm	nanometer
NO•	nitric oxide radical
NO_2	nitrogen dioxide
$O_2•^-$	superoxide anion radical
OGTT	oral glucose tolerance test
OH•	hydroxyl radical
$ONOO^-$	peroxynitrite
p14ARF	ARF tumor suppressor
PABA	para-aminobenzoic acid
PAD-4	peptidyl arginine deiminase-4
PAI-1	plasminogen activator inhibitor 1
PARP	poly ADP-ribose polymerase, a DNA repairing protein
PBMC	peripheral blood mononuclear cell
PBP	penicillin-binding proteins
PCR	polymerized chain reaction
PDGF	platelet-derived growth factor
PGE2	prostaglandin E2
P-gp	*P*-glycoprotein
Phe	phenylalanine
piRNAs	PIWI-interacting RNAs
PIWI	P-element induced wimpy testis
PKC	protein kinase C
PLC	phospholipase C
PLE	pressurized liquid extraction
PMSF	poly methyl fluoride
PPARγ	peroxisome proliferation activated receptor gamma
PR 3	proteinase 3
Pro	proline
PTMs	post-translational modifications
PTS	phosphotransferase system
PUFA	polyunsaturated fatty acids
PVDF	polyvinyl difluoride
RAAS	rennin angiotensin aldosterone system

RAGE	receptor for advanced glycation end products
RAS	rennin-angiotensin system
rasiRNAs	repeat-associated RNAs
Rb	retinoblastoma (oncogene)
RG	red ginseng
RNA	ribonucleic acid
RNS	reactive nitrogen species
ROO•	peroxyl radical
ROS	reactive oxygen species
RP-HPLC	reverse phase high pressure liquid chromatography
RSK2	ribosomal S6 kinase (a serine/threonine protein kinase)
RT-PCR	real-time polymerized chain reaction
SAM	S-adenosyl methionine
SC-CO$_2$	supercritical carbon dioxide
SDH	sorbitol dehydrogenase
SEM	scanning electron microscopy
SFE	supercritical fluid extraction
siRNAs	short interfering RNAs
SLE	systemic lupus erythematosus
SMN2	survival motor neuron
SOD	superoxide dismutase
TGF-β1	transforming growth factor beta-1
TKF	terminal kidney failure
TLC	thin layer chromatography
TMS1	target of methylation-induced silencing
TNF-alpha	tumor necrosis factor-alpha
TPC	total polyphenol
Trp	tryptophan
Tyr	tyrosine
UAE	ultrasound assisted extraction
USFDA	U.S. Food and Drug Administration
v/v	volume/volume
Val	valine
VEGF	vascular endothelial growth factor
VSMC	vascular smooth muscle cell
w/v	weight/volume
WHO	World Health Organization
λ_{max}	wavelength at which maximum absorption takes place

PREFACE 1

25 grams of soy protein a day,
as part of a diet low in saturated fat and cholesterol,
may reduce the risk of heart disease, [USFDA 21CFR101.82].
—Ramabhau Patil, PhD

To be healthy, it is our moral responsibility,
towards Almighty God, ourselves, and our family;
Eating fruits and vegetables makes us healthy,
Believe and have faith;
Reduction of food waste can reduce world hunger, and
can make our planet eco-friendly.
—Megh R. Goyal, PhD

We introduce this book volume under book series *Innovations in Plant Science for Better Health: From Soil to Fork*. This book mainly covers the current scenario of the research and case studies, and covers the importance of phytochemicals from plant-based sources in therapeutics, under three parts: Part I–Extraction of Bioactive Compounds from Plants; Part II–Plant Based Drugs; and Part III–Innovative Use of Plant-Based Drugs for Human Health.

This book volume sheds light on the potential of plants for human health from different technological aspects, and it contributes to the ocean of knowledge on food science and technology. We hope that this compendium will be useful for students and researchers of academia as well as for those working with the food, nutraceuticals, and herbal industries.

The contributions by the contributing authors to this book volume have been most valuable in the compilation. Their names are mentioned in each chapter and in the list of contributors. We appreciate you all for having patience with our editorial skills. This book would not have been written without the valuable cooperation of these investigators, many of whom are renowned scientists who have worked in the field of plant science and food science throughout their professional careers.

I am glad to introduce my new editors, Dr. P. P. Joy and Dr. Hafiz Ansar Rasul Suleria, who bring their expertise and innovative ideas on pharmaceutical sciences in this book.

The goal of this book volume is to guide the world science community on how plant-based secondary metabolites can alleviate us from various conditions and diseases.

We request that readers offer their constructive suggestions that may help to improve the next edition.

We express our admiration to our families and colleagues for understanding and collaboration during the preparation of this book volume.

As an educator, there is a piece of advice to one and all in the world: *"Permit that our almighty God, our Creator, provider of all and an excellent Teacher, feed our life with Healthy Food Products and His Grace; and Get married to your profession."*

—**Megh R. Goyal, PhD, PE,**
Senior Editor-in-Chief

PREFACE 2

If wealth is lost, nothing is lost,
If health is lost, something is lost, and
If character is lost, everything is lost.
—P. P. Joy, PhD
https://www.researchgate.net/profile/Pp_Joy

The tenet "Let food be thy medicine and medicine be thy food," espoused by Hippocrates nearly 2500 years ago, is receiving renewed interest in terms of functional foods, which are processed foods containing ingredients that aid specific bodily functions in addition to being nutritious. Functional food is a food containing health-giving additives. Functional foods can be considered to be those whole, fortified, enriched or enhanced foods that provide health benefits beyond the provision of essential nutrients (e.g., vitamins and minerals) when they are consumed at efficacious levels as part of a varied diet on a regular basis. Use of dietary supplements, functional foods, and nutraceuticals is increasing as the industry is responding to consumers' demands.

This book volume, *Plant Secondary Metabolites for Human Health: Extraction of Bioactive Compounds,* deals with recently advanced research in medical and nutrition sciences, natural products, and health-promoting foods that could possibly reduce the risk of diseases while enhancing overall well-being. Nowadays, this topic has been receiving extensive attention from both health professionals and the public.

Part I: Extraction of Bioactive Compounds from Plants describes the advances in the extraction of bioactive-compounds from plants. Advanced extraction techniques such as enzyme assisted, microwave-assisted, ultrasound-assisted, pressurized liquid extraction, and supercritical extraction techniques were used for the purpose that is described in detail. These compounds have been utilized for the production of pharmaceutical supplements and, more recently, as food additives to increase the functionality of foods. This section describes the therapeutic activities of natural resources. It discusses functional foods and plant extracts in human health, the vitality of phytochemicals in cell signaling and biological assays. Plants are inevitably the largest suppliers of drugs or compounds that can serve as lead compounds for the manufacture of drugs.

Part II: Plant-Based Drugs covers plant products, their health-promoting potential, and natural remedies for lifestyle diseases. These natural products and secondary metabolites are increasingly becoming significant in preventive and therapeutic medication. The incorporation of any functional plant food in the daily diet is a better endeavor to prevent the progression of chronic disorders. It also provides *Eugenia uniflora's* botany, physical characteristics, uniqueness, uses, distribution, importance, phytochemistry, traditional importance, nutritional importance, bioactivities, and future trends. It also discusses plant biotechnological interventions for bioactive secondary metabolites, epigenetics, and functional foods. Epigenetics involves the heritable changes in gene expression without any change in the underlying DNA sequence. Functional foods, beyond providing basic nutrition, may offer a potentially positive effect on health by acting as epigenetic modulators that cure various disease conditions, such as metabolic disorders, cancer, and chronic inflammatory reactions.

Part III: Innovative Use of Plant-Based Drugs for Human Health is about therapeutic activities of natural resources as xylitol and aldose reductase inhibitors. The potential of xylitol as an immunomodulator is discussed. Xylitol is found to be beneficial in the treatment of diabetes, pulmonary infection, otitis media, and osteoporosis. The role of aldose reductase inhibitors in combating hyperglycemia and glycation, oxidative stress and immune functions, ROS production, extracellular trap formation, etc. are discussed with reference to outcomes from different experimental studies.

This book volume is a treasure house of information and an excellent reference for researchers, scientists, students, growers, traders, processors, industries, dieticians, medical practitioners, and others.

—**P. P. Joy, PhD**
Editor

PREFACE 3

In the recent era, along with technological advancement and food-based strategies, changes in dietary patterns and nutritional awareness are becoming a foremost topic of concern. Globally, the consumption of therapeutic foods, including functional foods, is rising, as they are a vital component of dietary interventions for health promotion and disease management. Presently, due to the increased burden of diseases, people are more inclined towards the consumption of foods that provide additional health benefits along with fulfillment of nutritional requirements. Functional foods encompass physiologically active components, which may or may not have been modified to enhance their bioactivity and provide health benefits, in addition to basic nutrition. These foods may help to decrease the risk of diseases, prevent disease development, and enhance an individual's health. Owing to increased demand, the functional food market is growing and expanding at a strange level.

Plant-based functional foods are known to contain compounds (also referred to as phytochemicals) in the leaves, stems, flowers, and fruits that can help to promote human health. Therefore, plant products are drawing the attention of researchers and policymakers because of their demonstrated beneficial effects against diseases with high global burdens, such as diabetes, hypertension, cancer, and neurodegenerative diseases. The side effects associated with conventional medicine have awakened the interest of researchers to explore these plants as an alternative or complementary medicine. Various plants like onion, garlic, ginger, citrus, flaxseed, broccoli and other cruciferous vegetables, and green and black tea are utilized to develop functional foods that are effectively used to improve human nutrition. These plants were discovered by traditional healers to have activities against certain diseases mostly by chance or by testimonies from other users. Hence, there is a need for substantial scientific evidence in terms of efficacy, dosage, and safety in order for traditional herbs to have a place in modern medicine.

This book presents scientific reports on the therapeutic values of different plants against diseases. It aims to further encourage the need for the development of plant-based drugs through innovative and groundbreaking research studies and, thus, will help to promote the health and economic well-being of people around the world. The understanding of the therapeutic values of these plants will also help to improve their sustainability, as people and governments will be encouraged to preserve and conserve the plants for future generations. The book covers the phytochemistry and health-promoting potentials of plants against different ailments.

I thank our mentor, Dr. Megh R. Goyal, for his leadership qualities for inviting us to join his team; and for motivating us to get in love with our profession.

—Hafiz Ansar Rasul Suleria, PhD
Editor

PART I

Extraction of Bioactive
Compounds from Plants

CHAPTER 1

EXTRACTION METHODS FOR BIOACTIVE COMPOUNDS FROM PLANTS: AN OVERVIEW

A. SANGAMITHRA*, V. CHANDRASEKAR, and
SWAMY GABRIELA JOHN

*Corresponding author. E-mail: asokmithra@gmail.com

ABSTRACT

Consumer's increased awareness on health is fueling the industry to make an alteration on the processing and food products. Bioactive compounds are secondary metabolites that are produced within the plants. These bioactive compounds may produce either toxicological or pharmacological effects in man and animals based on their concentrations and purity. Such compounds have been used as a part of traditional medicines to cure many ailments and diseases. Thus, the efficient production and purification of plant-based extracts is a challenging task to the industry and academia. This chapter discusses conventional and advanced methods to extract bioactive compounds.

1.1 INTRODUCTION

The consumption of unbalanced food contributes to energy excess, overweight, obesity, and pave the way to many chronic diseases. However, at the same time, consumer's pursuit for health-enhancing food products has raised a remarkable progress in the food process industries towards the expansion and production of plant-based functional food products. According to the World Health Organization (WHO), around 80% of the global population depends on traditional medicine systems for their health needs [88]. Numerous epidemiological studies have shown a reduced risk of cancer in people consuming fruits and vegetables regularly [11]. Natural biocompounds from plants can be used as drugs, functional food ingredients, fragrances, food additives, and pigments. Most notably, plant products (such as fruits, vegetables, whole grains, and legumes) share the properties of healthy dietary patterns. Thus, the production and purification of plant-based extracts is an area of interest to the industry and academia. Significant epidemiologic evidence demonstrates a protective role in diets rich in fruits and vegetables, whole grains, and legumes on various chronic diseases.

This chapter presents a comprehensive review of the techniques/methods used for the plant-based bioactive component extraction. It elaborates both the conventional and advanced extraction techniques so far used by various researchers.

1.2   BIOACTIVE COMPOUNDS

The term "bioactive" is composed of two words: bio and active. As per etymology, the word 'bio' was derived from Greek word '*bios*,' which means life; and the word 'active' was from Latin 'activus' means 'capable of acting' or 'energetic' or 'lively' [26]. As per medical dictionary, bioactive components are defined as a substance having an effect on, that cause a reaction, or trigger a response in the living tissue [27, 49, 50]. Bioactive compounds are secondary metabolites produced by plants, which have a direct positive or negative effect on humans and animals depending on the dose or the bioavailability. Secondary metabolites are considered as end products of primary metabolism and may help plants to enhance the survival potential and to overcome the challenges, by letting them to interact with the environment [24]. These biocompounds act as defense chemicals, and their absence does not cause any bad effects in the plants. These compounds may promote good health in humans. Numerous epidemiologic evidence and experimental studies on plant-based diets have shown protective effects [76] on cardiovascular disease, cancer [36], osteoporosis, breast cancer [29], and other chronic diseases [74].

The bioactive compounds are secreted in every part of the plant cell and accumulated in a particular secretary tissue. The collections of cells are able to create greater amounts of specific biocomponents of varying chemical properties. These compounds are synthesized during the growth phase, or under specified seasons or conditions, which make their extraction and purification, quite difficult [84]. The major steps involved in the utilization of bioactive compounds from plant sources are: extraction, screening, isolation, and characterization, toxicological, and clinical evaluation [73].

Depending on the origin, plant-based secondary metabolites are classified into three major groups [31]: alkaloids, phenolics, and terpenoids. These major groups are further classified into subgroups (see, Figure 1.1). These secondary metabolites have also been recommended as additives in food products to improve the functionality of foods [5].

1.3 EXTRACTION METHODS

Extraction is an essential primary phase in the isolation of bioactive compounds to extract the components of interest from the plant materials for further separation and characterization. The major steps include: cleaning, prewashing, drying, and grinding to attain a homogenous sample. Size reduction of plant material often increases the contact surface of the sample in the solvent system and improves the kinetics of extraction. The standardization of specific extraction method is to get most of the therapeutically desired components and to eliminate the inert material by treating with an appropriate solvent.

1.3.1 CONVENTIONAL EXTRACTION METHODS

Most of the traditional extraction methods are centered on the extracting ability of the selected solvents and involves heating/mixing. These methods utilize the ability of the solvent to solvate certain bioactive components from a solid. The conventional methods are time-consuming and require relatively enormous quantities of solvents [20]. Plant-based bioactive compounds can be extracted using the following conventional extraction methods: maceration, infusion, percolation, and soxhlation.

A defined solvent or mix of solvent is referred to as the "menstruum." The plant material remaining after the extraction of the soluble components is called as marc. The menstruum selection depends on the nature of the target compound to be extracted. A variety of solvents are employed to separate the desired components from various natural sources (Table 1.1).

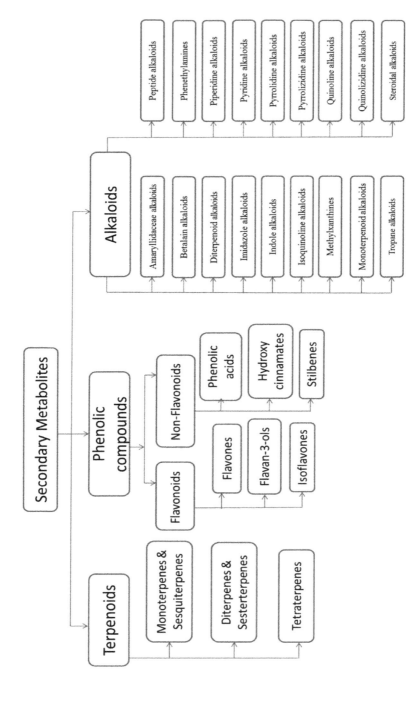

FIGURE 1.1 Classification of plant-based secondary metabolites.

Solvents such as ethanol, methanol, or ethyl-acetate are preferred for extraction of hydrophilic compounds; whereas, dichloromethane or a mixture of methanol and dichloromethane is preferred for the extraction of hydrophobic compounds. In a few cases, extraction with hexane is used to remove chlorophyll [17].

TABLE 1.1 Types of Solvents Used for Bioactive Component Extraction

Type of solvent	Bioactive compound
Acetone	Flavonols
Chloroform	Terpenoids, Flavonoids
Dichloromethanol	Terpenoids
Ethanol	Tannins, Polyphenols, Flavonol, Terpenoids, Alkaloids
Ether	Alkaloids, Terpenoids
Methanol	Anthocyanins, Terpenoids, Saponins, Tannins, Flavones, Polyphenols
Water	Anthocyanins, Tannins. Saponins, Terpenoids

1.3.1.1 MACERATION

Maceration has been practiced on a small scale and for homemade preparation of extracts. It is one of the cheapest methods to extract bioactive compounds from plant materials. Maceration is carried out by steeping the plant material in an organic solvent in a closed container generally at room temperature for a specified period of time. The steps involved in maceration are shown in Figure 1.2. Before extraction, the plant materials are grounded to improve the surface area for efficient mixing with solvent. The plant material and extracting solvent can be in the ratio of 1:10 w/v [80].

Frequent stirring of the solution increases the diffusion rate and also reduces the concentration of solution at the sample surface. The container is kept for a specific period of the interval until the cell structure is softened and penetrated by the solvent. The diffusion of solvent dissolves the soluble constituents that are extracted out. Finally, the extract is filtered to remove the solid residue. The solid residue is also pressed to recover the remaining occluded solution. The extraction process can be repeated several times to ensure complete extraction of desired compounds from the plant material. Maceration process can take from hours to weeks at room temperature [75]. Repeated maceration is more efficient than single maceration in cases, where active constituents are left behind. Triple maceration is performed in case, where the marc cannot be pressed.

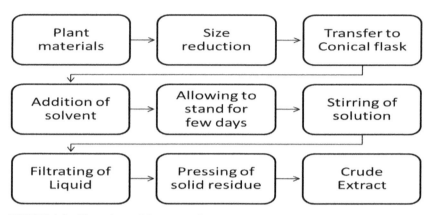

FIGURE 1.2   Flow chart of the maceration process.

1.3.1.2   INFUSION

Infusion is the process of creating a new substance by steeping another substance in a liquid, generally water. Infusion type of extraction is similar to maceration, but the process is carried out at a desired or elevated temperature usually higher than room temperature (up to 100°C) for a specific period of time. This process is highly suitable for readily-soluble constituents from the plant material. Infusion is generally made using boiling water as the extracting solvent. The most common example of infusion is tea or coffee preparation. The plant material is left to remain suspended in the solvent and finally filtered to remove the plant material from the extract [25].

1.3.1.3   PERCOLATION

Percolation process is frequently used to extract the essence for the production of tinctures and fluid extracts. It is a downward displacement of the solvent through the bed of raw materials to get the extract.

Percolation process is performed in a device called percolator. A percolator is a cone-shaped arrangement open at both ends (Figure 1.3).

The raw materials may be size reduced initially and transferred to the percolator. The materials are moistened with the required quantity of selected menstruum, and the mass is allowed to rest for approximately four hours. The moistened materials are then packed in the percolator. Fresh menstruum is further added to saturate the material and allowed to stand for about 24 hours. The stopcock valve of the percolator is then opened, and the liquid is allowed to drip slowly. The limitation of percolation is that, if the raw material is not uniformly distributed, then the solvent may not reach all the areas, which lead to incomplete extraction. A huge quantity of solvent is required, and also the process is time-consuming.

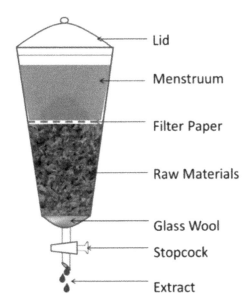

FIGURE 1.3 **(See color insert.)** Schematic diagram of percolator.

1.3.1.4 SOXHLATION

Soxhlation is a continuous hot extraction process using Soxhlet extractor. Soxhlet extraction is one of the standard procedures to relate the success of advanced extraction methodologies [6]. In this extraction process, the finely ground sample is packed in a thimble prepared from strong filter paper and placed in the body of Soxhlet apparatus (Figure 1.4).

The extracting solvent is placed in the flask, and it is heated. The liquid form of solvent is slowly converted into vapor form. These vapors enter into the condenser through the adjacent tube and get condensed into a hot liquid, which flows on the thimble containing the raw material and extracts the contents. This is a continuous process and is carried out until a drop of solvent from the siphon tube does not leave a residue. The benefit of this method is that huge quantities of the drug can be extracted with a much lesser quantity of solvent. This method can be used in both small scale and large scale, but seems to be highly economical and viable when improved into a continuous extraction process on a medium or large scale.

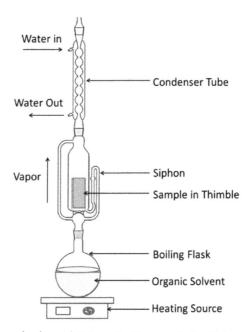

FIGURE 1.4 (**See color insert.**) Schematic diagram of a Soxhlet apparatus.

1.3.2 NOVEL EXTRACTION METHODS

Escalating demand for green processes and healthy products has given upswing to alternative extraction processes that are eco-friendly, mild, effective, and reduced by-products [65]. Novel extraction techniques can result in an increased yield in a shorter duration [15]. The improved extraction techniques overcome the limitations of conventional solvent extraction process such as the requirement of huge quantity of solvent, high energy to separate solute, and possible degradation of thermolabile compounds. The advanced methods for the extraction of bioactive compounds from plants are discussed in this section.

1.3.2.1 ENZYME-ASSISTED EXTRACTION

Enzyme-aided extraction of plant-based biologically active compounds is a feasible substitute to traditional solvent-based extraction techniques. This method is eco-friendly and employs low quantity of solvent. Enzymes used for the extraction process may be from bacteria, fungi, animals, or plant extracts. They are classified as: hydrolyzing enzymes, ligases, oxidation-reduction enzymes, group transfer enzymes, carboxylation enzymes, and isomerizing enzymes.

Enzyme-based extraction is a high yielding technology, which also removes the undesired components from cell walls and offers the benefit of high catalytic efficiency and preserves original value of the natural products to a high degree [34]. The efficiency of enzyme-assisted extraction process depends on the reaction temperature, time of extraction, enzyme concentration, pH of the system, and particle size of the substrate. Enzyme-assisted extraction has been used to extract polysaccharides, oils, natural pigments, flavors, and medicinal compounds (Table 1.2). Based on the catalytic property, a particular enzyme acts on a specific substrate [16]. Compared to traditional methods, this technique has numerous benefits such as: high extraction yield, lower cost of investment and energy requirements, high reproducibility at shorter duration, and simplified manipulation [23, 56, 69].

Few of the bound phytochemical substances, which are dispersed in the cytoplasm, are retained in the polysaccharide-lignin network, and which are associated with plant cell wall polymers are not easily reached by solvents. Enzymes disrupt the cell wall membranes, increases the permeability of the cell wall and enable better release of bioactive compounds [62].

Cell wall degrading enzymes (cellulase, protease, pectinase, hemicellulase, and phospholipase) can be employed before extraction to collapse

TABLE 1.2   Experiments on Enzyme-Assisted Extraction of Different Raw Materials

Material	Bioactive compound	Enzymes	Remarks	Reference
Apple skin	Epicatechin, Procyanidin B2, Rutin, Chlorogenic acid, Phloridzin, a Phloridzin derivative	Pectinase, Cellulase, Protease	• Increased in Fick's module values • Enhanced dosage and degradation of pectin improved phenol transfer	[63]
Bay leaves	Essential oil	Cellulase, hemicellulase, xylanase, and a ternary mixture of enzymes	• Yield was high in individual enzyme than a mixture of enzymes • Improved extraction by cellulolytic and hemicellulolytic activities of enzymes	[12]
Black Pepper and Cardamom	Essential oils β-caryophyllene, α-terphenyl acetate	A mixture of cellulose, β-glucanase, pectinase, and xylanase	• Improved pore formation, pore size, porosity • Yield of the β-myrcene and β-caryophyllene increases on enzyme pre-treatment • Superior quality oil	[14]
Chinese licorice, Chinese skullcap	Flavonoids	*T. viride* cellulase, *P. decumbens* naringinase	• Total flavonoid yield of about 53.23 mg/g obtained with 2 mg/ml cellulase	[90]
Citrus peels	Phenolics	Cellulase	• Highest recovery • Variation of the phenolic contents depended on the extraction conditions	[41]
Ginger	6-Gingerol	α-amylase, Viscozyme, Cellulase, Protease, and Pectinase	• Enzyme treatment followed by acetone extraction — oleoresin and gingerol	[45]
Lemon balm (*Melissa officinalis*)	Rosmarinic acid	Cellulase, Endo-β-1,4 xylanase, pectinase	• Enhanced total phenolic and antioxidant activity compared to the non-enzymatic sample	[51]

TABLE 1.2 *(Continued)*

Material	Bioactive compound	Enzymes	Remarks	Reference
SilybumMarianum	Silybin	Cellulase	• Higher yield than ethanol extraction • SEM and TEM revealed ruptured cell wall	[43]
Tomato waste	Total carotenoid and lycopene	Pectinase and cellulase	• 6-fold and a 10-fold increase of carotenoid and lycopene in the enzyme-treated sample • Maximum carotenoid yield at 70 U/g pectinase and 122.5 U/g cellulase	[81]

the cell wall integrity and to enhance the release of bounded compounds. Enzyme-based extraction may be classified into two types such as:

- **Enzyme-assisted aqueous extraction** (EAAE) method is used to break the cell walls by rupturing polysaccharide protein colloid. This method is mainly extended for the extraction of lipophilic compounds [6].
- In **Enzyme-assisted cold pressing** (EACP) technique, enzyme facilitates hydrolysis of the cell wall, as the polysaccharide-protein colloid is not available in the system. The cold pressing technique is an alternate solution for bioactive compound extraction from oils due to its non-toxic and non-flammable properties [38].

1.3.2.2   MICROWAVE-ASSISTED EXTRACTION (MAE)

Microwaves are considered as very high-frequency radio waves. The electromagnetic radiation with the wavelength ranges between 100 cm and 0.1 cm, and a frequency range between 300 MHz and 300 GHz (Figure 1.5). They are the combination of the oscillating electric and magnetic field. Microwaves are applicable to many industrial processes like cooking, thawing, sterilization, and disinfestations. Microwaves are employed to extract principle compounds like carotenoids, as terpenoids, alkaloids, and saponins from biological materials [91]. Compounds extracted using microwave exhibit better functional and antioxidant properties [42, 79]. Microwave-assisted extraction (MAE) consumes less solvent and less process time during extraction.

MAE involves following three sequential steps [4]:

- Extraction of solutes from the cells, generally under high temperature and pressure.
- Diffusion of solvent across the sample.
- Release of solutes from the sample to solvent.

The principle of microwave heating is based on its direct impact on polar materials [40]. The conversion of electromagnetic energy into heat energy is due to two principles namely ionic conduction and dipolar polarization or dipolar rotation [32]. During the mechanism of ionic conduction, due to the collision of charged particles with neighboring atoms or molecules, the

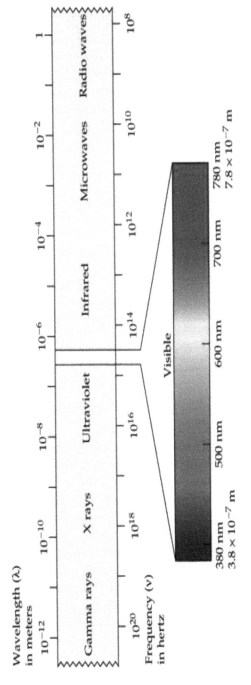

FIGURE 1.5 (See color insert.) Electromagnetic spectrum.

internal resistance causes the heat generation. Alternatively, during dipole rotation, ions tend to align with continuously varying electric field. The frequent changes in orientation produce molecular collision and friction, which produces heat [91]. Figure 1.6 represents the schematic diagram of MAE. Heat generated in the solid matrix evaporates the moisture content and builds up the pressure in the sample. Due to the vapor pressure difference, water vapor escapes from the sample and disrupts the cell wall improves the porosity of the sample matrix. Extracting solvent penetrates through the pores of the sample and interacts with the analyte; thus improves the efficiency of extraction.

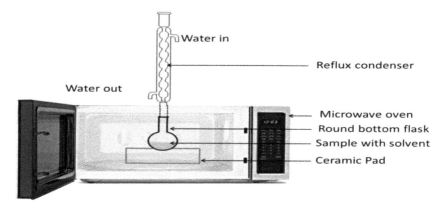

FIGURE 1.6   **(See color insert.)** Schematic diagram of microwave-assisted extraction.

MAE efficiency is influenced by factors such as: microwave power, frequency, dielectric constant, loss factor, type of sample, sample size, moisture content, the composition of solvent, solubility, solid to feed ratio, extraction temperature, process time, extraction pressure, and a number of extraction cycles [46]. Polar solvents pair well with microwaves and reach elevated temperatures within a short time. However, non-polar solvents are transparent to microwaves. Polar solvents absorb more microwave energy than non-polar solvents as dielectric constant extracts more analyte from the sample matrix [86]. Higher dissipation factor solvent raises the temperature and pressure of the sample quickly whereas it rises slowly at a lower dissipation factor [2]. MAE has been carried out by numerous researchers to extract anthocyanin, phenolics, alkaloids from herbs, black currant, tealeaves, mulberry fruits, etc. (Table 1.3).

TABLE 1.3 Studies on Microwave - Assisted Extraction of Different Raw Materials

Materials	Bioactive compounds	Extraction parameters	Remarks	Reference
Black current marc	Anthocyanin	Single mode cavity resonator, at a magnetron frequency of 2.45 GHz Solvent: Aqueous hydrochloric solution Microwave power: 140–700 W pH of solvent: 2 and 7 Solid to liquid ratio: 1:10–1:20 Extraction time: 10–30 min	• Maximum yield of 20.4 mg/g anthocyanin at pH 2, with an extraction time of 10 min, microwave power of 700 W, solid to liquid ratio of 1:20 in MAE • Anthocyanin concentration in solvent phase of MAE increased by 20% of CSE	[60]
Fruit hulls of *Camellia oleifera*	Total polyphenol Total flavonoid	Microwave extraction (Sineo Microwave Equipment Co., Ltd., Shanghai, China) – 2450 MHz Liquid to solid ratio: 11.59–28.41ml/g Extraction time: 8.18–41.82 min Extraction temperature: 53.18–86.82°C	• Optimum conditions for extraction: 15.33:1 (ml/g) liquid to solid ratio for an extraction time of 35 min, with extraction temperature of 76°C • Polyphenol yield was about 15.05 ± 0.04% • Total flavonoid in the extract was 140.06 mg/g	[92]
Fruits of Mulberry	Anthocyanin	Galanz (800W, 2450 MHz) microwave oven (Guangzhou, China) Methanol concentration: 30–70% Microwave power: 320–480 W Extraction Time: 80–160 s Fixed liquid-to-solid ratio: 25:1	• Optimized conditions: Acidified methanol of 59.6%, 425 W Microwave power, and 132 s of extraction time • Anthocyanins of about 54.72 mg was obtained • MAE was more rapid and efficient than CSE	[94]

TABLE 1.3 *(Continued)*

Materials	Bioactive compounds	Extraction parameters	Remarks	Reference
Tea leaves	Tea poly-phenols and caffeine	Type of solvents: Ethanol, methanol, and acetone Solvent concentration: 0–100%, v/v Time for Pre-leaching: 0–90 min Time for Extraction: 0.5–8 min Liquid to solid ratio: 10:1–25: 1 ml /g	• 20:1 (ml /g) was sufficient to obtain higher extraction • 30% tea polyphenols and 4% tea caffeine with MAE for 4 min • Acetone gave higher extraction • At 90 min of pre-leaching time, polyphenols increased from 28.06 to 29.59%, and caffeine increased from 3.55 to 4.04% • MAE greatly reduced the extraction time than conventional methods	[59]
Myrtus communis leaves	Total phenolic compounds	Domestic microwave oven system (2450 MHz, Samsung Model NN-S674MF, Kuala Lumpur, Malaysia) Microwave power: 400–600 W Time of Extraction: 30–90 s Liquid to solid ratio: 20–40 ml/g Ethanol: 20–100%	• Optimum conditions obtained at 42% Ethanol, 500W Microwave power for a 62 s with a liquid to solid ratio of 32 ml/g • TPC was 162.49 ± 16.95 mg Gallic acid equivalent/g • Higher extracts of Tannins, flavonoids, and antioxidant activities in MAE than CSE and UAE	[18]
Stephania Sinica Diels	Alkaloids	MDS–8 Microwave Workstation (Shanghai Sineo Microwave Chemical Technology Co.) Microwave power: 150–400 W Extraction temperature: 40–100°C Extraction time: 15–240 s Liquid-solid ratio: 10–40 ml/g Ethanol concentration: 20–95%	• Maximum yield of alkaloid was achieved at the liquid-solid ratio of 24:1 (ml/g), with 65% of ethanol at 60°C and microwave power of 150 W in the 90s • MAE showed higher extraction efficiency and shorter processing time than soxhlet and UAE	[89]

1.3.2.3   ULTRASOUND ASSISTED EXTRACTION (UAE)

Ultrasound is a sound wave at a frequency of above 20 kHz, which is above the threshold limit of human hearing. Ultrasonic waves penetrate through the material at a characteristic speed of waves [35]. Depending on the frequency range, ultrasound may be classified into low frequency and high-frequency waves. High-frequency ultrasound waves are used for non-destructive analytical technique in food processing whereas low-frequency range ultrasound is used for wide applications like altering physio-chemical properties of food material [48]. Ultrasound-assisted extraction (UAE) is one of the green technologies for extracting analytes. The process is non-toxic, reduces the consumption of solvent and energy required for extraction, and yields high purity of final product.

Ultrasound causes cavitation while it is passing through the fluid of food and it is propagated by continuous compression and rarefaction of waves (Figure 1.7).

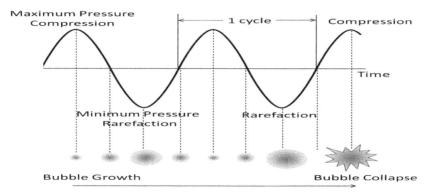

FIGURE 1.7   Ultrasonic cavitation.

Rarefaction cycle of ultrasound wave exceeds at a high power input and form cavitation bubbles from gas nuclei present in the fluid which causes shear forces. Cavitation bubbles generate the turbulence and inter-particle collisions in the solvent which intensifies the eddy diffusion and internal diffusion [33]. This cavitation is distributed throughout the fluid of the food material and grows continuously to a critical size. Bubbles are instable formed after critical size and imploded suddenly [7, 77]. The implosion of cavitation increases temperature to about 5000 K and pressures up to

1000 atm and accumulates temperature in hotspots which produce high shear energy waves and turbulence in the fluid [85]. Furthermore, at the surface of the sample matrix cavitation creates a microjet force to impinge the solvent to the sample matrix which results in more surface area of the sample. The strong microstreaming current creates bubble size variation and implodes subsequently.

Moreover, breaking of water molecules present in the food material generates highly reactive free radicals [68]. This phenomenon results in the increased mass transfer coefficients. Due to this, perturbation of solvents in micropores of sample matrix occurs. In simpler terms, the mechanism of ultrasonic extraction is by breaking the vegetal cells and improving diffusion and osmotic processes. Thus, it enhances the solvent diffusion into the sample matrix and removes the analytes from the active site [47]. Figure 1.8 represents the laboratory scale extraction using the ultrasonic probe. The jacketed vessel controls the extraction temperature by circulating cooling water through the system. A transducer is joined to the horn and a probe. The sonotrode probe is immersed into the sample. UAE recently used it for isolating various bioactive components from plant materials, and results are reported in Table 1.4.

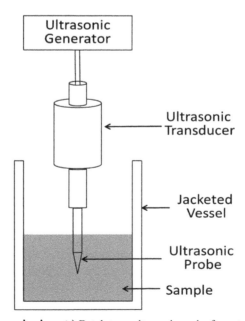

FIGURE 1.8 (**See color insert.**) Batch type ultrasonic probe for extraction.

TABLE 1.4 Experiments on Ultrasound-Assisted Extraction of Different Raw Materials

Plant material	Bioactive compound	Extraction parameters	Remarks	Reference
Boldo Leaves (*Peumusboldus* Mol.)	Boldine	Sonotrode (BS2d34, Hielscher) Solvent: Water Extraction time: 10–40 min Ultrasonic Intensity: 10–23 W/cm^2 Temperature: 10–70°C	• Optimum conditions: Sonication power of 23 W/cm^2, 40 min extraction time, and temperature of 36°C • Higher yield and less time compared to conventional extraction • UAE took 30 min whereas conventional maceration took 120 min	[61]
Mulberry	Anthocyanins	Ultrasonic device (KJ1004B, Kejin Instrument Company, China), 200 W, 40kHz Methanol concentration: 10–50% Extraction temperature: 20–100 min Liquid to solid ratio: 5: 1–20: 1	• Optimized conditions: Methanol of 63.8%, temperature of 43.2°C, liquid to solid ratio of 20: 1 (v/w) for 40 min • Maximum yield of 64.70 ± 0.45 mg/g was obtained • Anthocyanin yield decreased when temperature from 40 to 60°C due to degradation	[93]
Olive leaves	Polyphenols	Ultrasonic bath (Protech, 220 V and 50 Hz) at 25°C Solvent concentration: 0–100% of ethanol Solid to liquid proportion: 500/10–20 mg/ml Extraction time: 20–60 min	• Optimum conditions: solid to liquid proportion of 500mg/10ml, extraction time of 60 min and 50% ethanol composition • Solvent concentration was found to be the most significant parameter	[71]
Potato Peel Waste	Steroidal alkaloids	1500W ultrasonic system (VC 1500, Sonics, and Materials Inc., Newtown, USA) with a 19 mm diameter probe Solvent: Methanol	• Optimum Conditions: 61 μm amplitude, for 17 min of extraction time yielded 1102 μg of alkaloids/g of dried potato peel	[30]

TABLE 1.4 *(Continued)*

Plant material	Bioactive compound	Extraction parameters	Remarks	Reference
		Amplitude: 24.40, 30.5, 42.70, 54.9, 61.0 μm Ultrasound intensity: 9.24, 10.16, 13.28, 17.17, 22.79 W/cm² Processing time: 3, 5, 10, 15, 17 min Durations of pulse: 5s on and 5s off	• In particular, 273, 542.7, 231 and 55.3 μg/g of a-solanine, α-chaconine, solanidine, and demissidine respectively are obtained from dried potato peel by UAE	
Rice Bran	Polyphenols and antioxidants	Ultrasonic bath RK103H (Bandelin Sonorex, Germany) with 35 kHz, 140 W Solvent: Ethanol, ethyl acetate, methanol, and n-hexane Sonication time: 45°C for 30 min Liquid to solid ratio: 10:1, 20:1, 40:1, 80:1	• Ethanol found to be significant on all response • 65–67% ethanol, 51–54°C, 40–45 min was found to be the optimum condition • Extraction yield ranged from 11 to 20.2% • UAE offers reduced usage of solvents, temperature, time, and water consumption for extraction	[82]
Tomato Waste	Lycopene	High-intensity ultrasonic probe of 200 W and 24 kHz (Model UP 400S, Dr. Hielscher, Germany) with a H14 sonotrode Power: 50, 65 and 90 W Extraction time: 1–30 min Solvent: Mixture of hexane: acetone: ethanol in ratio 2:1:1 Liquid to Solid ratio: 50:1, 35:1 and 20:1	• Results compared with CSE • Optimum CSE: Solvent solid ratio of 50:1 at 60°C for 40 min • Optimum conditions was obtained for liquid to solid ratio of 35:1 (v/w), for 90 W ultrasonic power for 30 min • Lycopene extracted -reduced time, temperature, and solvent than CSE	[37]

1.3.2.4 PRESSURIZED LIQUID EXTRACTION (PLE)

Pressurized liquid extraction (PLE) is called as accelerated solvent extraction or pressurized solvent extraction or enhanced solvent extraction or subcritical extraction technology. It is mentioned as pressurized hot water extraction or sub-critical water extraction or superheated water extraction, if water is utilized as a solvent. Currently, several bioactive compounds are extracted from natural sources using PLE. Food grade compounds can be extracted by PLE by using water or other GRAS solvents [64].

This extraction technique is appropriate for a broad category of solutes, from polar to non-polar based on the solvent type, process temperature, time for extraction, particle size, and water content of the sample. In PLE, liquid solvent extraction is taking place at elevated pressure and temperature to ensure rapid extraction of compounds [21]. During the whole extraction process, pressure, and temperature are maintained so that the solvent should remain in liquid state. Wide ranges of temperatures can be adopted in PLE; generally, it ranges from ambient temperature to higher temperature of about 200°C with pressure ranging from 35 to 220 bars to extract the analyte from the sample matrix [54]. At high pressure, solvents are forcefully and rapidly filled in the solid matrix, and at high temperature, the polarity of the solvent is adjusted to match with the polarity of the compounds to be extracted [21, 52]. Moreover, at a temperature higher than the boiling point, solubility, selectivity, and diffusivity of solvents are enhanced. Thus, more compounds are extracted with less quantity of solvents in a shorter duration.

PLE method can be static or dynamic. In static PLE method (Figure 1.9), the solvent is pumped to the extraction chamber for only once or in cycles. After the cycles are completed, solvent along with the compounds are removed from the sample vessel and filtered through filter cloth. Filtered compounds are dried using dehydrating agent such as inert diatomaceous earth [72]. Extraction efficiency might be less due to less quantity of solvent filled in the extraction chamber [57, 83].

In dynamic PLE method (Figure 1.10), the solvent is pumped continuously to the extraction chamber where the sample is kept. Extraction pressure and temperature are preset.

Once the preset pressure is reached in the extraction chamber, solvent at preset temperature is pumped to the extraction chamber. Time taken for the extraction varies between 5–15 min. The solvent continuously extracts the compound and collected in a filter to separate the compounds and the solvent. Eventually, after extraction process, the sample is removed from the extraction chamber. After extraction, solvent can be recycled and used for another extraction. Dynamic PLE method is useful for higher capacity extraction, and automation is also possible in dynamic PLE. However, no commercial dynamic PLE system is available.

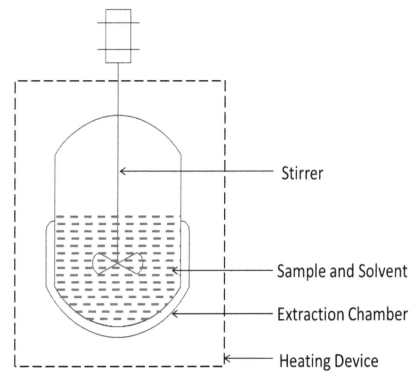

FIGURE 1.9 Experimental set up of static pressurized liquid extraction.

Pressure, temperature, solid matrix, and particle size are major factors affecting the PLE. Extraction time is significantly less compared to conventional extraction methods. Increasing the temperature and reducing the particle size will reduce the long extraction times. High temperature enhances the interaction between sample matrix and analyte by creating Vander Waals forces, dipole attraction and hydrogen bonding [67]. Further, it decreases the cohesive and adhesive interaction between analyte and

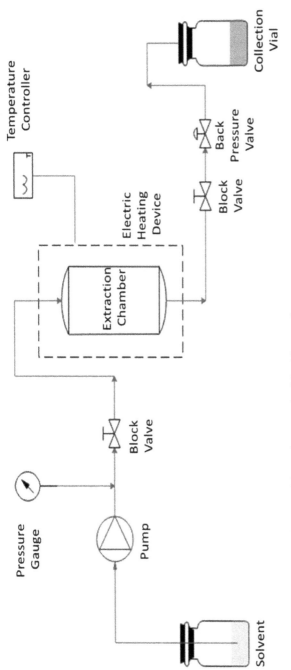

FIGURE 1.10 Experimental set up of dynamic pressurized liquid extraction.

sample matrix by decreasing activation energy for desorption of analyte from the sample matrix. High temperature reduces the solvent viscosity, thereby enhancing its penetration inside the matrix [58]. However, longer heating time may affect the quality of heat sensitive bio-compounds by disintegration and hydrolytic degradation [22, 53]. Hence, it is necessary to maintain the optimized time for the extraction [13, 55]. Selectivity of the solvent is important as that high temperature of solvent may co-extract analytes other than desired compounds.

After extraction, method of separation of compounds from the solvents affects the extraction efficiency. Apart from that, the entire process is affected by the mass transfer of molecules, analyte, and solvent [57]. Higher diffusion of the solvent increases the extraction efficiency [58] and thus decreases the amount of solvent required [10, 67]. The moisture content of sample, pH of the sample and solvent, presence of additives such as anti-coagulant agent, moisture scrubber, dispersing agents, surfactants, and antioxidants also affect the extraction process. Moisture content of the sample affects the extraction efficiency due to its polarity. Dispersing agents are used to absorb the moisture in sample matrix. These dispersing agents fill the cell pores of the sample matrix. Reducing cell volume of sample matrix reduces the consumption of solvents and thus increases the extraction efficiency. At elevated temperature, solvents may degrade the quality of the compounds to be extracted. This method cannot be used for heat-sensitive compounds [2], and high temperature of the liquid causes damage to structure and functional activity of extracted compound. Since the process involves high pressure and temperature, safety measures must be considered.

Sub-critical water extraction is a kind of PLE technology which is based on the same principles but using different solvents to extract the compounds. In this method, the water is used at high temperature exceeding the boiling point (100°C) and is compressed below its critical temperature (374°C). Generally, pressure range of 50–100 bars is usually employed to retain the liquid state. Figure 1.11 represents the phase diagram of water. At sub-critical condition, water possesses two distinguishing properties: low dielectric constant and high ionized state of the water [1]. The rise of liquid water temperature weakens the hydrogen bonds, leading to a lower dielectric constant. The physicochemical characteristics of water may be varied by changing and monitoring temperature and pressure within sealed systems. Due to reduced viscosity, surface tension and disassociation constant of sub-critical water, it can easily solubilize organic compounds from polar to non-polar like phytochemicals, which are generally insoluble in water [66].

The major factors affecting the extraction efficiency in sub-critical water extraction are: extraction temperature, pressure, time, flow rate, and particle size of sample. The increased temperature increases the solubility of the substances in the sample matrix, improves mass transfer as well decreases the surface tension of water which facilitates better diffusion into the sample matrix. A suitable flow rate is necessary to permit contact between sample and solvent to solubilize the desired compounds. At the same time, the flow rate should not be too high so that the extract is not too diluted. Appropriate particle size is necessary to maximize the contact surface.

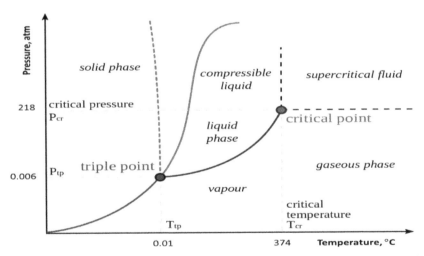

FIGURE 1.11 **(See color insert.)** Water phase diagram as a function of temperature and pressure.

1.3.2.5   SUPER CRITICAL FLUID EXTRACTION (SFE)

Supercritical fluid extraction (SFE) is one of the non-conventional, efficient, and environment-friendly extraction technologies to extract compounds from the sources like plants, spices, agro wastes, algae, and microalgae [28]. It is referred to as supercritical water extraction, if water is employed as the solvent. Currently, SFE is employed in numerous large-scale applications such as: fatty acid refining, decaffeination, essential oil and flavor extraction from natural sources [19, 86]. In SFE, a variety of solvents (such as ethylene, xenon, methane, fluorocarbons, nitrogen, and carbon dioxide) are used for extraction of various compounds [78]. However, most of the extraction processes use carbon dioxide, as it readily solubilizes the lipophilic substances and it is also Generally Recognized as Safe by USFDA and EFSA. Moreover, carbon dioxide is non-explosive, non-toxic, low cost, environment-friendly and easily separated from the final products [70, 87]. All the solvents exhibit fluid properties such as: density, viscosity, diffusivity, and dielectric constant. The nature of a fluid change depending on pressure and temperature. At high pressure and temperature, fluids become supercritical fluid where its properties lie between gas and liquid [86]. The density of the supercritical fluid is comparable to that of liquid whereas, the viscosity of the supercritical fluid is similar to that of the gas [78, 87].

The SFE system (Figure 1.12) consists of a reservoir of mobile phase, generally CO_2, a pump to pressurize the gas, vessel, and pump for co-solvent, a heating device, extraction chamber, a pressure controller, and separator. Flow meters and gas meters are also attached to the system. Raw materials are placed in the extraction vessel. A high-pressurized fluid with higher temperature is pumped to the extraction vessel. Once the analyte from the sample matrix diffused to the supercritical fluid, then the fluid along with dissolved analyte is transported to the separators. After separation of analyte from the fluid, fluids are recycled or released to the environment [78]. Several factors influencing the extraction efficiency are: supercritical fluids, co-solvents, extraction pressure, temperature, moisture content and particle size of raw materials, carbon dioxide flow rate, solvent to solid ratio and extraction time.

The advantage of using carbon dioxide as solvent in SFE is its gaseous state at room temperature and pressure that makes solvent free recovery of compound. The limitations of SC-CO_2 is less efficiency in extracting polar compounds due to its low polarity nature [28]. Hence, co-solvent is added in small quantities along with carbon dioxide to enhance the solubility of analyte and the selectivity of the process [78]. Hexane, methanol, ethanol, isopropanol, acetonitrile, and dichloromethane are employed as co-solvent. Among these, ethanol is widely used co-solvent in supercritical fluid extraction due to its less toxicity and miscibility in CO_2 [39, 44]. At pressure range from 7 to 21 MPa, relative permittivity of supercritical CO_2 is 1.4, so it effectively extracts the lipids and other non-polar organic compounds.

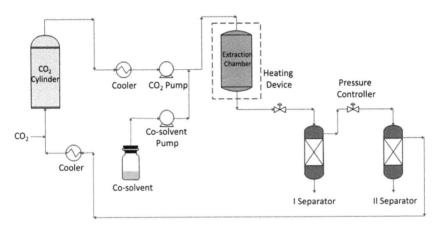

FIGURE 1.12 **(See color insert.)** Experimental set up of supercritical fluid extraction.

1.3.2.6 INSTANT CONTROLLED PRESSURE DROP ASSISTED EXTRACTION (DIC)

Instant controlled pressure drop assisted extraction (DIC) is a thermos-mechanical process, which overcomes the difficulty exerted in steam distillation, i.e., partial transfer of vapor towards the core of raw material. In DIC method, the raw material is subjected to high pressure/high temperature for a shorter period and immediately followed by sudden pressure drop of 5kPa. The high pressure/high temperature is achieved by exposing the raw material to high temperature compressed air, saturated steam pressure, high-pressure microwaves, etc. Depending on the raw material, the high pressure applied ranges between 0.1 and 1 MPa with a temperature increase from 100 to 180°C, when saturated steam is used.

The sudden drop in pressure triggers vaporization of volatile compounds, instantaneous cooling of raw material and swelling or rupturing of the cells. The instant cooling of raw material prevents thermal degradation of raw material which is common in conventional high-temperature extraction process. Due to auto-vaporization of volatile compounds, this method is found to be suitable for direct extraction of volatile compounds [8]. For the extraction of non-volatile substances like flavonoids, DIC method is utilized as a pre-treatment for extraction [3].

DIC allows swelling of cell structure with more effective washing stage and higher diffusivity. However, severe DIC conditions may involve breaking of cell walls, which allows the valuable compounds to be easily accessible for extraction. DIC combined with solvent extraction can be done very efficiently in a shorter duration with less solvent [9].

1.4   FUTURE PERSPECTIVES

Previous research studies have reported that only very little of plant biodiversity has been identified for its bioactive compounds. There is greater scope for identification and screening of bioactive compounds from untapped plant sources, which may provide an enhanced counterpart of synthetic drugs. Also, the rising trend of consumers towards health-conscious eating will continue to fuel concern about benefits of functional foods. Therefore, proper identification and extraction of active ingredients are to be made by the processors. The most common method of extraction is the conventional solvent extraction technique, which utilizes enormous quantities of organic solvents. Those solvents are reported to be volatile and flammable. The growing technological and economic demands can be met by environment-friendly green extraction technology, which can operate at relatively low cost and also easy to be scaled-up to industrial scale. Numerous studies on advanced extraction techniques have reported the optimized process parameters to yield maximum bioactive compounds, but still, the scale-up of these techniques remains a challenge, which needs further development.

1.5   SUMMARY

Extraction is the essential primary step in the isolation of bioactive compounds, because it is essential to extract the required bioactive components. The most commonly used conventional methods of extraction are: maceration, infusion, percolation, and soxhlation. As an alternative to conventional methods, advanced extraction techniques (such as enzyme assisted, microwave assisted, PLE, ultrasound assisted and supercritical extraction techniques) have resulted in increased yield within a shorter duration. More focus on research studies related to the identification and extraction of biologically active compounds from untapped natural sources are required to meet the escalating global demands for the functional ingredients. Also, the scale-up of the advanced techniques remains a challenge for further research and development.

KEYWORDS

- **enzyme assisted extraction**
- **instant controlled pressure drop assisted extraction**
- **maceration**
- **microwave assisted extraction**
- **soxhlation**
- **supercritical fluid extraction**

REFERENCES

1. Adachi, S., (2009). *Properties of Subcritical Water and its Utilization* (p. 112). Division of food science and biotechnology, Graduate School of Agriculture, Kyoto University, Japan.
2. Ajila, C., Brar, S., Verma, M., Tyagi, R., Godbout, S., & Valero, J., (2011). Extraction and analysis of polyphenols: Recent trends. *Critical Reviews in Biotechnology, 31*(3), 227–249.
3. Allaf, T., Tomao, V., Ruiz, K., & Chemat, F., (2013). Instant controlled pressure drop technology and ultrasound assisted extraction for sequential extraction of essential oil and antioxidants. *Ultrasonics Sonochemistry, 20*(1), 239–246.
4. Alupului, A., Călinescu, I., & Lavric, V., (2012). Microwave extraction of active principles from medicinal plants. *UPB Science Bulletin, Series B., 74*(2), 1454–2331.
5. Ayala-Zavala, J., Rosas-Domínguez, C., Vega-Vega, V., & González-Aguilar, G., (2010). Antioxidant enrichment and antimicrobial protection of fresh-cut fruits using their own byproducts: Looking for integral exploitation. *Journal of Food Science, 75*(8), R175–R181.
6. Azmir, J., Zaidul, I., Rahman, M., Sharif, K., Mohamed, A., Sahena, F., et al., (2013). Techniques for extraction of bioactive compounds from plant materials: A review. *Journal of Food Engineering, 117*(4), 426–436.
7. Barbosa-Cánovas, G., & Rodriguez, J., (2002). Update on nonthermal food processing technologies: Pulsed electric field, high hydrostatic pressure, irradiation, and ultrasound. *Food Australia, 54*(11), 513–520.

8. Berka-Zougali, B., Hassani, A., Besombes, C., & Allaf, K., (2010). Extraction of essential oils from Algerian myrtle leaves using instant controlled pressure drop technology. *Journal of Chromatography A, 1217*(40), 6134–6142.

9. Besombes, C., Berka-Zougali, B., & Allaf, K., (2010). Instant controlled pressure drop extraction of Lavandin essential oils: Fundamentals and experimental studies. *Journal of Chromatography A, 1217*(44), 6807–6815.

10. Björklund, E., Nilsson, T., & Bøwadt, S., (2000). Pressurized liquid extraction of persistent organic pollutants in environmental analysis. *TRAC Trends in Analytical Chemistry, 19*(7), 434–445.

11. Block, E., (1992). The organosulfur chemistry of the genus Allium–implications for the organic chemistry of sulfur. *Angewandte Chemie International Edition in English, 31*(9), 1135–1178.

12. Boulila, A., Hassen, I., Haouari, L., Mejri, F., Amor, I. B., Casabianca, H., & Hosni, K., (2015). Enzyme-assisted extraction of bioactive compounds from bay leaves (*Laurus nobilis L.*). *Industrial Crops and Products, 74*, 485–493.

13. Camel, V., (2001). Recent extraction techniques for solid matrices-supercritical fluid extraction, pressurized fluid extraction, and microwave-assisted extraction: Their potential and pitfalls. *Analyst, 126*(7), 1182–1193.

14. Chandran, J., Amma, K. P. P., Menon, N., Purushothaman, J., & Nisha, P., (2012). Effect of enzyme assisted extraction on quality and yield of volatile oil from black pepper and cardamom. *Food Science and Biotechnology, 21*(6), 1611–1617.

15. Chemat, F., Tomao, V., & Virot, M., (2008). Ultrasound-assisted extraction in food analysis. In: *Handbook of Food Analysis Instruments* (pp. 85–103). CRC Press.

16. Cheng, X., Bi, L., Zhao, Z., & Chen, Y., (2015). Advances in enzyme assisted extraction of natural products. In: Yarlagadda, P., (ed.), *AER-Advances in Engineering Research* (3rd edn., pp. 371–375).

17. Cos, P., Vlietinck, A. J., Berghe, D. V., & Maes, L., (2006). Anti-infective potential of natural products: How to develop a stronger *in vitro* 'proof-of-concept.' *Journal of Ethnopharmacology, 106*(3), 290–302.

18. Dahmoune, F., Nayak, B., Moussi, K., Remini, H., & Madani, K., (2015). Optimization of microwave-assisted extraction of polyphenols from *Myrtus communis L.* leaves. *Food Chemistry, 166*, 585–595.

19. Daintree, L., Kordikowski, A., & York, P., (2008). Separation processes for organic molecules using SCF Technologies. *Advanced Drug Delivery Reviews, 60*(3), 351–372.

20. De Castro, M. L., & García-Ayuso, L., (1998). Soxhlet extraction of solid materials: An outdated technique with a promising innovative future. *Analytica Chimica Acta, 369*(1), 1–10.

21. Dunford, N. T., Irmak, S., & Jonnala, R., (2010). Pressurized solvent extraction of policosanol from wheat straw, germ and bran. *Food Chemistry, 119*(3), 1246–1249.

22. Fernández-González, V., Concha-Graña, E., Muniategui-Lorenzo, S., López-Mahía, P., & Prada-Rodríguez, D., (2008). Pressurized hot water extraction coupled to solid-phase microextraction–gas chromatography–mass spectrometry for the analysis of polycyclic aromatic hydrocarbons in sediments. *Journal of Chromatography A, 1196*, 65–72.

23. Fleurence, J., Massiani, L., Guyader, O., & Mabeau, S., (1995). Use of enzymatic cell wall degradation for improvement of protein extraction from *Chondrus crispus*, *Gracilaria verrucosa* and *Palmaria palmata*. *Journal of Applied Phycology, 7*(4), 393–397.

24. Harborne, J. B., (1993). *Introduction to Ecological Biochemistry* (p. 384). Academic Press, Elsevier: London.

25. Harbourne, N., Marete, E., Jacquier, J. C., & O'Riordan, D., (2013). Conventional extraction techniques for phytochemicals. In: Tiwari, B. K., Brunton, N. P., & Brennan, C., (eds.), *Handbook of Plant Food Phytochemicals: Sources, Stability, and Extraction* (pp. 397–411). John Wiley & Sons Ltd: UK.

26. Harper, D., (2016). *Online Etymology Dictionary*. http://www.etymonline.com. (accessed on November 11).

27. Harris, P., Nagy, S., & Vardaxis, N., (2014). *Mosby's Dictionary of Medicine, Nursing and Health Professions-Australian & New Zealand Edition* (p. 2040). Elsevier Health Sciences.

28. Herrero, M., Cifuentes, A., & Ibanez, E., (2006). Sub-and supercritical fluid extraction of functional ingredients from different natural sources: Plants, food-by-products, algae, and microalgae: A review. *Food Chemistry, 98*(1), 136–148.

29. Hooper, L., & Cassidy, A., (2006). A review of the health care potential of bioactive compounds. *Journal of the Science of Food and Agriculture, 86*(12), 1805–1813.

30. Hossain, M. B., Tiwari, B. K., Gangopadhyay, N., O'Donnell, C. P., Brunton, N. P., & Rai, D. K., (2014). Ultrasonic extraction of steroidal alkaloids from potato peel waste. *Ultrasonics Sonochemistry, 21*(4), 1470–1476.

31. Irchhaiya, R., Kumar, A., Yadav, A., Gupta, N., Kumar, S., Gupta, N., et al., (2015). Metabolites in plants and its classification. *World Journal of Pharmacy and Pharmaceutical Sciences, 4*(1), 287–305.

32. Jain, T., Jain, V., Pandey, R., Vyas, A., & Shukla, S., (2009). Microwave assisted extraction for phytoconstituents-an overview. *Asian Journal of Research in Chemistry (AJRC), 2*(1), 19–25.

33. Ji, J. B., Lu, X. H., Cai, M. Q., & Xu, Z. C., (2006). Improvement of leaching process of Geniposide with ultrasound. *Ultrasonics Sonochemistry, 13*(5), 455–462.

34. Kim, S. K., (2011). *Handbook of Marine Macroalgae: Biotechnology and Applied Psychology* (p. 592). John Wiley & Sons, New York.

35. Knorr, D., Zenker, M., Heinz, V., & Lee, D. U., (2004). Applications and potential of ultrasonics in food processing. *Trends in Food Science & Technology, 15*(5), 261–266.

36. Kris-Etherton, P. M., Hecker, K. D., Bonanome, A., Coval, S. M., Binkoski, A. E., Hilpert, K. F., Griel, A. E., & Etherton, T. D., (2002). Bioactive compounds in foods: Their role in the prevention of cardiovascular disease and cancer. *The American Journal of Medicine, 113*(9), 71–88.

37. Kumcuoglu, S., Yilmaz, T., & Tavman, S., (2014). Ultrasound assisted extraction of lycopene from tomato processing wastes. *Journal of Food Science and Technology, 51*(12), 4102–4107.

38. Latif, S., & Anwar, F., (2009). Physicochemical studies of hemp *(Cannabis sativa)* seed oil using enzyme-assisted cold-pressing. *European Journal of Lipid Science and Technology, 111*(10), 1042–1048.

39. Lemert, R. M., & Johnston, K. P., (1990). Solubilities and selectivities in supercritical fluid mixtures near critical endpoints. *Fluid Phase Equilibria, 59*(1), 31–55.
40. Letellier, M., & Budzinski, H., (1999). Microwave assisted extraction of organic compounds. *Analysis, 27*(3), 259–270.
41. Li, B., Smith, B., & Hossain, M. M., (2006). Extraction of phenolics from citrus peels: II. Enzyme-assisted extraction method. *Separation and Purification Technology, 48*(2), 189–196.
42. Li, Y., Skouroumounis, G. K., Elsey, G. M., & Taylor, D. K., (2011). Microwave-assistance provides very rapid and efficient extraction of grape seed polyphenols. *Food Chemistry, 129*(2), 570–576.
43. Liu, H., Du, X., Yuan, Q., & Zhu, L., (2009). Optimization of enzyme assisted extraction of silybin from the seeds of *Silybum marianum* by Box–Behnken experimental design. *Phytochemical Analysis, 20*(6), 475–483.
44. Liza, M., Rahman, R. A., Mandana, B., Jinap, S., Rahmat, A., Zaidul, I., & Hamid, A., (2010). Supercritical carbon dioxide extraction of bioactive flavonoid from *Strobilanthes crispus* (Pecah Kaca). *Food and Bioproducts Processing, 88*(2), 319–326.
45. Manasa, D., Srinivas, P., & Sowbhagya, H., (2013). Enzyme-assisted extraction of bioactive compounds from ginger (*Zingiber officinale Roscoe*). *Food Chemistry, 139*(1), 509–514.
46. Mandal, V., Mohan, Y., & Hemalatha, S., (2007). Microwave assisted extraction—an innovative and promising extraction tool for medicinal plant research. *Pharmacognosy Reviews, 1*(1), 7–18.
47. Mason, T., Paniwnyk, L., & Lorimer, J., (1996). The uses of ultrasound in food technology. *Ultrasonics Sonochemistry, 3*(3), S253–S260.
48. McClements, D. J., (1995). Advances in the application of ultrasound in food analysis and processing. *Trends in Food Science & Technology, 6*(9), 293–299.
49. Mifflin, H., (2007). *The American Heritage Medical Dictionary* (p. 909). Houghton Mifflin Company, New York.
50. Miller, B. F., & Keane, C. B., (2005). *Miller-Keane Encyclopedia and Dictionary of Medicine, Nursing, and Allied Health* (7th edn., p. 2344). Saunders Limited, New York.
51. Miron, T., Herrero, M., & Ibáñez, E., (2013). Enrichment of antioxidant compounds from lemon balm (*Melissa officinalis*) by pressurized liquid extraction and enzyme-assisted extraction. *Journal of Chromatography A, 1288*, 1–9.
52. Miron, T., Plaza, M., Bahrim, G., Ibáñez, E., & Herrero, M., (2011). Chemical composition of bioactive pressurized extracts of Romanian aromatic plants. *Journal of Chromatography A., 1218*(30), 4918–4927.
53. Moreno, E., Reza, J., & Trejo, A., (2007). Extraction of polycyclic aromatic hydrocarbons from soil using water under subcritical conditions. *Polycyclic Aromatic Compounds, 27*(4), 239–260.
54. Mustafa, A., & Turner, C., (2011). Pressurized liquid extraction as a green approach in food and herbal plants extraction: A review. *Analytica Chimica Acta, 703*(1), 8–18.
55. Namiesnik, J., & Gorecki, T., (2000). Review-sample preparation for chromatographic analysis of plant material. *Journal of Planar Chromatography-Modern TLC, 13*(6), 404–413.

56. Nyam, K. L., Tan, C. P., Lai, O. M., Long, K., & Man, Y. B. C., (2009). Enzyme-assisted aqueous extraction of Kalahari melon seed oil: Optimization using response surface methodology. *Journal of the American Oil Chemists' Society, 86*(12), 1235–1240.

57. Ong, E. S., Cheong, J. S. H., & Goh, D., (2006). Pressurized hot water extraction of bioactive or marker compounds in botanicals and medicinal plant materials. *Journal of Chromatography A, 1112*(1), 92–102.

58. Palma, M., Piñeiro, Z., & Barroso, C. G., (2002). In-line pressurized-fluid extraction–solid-phase extraction for determining phenolic compounds in grapes. *Journal of Chromatography A., 968*(1), 1–6.

59. Pan, X., Niu, G., & Liu, H., (2003). Microwave-assisted extraction of tea polyphenols and tea caffeine from green tea leaves. *Chemical Engineering and Processing: Process Intensification, 42*(2), 129–133.

60. Pap, N., Beszédes, S., Pongrácz, E., Myllykoski, L., Gábor, M., Gyimes, E., Hodúr, C., & Keiski, R. L., (2013). Microwave-assisted extraction of anthocyanins from black currant marc. *Food and Bioprocess Technology, 6*(10), 2666–2674.

61. Petigny, L., Périno-Issartier, S., Wajsman, J., & Chemat, F., (2013). Batch and continuous ultrasound assisted extraction of boldo leaves (*Peumus boldus* Mol.). *International Journal of Molecular Sciences, 14*(3), 5750–5764.

62. Pinelo, M., Arnous, A., & Meyer, A. S., (2006). Upgrading of grape skins: Significance of plant cell-wall structural components and extraction techniques for phenol release. *Trends in Food Science & Technology, 17*(11), 579–590.

63. Pinelo, M., Zornoza, B., & Meyer, A. S., (2008). Selective release of phenols from apple skin: Mass transfer kinetics during solvent and enzyme-assisted extraction. *Separation and Purification Technology, 63*(3), 620–627.

64. Plaza, M., Amigo-Benavent, M., Del Castillo, M. D., Ibáñez, E., & Herrero, M., (2010). Neoformation of antioxidants in glycation model systems treated under subcritical water extraction conditions. *Food Research International, 43*(4), 1123–1129.

65. Puri, M., Sharma, D., & Barrow, C. J., (2012). Enzyme-assisted extraction of bioactives from plants. *Trends in Biotechnology, 30*(1), 37–44.

66. Ramos, L., Kristenson, E., & Brinkman, U. T., (2002). Current use of pressurized liquid extraction and subcritical water extraction in environmental analysis. *Journal of Chromatography A., 975*(1), 3–29.

67. Richter, B. E., Jones, B. A., Ezzell, J. L., Porter, N. L., Avdalovic, N., & Pohl, C., (1996). Accelerated solvent extraction: A technique for sample preparation. *Analytical Chemistry, 68*(6), 1033–1039.

68. Riesz, P., & Kondo, T., (1992). Free radical formation induced by ultrasound and its biological implications. *Free Radical Biology and Medicine, 13*(3), 247–270.

69. Rosenthal, A., Pyle, D., & Niranjan, K., (1996). Aqueous and enzymatic processes for edible oil extraction. *Enzyme and Microbial Technology, 19*(6), 402–420.

70. Sahena, F., Zaidul, I., Jinap, S., Karim, A., Abbas, K., Norulaini, N., & Omar, A., (2009). Application of supercritical CO_2 in lipid extraction–a review. *Journal of Food Engineering, 95*(2), 240–253.

71. Şahin, S., & Şamlı, R., (2013). Optimization of olive leaf extract obtained by ultrasound-assisted extraction with response surface methodology. *Ultrasonics Sonochemistry, 20*(1), 595–602.

72. Saito, K., Sjödin, A., Sandau, C. D., Davis, M. D., Nakazawa, H., Matsuki, Y., & Patterson, D. G., (2004). Development of an accelerated solvent extraction and gel permeation chromatography analytical method for measuring persistent organohalogen compounds in adipose and organ tissue analysis. *Chemosphere, 57*(5), 373–381.

73. Sasidharan, S., Chen, Y., Saravanan, D., Sundram, K., & Latha, L. Y., (2011). Extraction, isolation and characterization of bioactive compounds from plants' extracts. *African Journal of Traditional, Complementary and Alternative Medicines, 8*(1).

74. Saura-Calixto, F., & Goni, I., (2009). Definition of the Mediterranean diet based on bioactive compounds. *Critical Reviews in Food Science and Nutrition, 49*(2), 145–152.

75. Seidel, V., (2012). Initial and bulk extraction of natural products isolation. In: *Natural Products Isolation* (pp. 27–41). Humana Press, New York.

76. Shahidi, F., (2009). Nutraceuticals and functional foods: Whole versus processed foods. *Trends in Food Science & Technology, 20*(9), 376–387.

77. Shukla, T., (1992). Microwave ultrasonics in food processing. *Cereal Foods World, 37*(4), 332–333.

78. Sihvonen, M., Järvenpää, E., Hietaniemi, V., & Huopalahti, R., (1999). Advances in supercritical carbon dioxide technologies. *Trends in Food Science & Technology, 10*(6), 217–222.

79. Simsek, M., Sumnu, G., & Sahin, S., (2012). Microwave assisted extraction of phenolic compounds from sour cherry pomace. *Separation Science and Technology, 47*(8), 1248–1254.

80. Stojanovi, J. B., & Veljkovi, V. B., (2007). Extraction of flavonoids from garden (*Salvia officinalis* L.) and glutinous (*Salvia glutinosa L.*) sage by ultrasonic and classical maceration. *J. Serb. Chem. Soc, 72*(1), 73–80.

81. Strati, I. F., Gogou, E., & Oreopoulou, V., (2015). Enzyme and high pressure assisted extraction of carotenoids from tomato waste. *Food and Bioproducts Processing, 94*, 668–674.

82. Tabaraki, R., & Nateghi, A., (2011). Optimization of ultrasonic-assisted extraction of natural antioxidants from rice bran using response surface methodology. *Ultrasonics Sonochemistry, 18*(6), 1279–1286.

83. Teo, C. C., Tan, S. N., Yong, J. W. H., Hew, C. S., & Ong, E. S., (2010). Pressurized hot water extraction (PHWE). *Journal of Chromatography A., 1217*(16), 2484–2494.

84. Verpoorte, R., Contin, A., & Memelink, J., (2002). Biotechnology for the production of plant secondary metabolites. *Phytochemistry Reviews, 1*(1), 13–25.

85. Villamiel, M., & De Jong, P., (2000). Influence of high-intensity ultrasound and heat treatment in continuous flow on fat, proteins, and native enzymes of milk. *Journal of Agricultural and Food Chemistry, 48*(2), 472–478.

86. Wang, L., & Weller, C. L., (2006). Recent advances in extraction of nutraceuticals from plants. *Trends in Food Science & Technology, 17*(6), 300–312.

87. Wang, L., Weller, C. L., Schlegel, V. L., Carr, T. P., & Cuppett, S. L., (2008). Supercritical CO 2 extraction of lipids from grain sorghum dried distillers grains with solubles. *Bioresource Technology, 99*(5), 1373–1382.

88. WHO, (2002). *Traditional Medicine Strategy: 2002–2005* (p. 74). World Health Organization, Geneva.

89. Xie, D. T., Wang, Y. Q., Kang, Y., Hu, Q. F., Su, N. Y., Huang, J. M., Che, C. T., & Guo, J. X., (2014). Microwave-assisted extraction of bioactive alkaloids from *Stephania sinica*. *Separation and Purification Technology, 130*, 173–181.

90. Xu, M. S., Chen, S., Wang, W. Q., & Liu, S. Q., (2013). Employing bifunctional enzymes for enhanced extraction of bioactives from plants: Flavonoids as an example. *Journal of Agricultural and Food Chemistry, 61*(33), 7941–7948.

91. Zhang, H. F., Yang, X. H., & Wang, Y., (2011). Microwave assisted extraction of secondary metabolites from plants: Current status and future directions. *Trends in Food Science & Technology, 22*(12), 672–688.

92. Zhang, L., Wang, Y., Wu, D., Xu, M., & Chen, J., (2011). Microwave-assisted extraction of polyphenols from *Camellia oleifera* fruit hull. *Molecules, 16*(6), 4428–4437.

93. Zou, T. B., Wang, M., Gan, R. Y., & Ling, W. H., (2011). Optimization of ultrasound-assisted extraction of anthocyanins from mulberry, using response surface methodology. *International Journal of Molecular Sciences, 12*(5), 3006–3017.

94. Zou, T., Wang, D., Guo, H., Zhu, Y., Luo, X., Liu, F., & Ling, W., (2012). Optimization of microwave-assisted extraction of anthocyanins from mulberry and identification of anthocyanins in extract using HPLC-ESI-MS. *Journal of Food Science, 77*(1), C46–C50.

CHAPTER 2

PLANT EXTRACTS AND FUNCTIONAL FOODS FOR BETTER HEALTH: VITALITY OF BIOACTIVE COMPOUNDS IN CELL SIGNALING AND BIOLOGICAL ASSAYS

K. M. THARA*

*Corresponding author. E-mail: tara_menon2003@yahoo.co.in

ABSTRACT

The bioactive components in plants are capable of controlling cell signal pathways either by triggering or by inhibiting the cell cycles/signals in human beings and have beneficiary effects in normalizing the pathways responsible for diseases. Food or combination of foods with required fiber content, useful flavonoids, vitamin C along with proper nutrients of vitamins, proteins, and carbohydrates and other functional components suitable to individual health can be designed. Functional food products are becoming more popular due to their effects in controlling the lifestyle diseases. Enriched or fortified food products are available in the market, and people are becoming more self-health oriented and turning to natural food products. Nutrigenomics (i.e., designing a food according to the genomic profile of a person or his health) is a developing branch of food science.

2.1 INTRODUCTION

The whole life on our planet depends on plants for existence and survival. Foods, especially plant-based, are not the only source of fortified or enhanced nutrition (such as carbohydrates, vitamins, proteins, etc.), but have curative effects in controlling ailments like obesity, diabetes, cancer, etc. The bioactive components are generally classified as: phenolics, flavonoids, terpenoids, glycosides, alkaloids, proteins, and enzymes depending on their structure. Some of them have antioxidant effects and remove or scavenge the highly reactive oxygen free radicals that are formed in our body. Some phytochemicals affect different components of cell cycle and development pathways of chronic diseases as triggering or inhibitory compounds. Plants like lemon, turmeric, garlic, ginger, green tea, aloe vera, cabbage, tomato, berry, grape, beetroot, moringa, banana, cucumber, etc.—that are common components of our daily diet—have proved to show different health supporting supplementary effects.

Plants are the main bioresource of traditional medicines, modern medicines, nutraceuticals, food supplements, folk medicines, pharmaceutical precursors and some basic components for synthetic drugs. It has been reviewed and reported by several investigators that a major percentage of higher plant species are utilized for medical use. After investigations on ethano-medicinal use of plants, pharmacologically active components are isolated and developed to a drug. These beneficiary effects combined with recent biotechnological advances have helped in developing genetically engineered food products such as: golden rice with β-carotene, iron rice with enriched iron content and food products with ω–3 fatty acids, vitamins, and other essential nutrients. It is very important to have knowledge of bioactive components for developing genetically engineered functional foods. Safety of the functional foods also should be scientifically proved. Several food products with dietary supplements are available nowadays in the market.

The major lifestyle diseases are mainly caused by toxic substances, high-fat content, etc. present in our daily diet and improper metabolism. The highly reactive oxygen species (ROS) thus generated in our body can damage the cellular components and normal cell cycles, resulting in these ailments. Plant-based biocompounds present in our daily food can reduce the risk of such diseases. Isolated compounds or purified plant extracts in required quantity can be added as food supplements. Ascorbic acid, α-tocopherol, flavonoids, polyphenolic compounds, proanthocyanidins can be recommended as food supplements. Phytochemicals act as triggering or inhibiting agents of molecular pathways of the cell cycle. This effect can be utilized in the development of the functional food. Table 2.1 presents some common foods, bioactive components present in it with their vitality in cellular pathways and beneficiary effects. Table 2.2 indicates the mechanism of catechin, a phytochemical, in triggering cell signals/

TABLE 2.1   List of Some Common Foods, Bioactive Components, Beneficiary Effects and its Action in Cellular Pathways

Food type	Functional component	Action/use	Mechanism of action*	Reference
Capsicum	Capsicin	Cholesterol-lowering. Antimicrobial	Inhibition of LDL oxidation, HMG3-hydroxy 5-methyl glutarate)-coA reductase inhibition, a decrease in adipogenesis, inhibition of PPRγ expression, inhibits DNA methylation (in an epigenetic level).	[169]
Chocolate	Flavonoids	Heart diseases	Angiogenesis, apoptosis, PI-3K' AKT pathways, MAPK, AMPK activation, inhibition of pre-adipocyte differentiation	[47]
Cranberry	Proantho-cyanidines	Urinary tract infections	Prevents biofilm formation of cell signaling and adhering on the epithelial cell wall.	[157]
Fiber-containing vegetable fruits, grains, cruciferous vegetables	Indoles, Glucosinolates Myricetin	Reduce the risk of a certain type of cancer, heart complaints, prevents lipids drops in adiposities	Inhibits tumor progression by regulating STAT3/ JAK–2 pathway/signaling, PPRγ activation	[169, 215]
Garlic	Allicin, Organosulphur compounds Ajoene	Antimicrobial, Anticancer, Lowering cholesterol and blood pressure	Antimitotic and microtubule binding and prevents cell proliferation, reduced expression of PPRγ, C/EBPA, adiponectin, adipocyte differentiation.	[106]
Grape	Polyphenolics, Resveratrol	Antioxidant, Anticancer	Triggering CD95 signaling-dependent apoptosis, MAPK, PI–3K' AKT pathways, AMPK activation in fatty acid oxidation, inhibition of pre-adipocyte differentiation, down regulates neuropeptide –Y associated with food intake, upregulation of tumor suppressor gene p21Cip1/ WAF1, Caspase signals, down-regulating antiapoptotic proteins surviving, BcL–2, Bcl-XL	[4, 20, 91, 157]

TABLE 2.1 *(Continued)*

Food type	Functional component	Action/use	Mechanism of action*	Reference
Green leafy vegetables	Lutein	Antioxidant	Oxidative damage of eye, cataract, age-related macular degeneration	[77, 106]
Green tea	Catechins, Epicatechin	Anti-obesity, Anticancer, Antioxidant	AMPK activation, downregulation of PPRγ, inhibition of lipase in adipogenesis, Anti-cancer –by inhibiting endotoxin-mediated Necrosis factor NF-K, inducing apoptosis and in controlling related cell signals, phase II detoxification of antioxidant enzyme by modulating Necrosis factor, Nrf2	[8, 11, 30, 36, 105]
Lemon and citrus fruits	Vitamin C	Antioxidant, Anticancer, Anti-inflammation	Chemo-preventive, protect DNA from oxidative damage	[68, 117]
	Quercetin	Antioxidant, Anti-inflammatory, Anticancer, Antimicrobial, Anti-allergic	Prevents adipogenesis, promotes apoptosis, cyclooxygenase, and lipoxygenase inhibition and prevention of leukotriene and thromboxane, the proinflammatory molecules, epigenetic inhibition of DNA methylation and histone activation	[66, 77, 101, 172]
	Kaempferol			
Mango	Lupeol	Anticancer, Antimicrobial	Fast mediated apoptosis Mitochondrial Akt/pKb and NF-Kappa signaling pathways.	[67, 110, 176, 181]
Oats	β-glucan	Control heart diseases	Reduce total cholesterol and LDL.	[169, 212]

TABLE 2.1 *(Continued)*

Food type	Functional component	Action/use	Mechanism of action*	Reference
Soybean	Genistein Proteins	Anti-obesity, Reduce total cholesterol	AMPK activation, promote apoptosis, MAPK, and P38 pathways in TNF and IL expression by blocking NF-Kb activity, Inhibit adipocyte differentiation or formation, decrease fatty acid synthase expression, epigenetic modulation of DNA by histone acetylation or methylation	[169, 208]
Tomato	Lycopene	Prostate Cancer, Antioxidant, Anti-inflammatory	ROS mediated DNA oxidative damage, inhibit COX–2 and LOX in a dose-dependent manner. LDL oxidation.	[27, 107, 183]
Turmeric	Curcumin	Anticancer, Antioxidant	Inhibiting NFk pathway Inhibiting protein kinase, c-myc mRNA expression, bcl–2MRNA expression, apoptosis by topoisomerase II inhibition, activate caspase 3 for mitochondrial death indifferent type of cells, triggering transcription factors TPA, neutralizing ROS	[85, 126, 127]

*PPARγ – peroxisome proliferation activated receptor gamma;

MAPK – mitogen-activated protein kinase;

AMPK – AMP-activated protein kinase; C/EBPF α– CCAT enhancer binding protein α;

COX – cyclooxygenase;

LOX – lipoxygenase;

HMG – 3-hydroxy 5-methyl glutarate-coA; and

NF – necrosis factor.

TABLE 2.2 Mechanism of Catechin, a Phytochemical in Triggering Cell Signals/Inhibition of Molecular Signals by Bioinformatics Study Using the Software Pass (Prediction of Activity Spectra for Substances) Prediction; [http://195.178.207.233/PASS/index.html]

Activity reported as in the cell signaling of	Point value activity calculated as a percentage	Activity reported	Point value activity
2-Dehydropantoate 2-reductase inhibitor	0.814	Anti-inflammatory	0.597
4-Coumarate-CoA ligase inhibitor	0.703	Estrogen receptor beta antagonist	0.519
Aldehyde oxidase inhibitor	0.795	Melanin inhibitor	0.559
Anticarcinogenic	0.801	Antifungal	0.583
Antihemorrhagic	0.926	27-Hydroxycholesterol 7-alpha-monooxygenase inhibitor	0.645
Antimutagenic	0.951	UGT1A7 substrate	0.635
Antineoplastic	0.681	Anaphylatoxin receptor antagonist	0.546
Antioxidant	0.828	Kidney function stimulant	0.610
Antiseborrheic	0.737	Cytostatic	0.539
Antiviral (Influenza)	0.692	CYP3A2 substrate	0.549
APOA1 expression enhancer	0.863	CYP1A2 substrate	0.607
Astringent	0.921	UGT1A8 substrate	0.626
Capillary fragility treatment	0.755	Antiprotozoal (Trypanosoma)	0.550
Cardioprotectant	0.647	3-Oxoacyl-[acyl-carrier-protein] synthase inhibitor	0.493
	—	TNF expression inhibitor	0.517
Caspase 3 stimulant	0.661	CF transmembrane conductance regulator agonist	0.501
Chemopreventive	0.791	CYP1A1 inhibitor	0.564
Chemoprotective	0.666	Laxative	0.501
Chlordecone reductase inhibitor	0.877	3-Dehydroquinate synthase inhibitor	0.600

TABLE 2.2 *(Continued)*

Activity reported as in the cell signaling of	Point value activity calculated as a percentage	Activity reported	Point value activity
Creatine kinase inhibitor	0.882	UGT1A10 substrate	0.608
CYP1A inhibitor	0.856	Beta-glucuronidase inhibitor	0.604
CYP1A substrate	0.888	UGT2B15 substrate	0.614
CYP1A1 substrate	0.927	CYP1B substrate	0.628
CYP2A4 substrate	0.704	Antitussive	0.524
CYP2C12 substrate	0.909	RELA expression inhibitor	0.617
CYP3A inducer	0.737	Hydroxysteroid dehydrogenase inhibitor	0.527
CYP3A4 inducer	0.759	CYP2I2 substrate	0.598
Fibrinolytic	0.959	Vasoprotector	0.652
Free radical scavenger	0.850	Glutathione S-transferase substrate	0.590
Glutathione-disulfide reductase inhibitor	0.779	CYP3A7 substrate	0.552
Hepatoprotectant	0.769	Glucan endo–1,6-beta-glucosidase inhibitor	0.576
HIF1A expression inhibitor	0.883	Antiulcerative	0.613
Histamine release inhibitor	0.791	CYP2J substrate	0.635
Histamine release stimulant	0.697	Estrogen antagonist	0.511
Histidine kinase inhibitor	0.848	1-Alkylglycerophosphocholine O-acetyl transferase inhibitor	0.594
HMOX1 expression enhancer	0.971	UGT1A3 substrate	0.646
JAK2 expression inhibitor	0.785	SMN2 expression enhancer	0.541
Kinase inhibitor	0.743	Nitric oxide antagonist	0.539

TABLE 2.2 *(Continued)*

Activity reported as in the cell signaling of	Point value activity calculated as a percentage	Activity reported	Point value activity
Lipid peroxidase inhibitor	0.908	Antihypercholesterolemic	0.631
MAP kinase stimulant	0.662	Nitrite reductase [NAD(P)H] inhibitor	0.509
Membrane integrity agonist	0.983	Aspulvinone dimethylallyl transferase inhibitor	0.705
Membrane permeability inhibitor	0.789	Sugar-phosphatase inhibitor	0.603
MMP9 expression inhibitor	0.722	Oxidoreductase inhibitor	0.571
Mucomembranous protector	0.962	GST M substrate	0.644
NADPH-ferrihemo protein reductase inhibitor	0.697	Antipyretic	0.516
Neurotransmitter antagonist	0.689	CYP2A5 substrate	0.508
P-benzoquinone reductase (NADPH) inhibitor	0.787	Feruloyl esterase inhibitor	0.589
Pectate lyase inhibitor	0.964	Aryl-alcohol dehydrogenase (NADP+) inhibitor	0.644
Peroxidase inhibitor	0.721	CYP19 inhibitor	0.525
Phosphatase inhibitor	0.708	Glucan 1,4-alpha-maltotriohydrolase inhibitor	0.538
Proliferative diseases treatment	0.695	Antiviral (Rhinovirus)	0.520
Reductant	0.952	CYP2B6 substrate	0.649
Sulfotransferase substrate	0.949	NOS2 expression inhibitor	0.631
SULT1A3 substrate	0.953	AR expression inhibitor	0.641
Testosterone 17-beta-dehydrogenase (NADP+) inhibitor	0.715	Vascular diseases	0.503

TABLE 2.2 *(Continued)*

Activity reported as in the cell signaling of	Point value activity calculated as a percentage	Activity reported	Point value activity
TP53 expression	0.959	Cytoprotectant	0.652
UDP-glucuronosyltransferase substrate	0.885	Iodide peroxidase inhibitor	0.612
UGT1A substrate	0.925	Beta-carotene 15,15'-monooxygenase inhibitor	0.632
UGT1A1 substrate	0.751	Morphine 6-dehydrogenase inhibitor	0.540
UGT1A6 substrate	0.957	Hepatic disorders treatment	0.642
UGT1A9 substrate	0.840	Fatty acid synthase inhibitor	0.577
UGT2B substrate	0.770	Ribulose-phosphate 3-epimerase inhibitor	0.584
UGT2B1 substrate	0.884	Hemostatic	0.611
UGT2B12 substrate	0.701	CYP3A5 substrate	0.540
Xanthine dehydrogenase inhibitor	0.689	Alopecia treatment	0.538

inhibition of molecular signals related to different cellular pathways using bioinformatics study by the software pass prediction.

This chapter discusses functional foods and plant extracts for better health; and vitality of phytochemicals in cell signaling and biological assays.

2.2   PLANT-DERIVED FUNCTIONAL FOODS AND BIOACTIVE COMPONENTS

Plant-based foods contain a wide range of bioactive components, which are mainly responsible for their biological activities to treat or prevent chronic as well as acute infectious diseases. These bioactive components also aid in the normal development of cells. Those activities may be due to a single entity or synergistic effect of different components present in it. Around 25% of total medicines in the market are derived from plants.

On a global basis, 130 drugs, all of the single entity, have been isolated, purified, and tested for efficacy [171]. These bioactive components are synthesized in plants as a result of metabolic activities. The secondary plant metabolites are generally classified according to its chemical nature (Figure 2.1). The structure of the compound is the basis of classification and its general biological activities.

Few of the main classes of plant secondary metabolites are discussed in this section.

2.2.1   PHENOLIC COMPOUNDS

Phenols are good antioxidant compounds [131, 221]. Some examples of simple phenols are: lignin, p-hydroxybenzoic acid, vanillic acid, gallic acid, salicylic acid, catechol, orcinol, hydroquinone, etc. Examples of phenylpropanoids are: coumarin, eugenol, myricetin, caffeic, and coumaric acid. Examples of natural polyphenolics [18] are: Clove, pepper, ginger, coriander, cardamom, etc. Fruits like strawberry, blackberry, blueberry, grapes, orange, lemon, etc. are also a good source of antioxidants such as polyphenolics. Vegetables like cabbage, cauliflower, avocado, etc. are also listed in such a group. Vitamin A, C, and E present in different fruits and vegetables are also antioxidants and help in preventing different diseases. It also has molecular triggering activity related to some cellular pathways such as sodium-dependent vitamin-C transporters (SVCTs) in blood, brain, and placental cells. Vitamin C also functions in nucleic acid and histone methylation and as a co-factor in hydroxylation reactions in the cells.

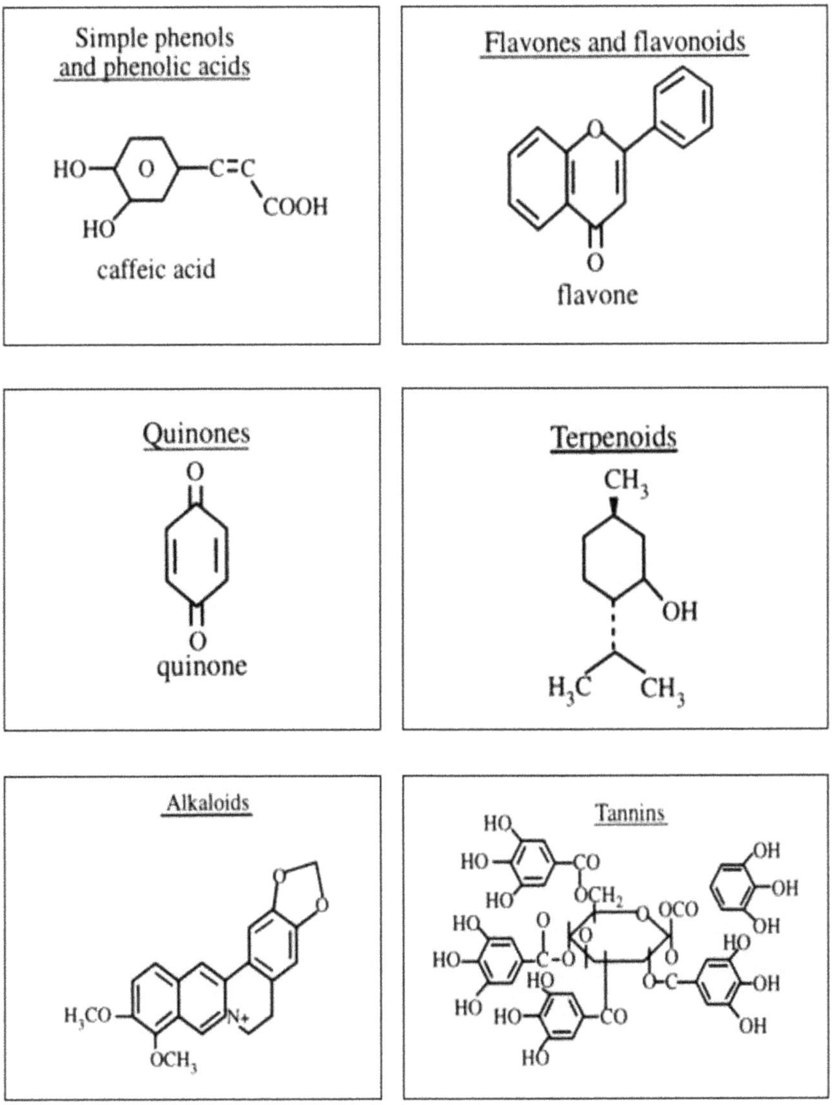

FIGURE 2.1   Structures of various secondary metabolites in plants, adapted from [44].

2.2.2 FLAVONOIDS

Flavonoids are mainly synthesized by the phenylpropanoid pathway. As the number of conjugated bonds increase, the intensity of the color also increases [212]. Such compounds are: flavones–acacetin, flavones–galanin, flavanones, anthocyanins, isoflavones–genistein, retinone, and chalcone–butein, phlorizin, and quercetin, catechins. These are present in different food items such as: green tea, coffee, and berries. Around 4000 flavonoids have been reported. It has a different action mechanism in a different type of cancer cells. It has actions like inhibition of DNA topoisomerase activity, regulation of heat shock protein expression, cell cyclear rest, activation of caspase, downregulation of Bcl–2, induction of differentiation and apoptosis, suppression of proliferation, etc., in a different type of cells.

2.2.3 TERPENES

Terpenes are produced by isopentyl diphosphate through Mevalonic acid pathway in plants. The presence of essential oils has been noticed and recorded from medieval age onwards. The fragrance of essential oil is due to the presence of terpenes [201]. Terpenes are classified as hemi-, mono-, di-, tri-terpenoids depending on the number of repeating isoprene units. Terpenoid is the common name given to a variety of compounds like terpenes. Around 30,000 members of terpenoids have been identified [173]. Terpenes have different functions such as: preventing the attack of insects, inhibiting the growth of competing plants, attracting pollinators and acting as plant hormones. It is found in found in fruits and vegetables. Some common examples of this class of compounds are: geraniol, menthol, lycopenes, carotenoids, and taxol. Terpenoids like taxol can inhibit cellular proliferation and activate differentiation and induce cell death in tumors by inhibiting cancer specific proteasome, NF-κB, an antiapoptotic protein, Bcl–2 in some hormone-related cancer cells like breast and prostate cancer.

2.2.4 ALKALOIDS

Alkaloids are nitrogen-containing cyclic compounds [42]. Examples are: vinblastine, resiprine, morphine, nicotine, cocaine, atropine, caffeine, quinine, Indole–3, acetic acid, camptothecin, berberine, papaverine, ajmaline, and ellipticine [30, 86, 102, 189, 192, 209]. Alkaloids like vincristine and vinblasine can inhibit microtubule formation in the cell cycle, activation of NF-κB cells, apoptosis, etc., in tumor cells.

2.2.5 TANNINS

A large number of polyphenolic groups are present in tannins. They can form complex compounds with proteins and can be precipitated. Hydrolyzable tannins found in pomegranate berries and in some other fruits can induce apoptosis in tumor cells. Cell signaling like down-regulation of cyclins A, upregulation of cyclin E, FAS independent apoptosis, activation of caspase–9, etc., are reported in different kind of cancer cells.

2.2.6 LIGNIN

Lignin is synthesized by shikimate and polyketide pathways. Several lignin compounds have antimicrobial properties. Some lignin derivatives are used as anticancer agents.

2.2.7 QUINONES

Quinones are mostly associated with plant respiration and photosynthesis. These are usually colored compounds and exist in yellow, orange, and red colors. Examples are: anthraquinone, benzoquinone, and naphthoquinone. Emodin and juglone are anticancer agents. It is an antioxidant, antiallergic, and anti-inflammatory agent. Celastrol is another quinone phytochemical and is reported to have an anticancer effect. It inhibits proteasomes, activates heat shock proteins, etc.

2.2.8 PROTEINS

Antimicrobial proteins and polypeptides are generally produced in plants by fighting against pathogens. Examples are: Lectins like concanavalin A (jack fruit) and Ricin.

2.2.9   NUCLEIC ACID

It generally exists as pyrimidine glycoside, which is known to have active biologicalactive properties. Examples are: vicine, convicine, methylated purines like theobromine.

Several plant-derived compounds have been marketed as therapeutics. Some of such products are summarized in Table 2.3 [76, 82, 220–232].

2.3   PLANT DERIVED COMPOUNDS: BIOLOGICAL ACTIVITIES AND MODES OF ACTION

Some of the experimentally proved biological activities of plant-derived products are: antimicrobial, anti-cardiogenic, anticancerous, anti-inflammatory, antidiabetic, bronchodilator, antitussive, antidiuretic, antipyretic, memory enhancing, digestive, immunomodulatory, etc. Around 75% of anti-infective and 60% of anticancer drugs are found to be developed from natural products [76].

2.3.1   ANTIMICROBIAL ACTIVITY

Drugs against multi-drug resistant strains have been developed from plant-derived biologically active compounds. As remedies for various ailments, plant extracts (like *Saraca indica, Azadirachta indica, Aloe vera, Piper longum, Santalum album, Ocimum sanctum*, etc.) are used traditionally.

Due to the presence of alkaloid content vasecine, *Adathoda vasica* is a well-known anti-depressant and antimicrobial agent. *Tinospora cordifolia* is used for diabetes, fever, ulcer, etc., and it contains berberine that is reported to be an antimicrobial agent [104]. Plants like *Curcuma longa* and *Allium sativum* are common constituents of Indian food [20]. These plant extracts are reported to contain curcumin and allicin, respectively that have anticancer, antioxidant, and antimicrobial properties. *Aloe vera* is a good source of antioxidants and is used as a cancer preventive agent [110]. The plants like *Bacopa monnieri* and *Phyllanthus amarus* are considered for their best contribution in the pharmaceutical field from India [161]. Phyllanthin is a compound isolated from Phyllanthus, and it is used in the treatment of jaundice. *Tylophora indica* is traditionally used as a bronchodilator [98]. Spices are also used as antimicrobial agents.

TABLE 2.3 List of Some Marketed Plant Products

Marketed product	Class of bioactive component	Plant source	Medical use	Reference
Atropine	Tropane alkaloid	*Atropa balladona*	Anticholinergic	http//en.m.wikipedia.org/wiki/atropine
Camptothecin	Quinoline alkaloid	*Camptotheca acuminata*	Anticancer, antineoplastic during S-phase of cell cycle	www.en.m.wikipedia/wiki/camptothecin
Chebulin	Anthraquinone	*Terminalia chebula*	Antihypertensive	www.ncbi.nih.gov/terminalia chebula
Digitalin and Digitoxin	Cardiac glycosides	*Digitalis purpurea*	Cardiac glycoside	https://en.wikipedia.org/wiki/Digitalis
Ginseng	Ginsenoside	*Panax ginseng*	Antidiabetic, anti-aging	www.drugs.com/npc/gengeng
Himpyrin	Terpenoid	*Cyperus rotundas*	Antibacterial, antipyretic	www.himalayawellness.com/himpyrin
Quinine	Alkaloid	*Cinchona officinalis*	Antimalarial	www.wikipedia.org//wiki/qunine
Reserpine and Serpentine	Cardiac glycoside	*Rauwolfia serpentine*	Antirheumatic, Antihypertensive	https://www.ncbi.nlm.nih.gov › NCB/reserpine
Tami flu	Shikimic acid	*Chinese starkness*	Anti H1N1	www.en.m.wikipedia.org/oseltamivir
Taxol	Diterpenoid	*Taxus brevifolia*	Anticancer (Ovarian, breast, lung, pancreas, etc.	www.en.m.wikipedia.org/wiki/paclitaxel
Vincristine and Vinblastine	Vinca alkaloids	*Vinca rosea*	Anticancer	www.en.m.wikipedia.org/wiki/vincristine

Even though a large number of plants were traditionally used as antimicrobial agents, the separation of the relevant active compounds from the plant extract and their commercial exploitation has not been satisfactorily carried out in many cases.

The possibility of the synergistic effect of different compounds in an extract or the cumulative effect of the extracts, when used in combination with antibiotics and other antimicrobial agents, also have to be seriously looked into, while searching for new strategies. Some of these extracts can be added in minimum quantity in our daily diet such as: tea, drinking water, etc. and found to prevent some microbial diseases effectively.

2.3.1.1   EMERGENCE OF MULTI-DRUG RESISTANCE AMONG MICROORGANISMS

Due to frequent and indiscriminate use of such drugs, an unanticipated situation has emerged, where microorganisms have gradually started to develop resistance to the drugs. It has become one of the major health threats of the era. Dreadful ailments such as diarrhea, meningitis, sexual, and respiratory infections and many types of infections, which are acquired from hospitals necessitate timely, perfect control, using effective medicines. The newly emerging and re-emerging diseases and resistance of their causative agents to the available drugs create one of the growing challenges in the medical field. This situation demands an intense search for more effective drugs to combat these new medical menaces [97, 113].

The development of resistant strains of microorganisms has reduced the action of common antimicrobial agents and hence increased the necessity of development of new plant-derived antimicrobial compounds. Antimicrobials developed from plants are found to be more effective against drug-resistant microbes due to its different structure and mode of action apart from common antibiotics. Some of the plant-derived compounds, which are reported to have antimicrobial properties against multidrug-resistant organisms are: rhein from rhubarb, plumbagin, resveratrol, gossypol, coumestrol, isoflavone, rhamnoside, berberine, etc. [16, 26, 65, 154]. Combating different mechanisms of resistance is a potential remedy for treating the microbial infections caused by multidrug-resistant pathogens. These compounds are present in some of our common foods.

While using drugs like Penicillin G, the resistant bacteria produces enzymes like β-lactamases, which destroy the drug in the drug inactivation or modification mechanism. Another mechanism is an alteration of target site by changes in structural or amino acid composition. In Penicillin Binding proteins (PBP), which are the target site of penicillin, alteration of amino acid composition makes the antibiotic (penicillin) ineffective. This leads to antibiotic resistance as seen in Multi-Drug Resistant *Staphylococcus aureus* (MRSA). In the mechanism related to the alteration of the metabolic pathway, Para amino benzoic acid (PABA) can be inactivated by sulphonamides and can prevent the bacterial growth. Some bacteria, which are sulphonamide resistant, do not require PABA for its growth. Instead, they directly utilize pre-formed folic acid in mammalian cells for its growth.

Reduced drug accumulation is another important drug resistance mechanism. Increase in active efflux or reducing the drug intake across the cellular membrane causes the drug accumulation [119]. Plant-derived compounds are effective in combating drug resistance by inhibiting the above active site of resistant organisms.

Some of the examples for multidrug-resistant pathogens are: resistant Penicillin *Staphylococcus aureus*, Methicillin-resistant *Staphylococcus aureus*, Vancomycin-resistant *Enterococci*, multiresistant *Salmonella*, and multi-resistant *Mycobacterium tuberculosis*. Development of microbial resistance to common antimalarial drugs also draws major attention as it causes resistance to HIV drugs.

Resistant organisms like MRSA carry a methicillin-resistant gene called methicillin-resistant gene (mecA), coding for a specific methicillin-resistant penicillin-binding protein. Genes like lysylphosphatdylglycerol synthetase gene (mpr F); and methionine sulphoxide reductase regulator gene (msrR) are also involved in the resistant mechanism of MRSA, which are closely related with membrane synthesis in MRSA [72]. A drug, which can block this gene, can prevent its action, i.e., microbial resistance. The increased resistance due to biofilm formation is also reported in *Pseudomonas aeruginosa*. *E. coli* is a common bacterium, which can infect through contaminated food.

2.3.1.2   *DRUG RESISTANCE IN MICROBES AND USE OF PHYTOCHEMICALS*

The major mechanisms by which microorganisms exhibit resistance are due to interference in the synthesis of the cell wall, protein, nucleic acid, and cell membrane and also in a metabolic pathway. These are either acquired or caused by mutations. Various types of ATP driven efflux pumps are reported in different organisms to explain the drug resistance [7].

Plant-derived compound berberine is an example for the use of plant products in combating drug resistance by inhibiting bacterial efflux pump (Figure 2.2). This can be achieved either by an increase in active pumping out or by a decrease in drug uptake across the cell membrane [120]. Some of the identified efflux pumps for antimicrobial resistance in the different microbial system are:

- ATP binding superfamily (ABC);
- Major facilitator superfamily (MFS);
- Multi antimicrobial extrusion family (MATE);
- Resistance nodulation cell division superfamily (RND); and
- Small multidrug resistance family (SMR).

Among these, ABC superfamily has only primary transporters. The second is secondary transporters using proton and sodium as an energy source, and MFS dominates in Gram-positive bacteria.

2.3.2 SYNERGISTIC EFFECT

Compounds having an antimicrobial effect can improve the antimicrobial activity of either antibiotics or extract when used alone. This is called synergistic effect. It will reduce the quantity of antibiotic to be consumed for treating the disease. Synergistic effect can be utilized in combating drug resistance development in microbes. In this case, plant extracts are used along with antibiotics [3]. Some of the combating strategies to fight against dreadful multidrug-resistant pathogens are use of multidrug potentiates or new plant products or combined use of plant products along with common drugs etc., which would provide synergistic effects. Berberine is a plant-derived compound, which has synergistic inhibitory effect with streptomycin on the growth of microorganisms like *Staphylococcus aureus* [188]. It has been experimentally proven that green tea polyphenols can effectively prevent the biofilm formation in certain organisms.

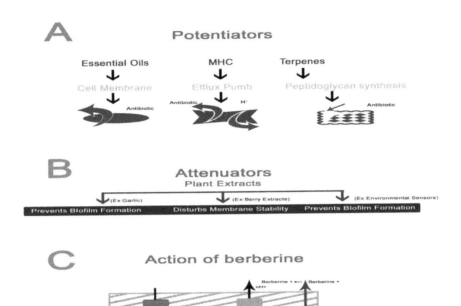

FIGURE 2.2   (See color insert.) Action of plant extracts on efflux pump of microbial cell wall synthesis pathway: (a) Potentiators; (b) Attenuators; and (c) Berberine; Adapted and modified from Ref. [188].

2.3.3   ANTIOXIDANT ACTIVITY OF DIFFERENT COMPOUNDS AND MECHANISMS OF ACTION

2.3.3.1   ANTIOXIDANT ACTIVITY

Antioxidants biocompounds can protect the cell from cellular oxidation and its damage by scavenging ROS or preventing the formation of ROS. Beneficial effect of antioxidants on promoting health is through several probable mechanisms, such as: directly neutralizing or scavenging free radicals, reducing peroxides, chelating with transition metals, and triggering the antioxidative defense enzymes [46]. It has also been reported that antioxidant activity of phytochemicals is mainly due to the component called phenolic compounds [82]. Many polyphenolic compounds are good antioxidant compounds against ROS [1, 45, 158]. Several workers have reported different antioxidant biocompounds from plant sources [18, 151, 154], such as: carnosol, quercetin, thymol, catechins, eugenol, carnosic acid, ellagic acid, hydroxytyrosol, rutin, gallic acid and its derivatives, tannins, morin, and rosemarinic acids. These have drawn attention because of their wide use in food preservation, dietary supplementation and in the treatment of different free radical-based diseases [70, 77].

Antioxidants can counter-balance the production of ROS. Enzymes such as superoxide (O_2^-) dismutase, glutathione peroxidase, and catalase are involved in endogenous cellular defense process of antioxidants. O_2^- dismutases catalyze the reduction of O_2^- anion to hydrogen peroxide (H_2O_2) and water. Glutathione peroxidases can scavenge most of H_2O_2, while catalase, located in peroxisomes, helps in elimination of high levels of H_2O_2 [47]. It is reported that ROS can generate single- or double-stranded DNA. These DNA damages can affect in stopping or induction of transcription, signal transduction pathways, genomic instability, and replication errors. Excess of free radicals of nitrogen and oxygen species (such as: hydroxyl radical (OH^-), H_2O_2, peroxynitrite (NO_2^-), and O_2^-) can cause cellular oxidative stress [124]. Such free radicals are the causative agents of cell proliferation, apoptosis inhibition angiogenesis, etc. in cancer cells. Many stages in cancer development (like cell transformation, proliferation, apoptosis resistance, metastasis, and angiogenesis) can be due to ROS. These ROS are also reported to make DNA aberrations, genetic changes, mutations, epigenetic changes, etc.

Few of the important signaling pathways of membrane-associated receptor such as tyrosine kinases are: cell-signaling kinases like membrane-associated protein kinases (MAPK), phosphatidylinositol 3-kinase (PI3K)/ Akt, and necrosis factor, NF-B (transcription factor). Some dietary antioxidants are useful in specifically targeting kinases and transcription factors in certain cellular pathways. Chemopreventive effects of dietary agents are mainly due to the signal transduction modulation in cellular pathways, which prevents carcinogenesis [8, 110]. It also inhibits DNA alteration or can also repair broken DNA. Some of the other effects are inhibition of antiapoptotic proteins, and induction of proapoptotic proteins. On exposure to any carcinogen, the quantitative equilibrium between detoxifying enzymes and carcinogen-activating enzymes determines the presence of carcinogenic moiety in the cell [110]. Table 2.4 indicates antioxidant activity of selected secondary metabolites from plants.

TABLE 2.4 Biological Activities of Different Plant Derived Compounds and Extracts

Plant species	Class of the bizoactive compound	Identified compound, if any	Biological activity/Action mechanism	Reference
Phyllanthus	Triterpenoids	Lupane	Cytotoxic	[39]
amarans			Amylase inhibitory	[12, 53]
Guoia villosa	Sesquiterpenes		Tyrosinase inhibitory	[54, 125, 208]
Spirostachys africana	Terpenoids		Cytotoxic, Antibacterial	[55, 134]
Curtisia dentata	Triterpenoid		Antibacterial, Antifungal	[180]
		Lupeol	Antiproliferative Anti-inflammatory. Antiprotozoal Antiproliferative, antiangiogenic, NF-κB activation, up-regulation of MAPK/P38 pathway, apoptosis. Akt/P13K pathway, etc.	[84, 177]
		Lupeol	Antitumor	[146]
		Pterostilbene	Antidiabetic	[87]
Premna tomentosa			Cytotoxic	[88]
	Triterpene	Lupeol	Anticancer, Anti-inflammatory	[177]
		Epicatechin	Antioxidant, DNA polymerase inhibitory	[11]
Amaranthus spinosus		-	Antioxidant, Antidiabetic	[84]
Vitis amurensis		Oligostilbene	Antioxidant, Lipid oxygenase inhibition	[63]
Citrullus colosyhes		Hesperidin	Antibacterial, Anticanadial	[64]
			Antioxidant, Antitumor. PPRγ activation, induce p53, activates NF-κB and apoptosis.	
Withania somnifera			Alzheimer's	[187]
Quassia amara, L.	Qouassioid		Antimalarial, Cytotoxic	[59]
Salvia plebea, R. Brown			Anti-inflammatory, Antiangiogenic, Antinociceptive	[89]

TABLE 2.4 *(Continued)*

Plant species	Class of the bizoactive compound	Identified compound, if any	Biological activity/Action mechanism	Reference
Boervahha diffusa. L.	—	—	Antiproliferative, Antiestrogenic	[186]
	Triterpenoid	—	Antimicrobial	[51]
Araucaria angustifolia	—	—	Antiviral	[65]
Clematis ganpiniana	Triterpenoid	—	Cytotoxic, Antibacterial	[25]
Juniperus communis L. (Cupressaceae)	Terpenoids	—	Antimycobacterial	[14]
Tamarindus Indica Linn.	Triterpene	Lupeol	Antidabetic, Antimicrobial	[72, 73]
Tamarindus indica Linn.	—	—	Antioxidant	[133]
Apples	—	—	Antioxidant	[30]
Apples	Triterpenes	—	Antioxidant	[75]
Pomegranate	Triterpene	Lupeol	Anti-inflammatory, Anticancer, Antimicrobial, Protease inhibitor, Topoisomerase inhibitor	[57, 204]
Etingera species (Zingiberaceae)	Terpenoids	—	Anticancer	[78, 177]
Andrographis paniculata	Diterpenoids	—	Platelet inhibition, Antimalarial, Anti-inflammatory	[38]
Bruguiera gymnorrhiza. J.	Triterpenes	—	Antioxidant, Antimicrobial	[96]
Rosmarinus officinalis	Diterpenoids	—	Antiplatelet, Antimicrobial	[34].
Centella asiatica.	—	—	Antithrombotic. Anticancer Antimicrobial. Modulatory	[9, 215]

TABLE 2.4 *(Continued)*

Plant species	Class of the bizoactive compound	Identified compound, if any	Biological activity/Action mechanism	Reference
Centaurea pullata:	Terpene	—	Antimicrobial	[52]
Shizegium aromaticum	—	Eugenol	Antifungal	[62]
	—	Curcumin	Antioxidant, Antimicrobial, Antimutagenic, Anti-inflammatory	[4, 26, 85, 126, 127]
	—	Lycopene	Anticancer, Anti-inflammatory	[107, 127, 165, 183]
Punica granatum (Pomegranate)	Anthocyanins		Anticancer, Antimicrobial, Antioxidant	[68, 117, 213]
Green tea	Poly phenols		Anticancer, Antioxidant, Antimicrobial, Antidiabetic	[1, 30, 36, 53, 90, 109, 156]
Quinones				
	Napthaquinone	Plumbagone, Juglone, Lawsone Shikonine	Antibacterial	[75, 99]
	Pterostilbene		Antioxidant, Antidiabetic	[6]
Osmitopsis asteris-coides (Asteraceae)	—		Antibacterial	[210]
	Anthraquinone	Emodin	Antibacterial, Cytotoxic	[22, 42]
Cassia fistula,	Anthraquinone		Antibacterial, Cytotoxic, Antidiabetic	[147]
Cinnamonum zeylanicum (cinnamon	—		Antidiabetic	[1]
Momordica charantia *Gymnema sylvestre*	—		Antidiabetic	[191]

TABLE 2.4 (Continued)

Plant species	Class of the bizoactive compound	Identified compound, if any	Biological activity/Action mechanism	Reference
Rumex crispus L. Rumex japoniccus	—	—	Antioxidant, Antibacterial	[58, 70]
Cetraria islandica (L) Ach.	—	—	Antioxidant	[74, 106]
Salvia tomentosa Miller (Lamiaceae)	—	—	Antibacterial, Antioxidant	[193]
Acanthus ilicifolius	—	—	Antioxidant, Hepatoprotective	[24]
Rhizophora mangle bark	—	—	Antioxidant	[178]
Ipomoea pescaprae	Quinic acid	—	Antimicrobial and anticancer Enzyme inhibitory	[194, 218]
Flavonoids				
Tagetes muta	Flavonoids	—	Antimicrobial, Anti-MRSA	[50, 195]
Hypericum species	—	—	Antimicrobial	[219]
Mosses	Flavonoids	—	Antimicrobial	[28]
Scutellaria barbata	Flavonoids	—	Antibacterial	[15]
Sophora exigua	Sophora flavones	—	Antibacterial	[195]
	Flavonoids	Quercetin	Enzyme inhibitor for topoisomerase IV	[212]
Bolusanthus spesiosus	Flavonoids	—	Antimicrobial	[69]
Cryptolepis sanguinolenta.	Flavonoid	—	Antioxidant, Antibacterial	[60]
Green tea	—	Quercetin	Antioxidant, Antimicrobial, Anticancer	[13]
	Stilbenes	Marchantins	Antiproliferative, Antifungal	[57]

TABLE 2.4 (Continued)

Plant species	Class of the bizoactive compound	Identified compound, if any	Biological activity/Action mechanism	Reference
Ximenia americana	—	—	Antimicrobial	[134]
Solidago chilensis Meyen	—	—	Antimicrobial Antimicrobial, Antiplatlet	[122, 132]
Solidago microglossa (Asteraceae) Allium cepa	Flavonoids	Quercetin	Anti-inflammatory, Antimicrobial	[163]
Ginseng	—	—	Nueroprotective, Anticancer Antimicrobial	[25, 144]
Thithonia diversifolia	—	Chlorogenic acid	Animicrobial (due to their polar propenoic side chain.)	[50]
Sophora flavescens	Flavonoid	Sophoroflavone G	Anti-inflammatory	[94]
Dorstenia barteri (Moraceae)	Flavonoids	—	Antimicrobial	[22]
Solidago herba.	Antioxidant	—	Anti oxidant	[206]
Solidago canadensis	Flavonoids	—	Glutathione *S. transferase* inhibitor	[17]
Galgnin	Flavonoids	—	Antimicrobial	[47]
Coffe	Flavonoids	—	Antibacterial, Antiplatelet	[13, 37]
Culcitium reflexum H. B	Flavonoid	Eupatorin	Antiproliferative	[15]
Culcitium reflexum H. B	Flavonoid	Apigenin	Anticancer	[164]
Culcitium reflexum	Flavonoid	Galangin	Anticancer, Antimicrobial, Anti-MRSA	[139]
Culcitium reflexum H. B	—	Quercetin	Antimicrobial, Anti-inflammatory	[32]

TABLE 2.4 (Continued)

Plant species	Class of the bizoactive compound	Identified compound, if any	Biological activity/Action mechanism	Reference
Hedera colchica. Pharmazie.	Flavonones	—	Anti MRSA	[1, 200]
Pithicellobium duke beneath., Erithinia latissima	—	Kaempferol	Antimicrobial, Antimicrobial, Anti-HIV	[130, 217]
—	Flavonoids	—	Antibacterial	[47]
Biotorentalis	—	Rutin	Antibacterial, Tyrosine inhibitor	[20]
	Flavonols		Antioxidant, Anti-inflammatory	[206]
Impatiens balsama L.	—	—	Antimicrobial	[214]
			Immunomodulatory, Anticancer	
Benincasa hispida,	—	—	Anti-inflammatory Ribosome inactivating, Anti ulcer	[114]
Glycyrrhiza uralensis Risch.	Flavonoid	Liquiritigenin	Antioxidant, Antibacterial, Xanthose oxidase inhibition, Anticancer	[114, 149]
Erythrina latissima	Flavonoid	—	Antioxidant, Antimicrobial, Anti HIV	[35]
Blumea balsamifera	Flavonoids	—	Xanthenes oxidase inhibitory	[151]
Pterocarpus indicus Aerva persica Sinofranchetia chinensis	—	Pterocarpol, Liquiritigenin, Isoliquiterigin	Antifungal	[56, 114]
Hydnophytum formicarum Jack	Flavnoid	Butein, Liquiterigin	Antioxidant, Antimicrobial, XO inhibitory, Antiproliferative	[212]
Berry	Phenolics	—	Antioxidant, Antimicrobial	[83]

TABLE 2.4 (Continued)

Plant species	Class of the bizoactive compound	Identified compound, if any	Biological activity/Action mechanism	Reference
	Flavonoid	Apigenin	Anticancer	[106]
	Flavonoid	Kaempferol	Anticancer Anti tuberculosis	[156]
	Isoflavone	Genistein	Anticancer, Antimicrobial	[95, [157]
	—	Resveratrol	Antimicrobial, Anticancer, Immunomodulatory, Antioxidant	[167, 176, 177]
Coffee	—	Caffeic acid	Antiulcer, Antidiabetic, Antioxidant	[162]
Coffee	Phenolic acid	—	Antimicrobial	[122]
Cocoa	—	Caffedyme	COX inhibitory	[173]
Swietenia mahagoni.	—	—	Protease inhibitory, Antimicrobial	[156]
	Polyphenols	Chlorogenicacid	Antifungal	[50]
	—	—	Antioxidant	[90]
Swietenia mahagoni.	Phenolics	—	Antioxidant	[101]
Spices	Phenolics	—	Antioxidant	[54]
Trichilla emetica	Phenolics	—	Antioxidant	[68].
Termalia bellerica. Ginseng	—	—	Antifungal, Antimalarial	[68]
Cydonia ablunga Miller	Polyphenols	—	Antimicrobial	[65]
	—	Catechin, epicatechin	Antimutagenic	[67]
Cocos nucifera	—	5-O-caffeoylquinic acid (chlorogenic acid), dicaffeoylquinic acid and caffeoylshikimic acid.	Antioxidant, Antimicrobial	[192]

TABLE 2.4 *(Continued)*

Plant species	Class of the bizoactive compound	Identified compound, if any	Biological activity/Action mechanism	Reference
Cratoxylum formosum Dyer)	—	Chlorogenic acid	Antioxidant, Antifungal, Anticancer	[87]
Hieracium pilosella L.	Flavonoids, Phenolics	Chlorogenic acid, apigenin—7-*O*-glucoside and umbelliferone		[77, 177]
Chamaecyparis obtusa var. formosana	—	—	Antioxidant	[130]
Hieracium pilosella L.	—	—	Antioxidant, Antimicrobial	[178]
Tribulus terrestris	Saponins	—	Antimicrobial, Antifungal	[115]
Eugenia jambolana	Saponin	—	Antibacterial	[193]
Juglans regia L.	Phenolic acids, tannins	—	Antioxidant, Antiproliferative, Antimicrobial	[106, 120]
Etingera species (Zingiberaceae)	—	—	Antioxidant, Antibacterial	[37]
Penstemon gentianoides	—	—	Anti-inflammatory	[129]
Murraya exotica	—	—	Anti-inflammatory	[112]
Coccinia grandis	Phenolic compounds	—	Antimicrobial	[166]
Ixora coccinea	—	—	Antioxidant	[175]
Klainedoxa gabonensis	—	—	Antimicrobial, Antioxidant	[93]
Cumum cymum	—	—	Antifungal	[177]
Alpinia galanga (L.) Wild	—	—	Antibacterial	[162]

TABLE 2.4 *(Continued)*

Plant species	Class of the bizoactive compound	Identified compound, if any	Biological activity/Action mechanism	Reference
		Alkaloids		
Cryptolepis sanguinolenta.	Alkaloid	Cryptolepine	Antibacterial, DNA intercalating	[71]
Hydrastis canadensis (goldenseal), Berberis vulgaris (barberry), and	Alkaloid	Berberine	Antimicrobial	[71]
Coptis chinensis Coptis or golden thread	Alkaloid	Berberine	Antimicrobial	[104]
Berberis aquifolium (Oregon grape),	Alkaloid	Berberine	Cholesterol-lowering	[114]
Berberis aristata	Alkaloid	Berberine	Antidiabetic	[145]

2.3.4 OBESITY AND FOOD HABITS

The high-calorie fat content in the food and modern lifestyle is mainly responsible for obesity. Different components in our food (like phenolic, flavonoids, terpenoids, etc.) have anti-adipogenic effects. Compounds like quercetin, Gallic acid, etc. can inhibit PPRγ gene in adipogenesis. Similarly honey, turmeric, and tomato are also good source of anti-adipogenic compounds. Capsaicin can induce apoptosis in adipocytes in a dose-dependent manner [46]. Compounds derived from plants (such as: catechin, catechin gallate, caffeic acid, resveratrol, allicin, berberine, butein, apigenin, capsaicin, celastrol, epigallocatechin gallate, curcumin, genistein, quercetin, and taxol) can cause modulation in cellular pathways. This results in angiogenesis, invasion, metastasis, proliferation, and survival in tumor. Most of these products are antioxidant and anti-inflammatory also. Thus these agents can modulate various cell signaling pathways.

In adipogenesis, there is an increase of mRNA expression levels of adipogenic transcription factors CCAT/enhancer binding protein a (C/EBPa) and proliferation activation receptor c (PPARc). This regulates mRNA expression levels of adipocyte-specific genes. Resveratrol and quercetin inhibit the PPRγ and C/EBPa and also enhance apoptosis in 3T3-L1 cells [94]. The green tea polyphenol EGCG can also induce apoptosis in preadipocytes (Figure 2.3). The apoptotic effects are dependent on Cdk2- and caspase 3 and could be resulted in the inhibition of cell mitogenesis [97]. Both in inhibition of adipocyte differentiation and apoptosis of mature adipocytes by capsaicin, EGCG, and genistein, AMPK is a significant and specific factor [105].

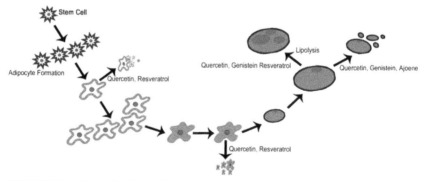

FIGURE 2.3 (See color insert.) Action of some plant-derived compounds in triggering/inhibiting adipogenesis pathways; modified and adapted from [105].

2.3.5   CYTOTOXICITY

A biological active compound should be less cytotoxic to human cells and organs. Then only it can be used as a drug to treat a particular disease. Many compounds have failed in the market due to its cytotoxicity and other side effects. World Health Organization (WHO) has described some guidelines to determine the cytotoxicity of a product.

2.3.6   ANTI-INFLAMMATORY EFFECT

Examples of bioactive anti-inflammatorybioactive components are: curcumin, kaempferol, resveratrol, quercetin, etc. Some of the different pathways, which trigger the inflammation, are cyclooxygenase (COX) and lipoxygenase (5-LOX) pathway. In COX, arachidonic acid is metabolized to prostaglandins and throbaxine. Lipoxygenase (5-LOX) is another pathway, in which leukotrienes (LTs) are produced from arachidonic acid. The metabolism of arachidonic acid has a major role in the series of events occurring in the mechanism of inflammation. Any compound, which can inhibit this LOX and COX, can reduce the formation of LTs and PGs and thus can prevent inflammation.

2.3.7   ANTICANCER ACTIVITY

Some examples of plant-derived anticancer agents are: phenols, plumbagin, vincristine, vinblastine, and coumarin. Coumarin can induce apoptosis in cancerous cells. Innate suicide mechanism in mammals is termed as apoptosis. It can also control cellular defense reactions, metamorphosis, morphosis, and tissue homeostasis. Apoptosis can be morphologically characterized by cell shrinkage, condensation of chromatin and cleavage of DNA to smaller 180bp fragments. This fragmentation can be visualized as DNA laddering [55]. Several plant-derived compounds (like resveratrol, lupeol, and genestine) can effectively induce apoptosis in myeloma cells by regulating the cell signaling pathways of cellular growth and proliferation [155]. Lupeol is an anticancer agent, and it affects by NF-κB activation, up-regulation of MAPK/P38 pathway, apoptosis, Akt/P13K pathway, etc. depending on type of cancer. Quercetin is another anticancer compound, which can upregulate Bax, resulting in apoptosis in cancer cells.

It can also suppress Bcl protein activity and stimulate DNA fragmentation. Resveratrol also can play an important role in apoptosis through activating caspase signal, down-regulation of anti-apoptotic proteins Bcl–2. It can upregulate tumor suppressor gene and p53.

Major biological activities of some phytochemicals are: antimicrobial, anticancer, antioxidant, anti-diabetic, anti-inflammatory, enzyme inhibitory, antihypertensive, etc. Some reports of important activities of biocompounds are summarized in Table 2.4.

2.3.8 BIOLOGICAL ACTIVITIES OF SOME PLANT DERIVED COMPOUNDS

2.3.8.1 ANTIMICROBIAL ACTIVITY OF GREEN TEA POLYPHENOLS

It is noticed that green tea has antimicrobial activity and it is a mixture of catechin compounds [44]. These compounds can inhibit the growth of different organisms like *Vibrio cholerae* O1 [33], *Streptococcus mutans*, *Shigella*. It has been experimentally proven that (−)-gallocatechin–3-gallate, (−)-catechin–3-gallate, (−)-epicatechin–3-gallate, (−)-epigallocatechin–3-gallate, theaflavin–3, 3′-digallate, theaflavin–3-gallate and theaflavin–3′-gallate are good antimicrobial agents at micromolar levels. These compounds can be effective than antibiotics like vancomycin, tetracycline at equivalent concentrations [92]. It was reported that flavonoids isolated from licorice showed growth inhibitory effect on methicillin-sensitive *Staphylococcus aureus*, methicillin-resistant *Staphylococcus aureus*, *Bacillus subtilis Escherichia coli*, *Klebsiella pneumonia* and *Micrococcus luteus*.

2.3.8.2 ANTIMICROBIAL ACTIVITY OF QUINONES

The antihemorrhagic activity of quinones is related to its ability of oxidation in body tissues. The antimicrobial activities of quinones by interaction with proteins and by enzyme inhibitions have also been reported. The antimicrobial and anti-depressant properties of *Hypericum perforatum* contain anthraquinone and Hyperisine. It is reported that an anthraquinone from *Cassiaitalica*, is bacteriostatic effect against *Bacillusanthracis*,

Corynebacterium pseudodiphthericum, and *Pseudomonasaeruginosa* and bactericidal against *Pseudomonas pseudomallei*. The antimicrobial activity of naphthoquinones from Brazilian plant, *Punica granatum*, has been reported. The antimicrobial activity was reported for salvipisone, the compound with p-naphthoquinone [128]. Antibacterial activity of an anthraquinone namely emodine was reported [22]. In microbial cells, the target of quinones is components like membrane-bound enzymes, polypeptides on the cell walls, adhesions exposed on the cell surface. As reported, quinones can form stable complex compounds with nucleophilic amino acids in proteins, resulting in inactivation and loss of the function of protein [44].

2.3.9 ANTICANCER EFFECT OF SOME PLANT DERIVED COMPOUNDS

2.3.9.1 GENISTEIN

Genistein (5, 7, 4-trihydroxyisoflavone) belongs to isoflavone class of phytochemicals and is present in soybean. This is a dietary component with protective and therapeutic action against cardiovascular diseases, cancers, and osteoporosis in humans. Genistein can inhibit the action of enzymes like topoisomerase II, tyrosine protein kinase and ribosomal S6 kinase in animal cell cultures [121, 210].

2.3.9.2 RESVERATROL

Resveratrol (3, 4, 5-trihydroxy-*trans*-stilbene) is a phytoalexin mainly found in grape skins, peanuts, and red wine. It has been reported to have a broad range of biological and therapeutic properties. Resveratrol can effect in three major stages of carcinogenesis like initiation, promotion, and progression [4, 157].

2.3.9.3 QUERCETIN AND KAEMPFEROL

Quercetin and kaempferol have anticancer and anti-inflammatory properties. The anti-inflammatory activity of quercetin is by inhibiting the production of pro-inflammatory cytokine called tumor necrosis factor-alpha (TNF-alpha) [66]. It has been proved that anti-inflammatory effect of quercetin and kaempferol is by the modulation of inducible COX–2 (Cyclooxygenase), nitric oxide synthase and CRP (C-reactive protein). Quercetin and kaempferol act by mechanism, which involves the upregulation of proinflammatory genes by the blockade of NF-kappa B activation. The anti-inflammatory activity of quercetin is due to the inhibition of enzymes like lipoxygenase, and by the inhibition of other inflammatory mediators. Quercetin and isoquercetin can suppress eosinophilic inflammation and therefore can be used for treating allergies [172]. Quercetin and kaempferol at 20–50uM reduced NO production, while PGE_2 (Prostaglandin E2) secretion was suppressed by kaempferol [83].

2.3.9.4 LUPEOL

Lup–20(29)-en–3h-ol (Lupeol) comes in the triterpene class of phyto-chemicals, and is generally present in fruits like: figs, grapes, mango, olive, and strawberry. It is also found in some vegetables and medicinal plants. It is a powerful anti-inflammatory, anticancer, antiarthritic, and antimalarial compound with activity in both *in vitro* and *in vivo* systems. Lupeol is a potent inhibitor of protein kinases, serine proteases and DNA topoisomerase II, the major targets in anticancer treatment [67]. It has also been reported to improve the epidermal tissue reconstitution, promote differentiation and stop the cell growth of melanoma cells [80]. It can induce cell cycle arrest at G1/S resulting in cyclin D. CDK2 activation in melanoma cells. Lupeol possesses antitumor enhancing effects on tumori-genesis model in mouse skin [176, 177].

Lupeolcan inhibit cell viability at an optimum level causing apoptotic death of prostate cancer cells. Lupeol can induce the cleavage of PARP (Poly ADP-ribose polymerase, a DNA repairing protein) and degradation of acnes protein with an increase in the expression of FAS-associated protein with domain (FADD) and Fatty acid synthase gene (FAS) receptors [176]. FAS-mediated apoptotic pathway can be specifically activated by lupeol in androgen-sensitive prostate cancer cells. Lupeol showed synergistic effect along with anti-FAS monoclonal antibody used against cancer treatment.

2.4 BIOLOGICAL ASSAYS

2.4.1 EXTRACT PREPARATION

Methods given in this section have mostly been described by earlier with some modifications [193]. Geographical locations, morphology, age, time of collection, etc. of the collected plant material have to be recorded. Herbarium of the collected material may be prepared and deposited. The plant has to be then identified. Plant materials should be washed in tap water, rinsed several times in distilled water and dried. The 20 g of this material is taken and powdered in a warring blender. This powder is extracted using suitable solvents [46, 79, 148, 152].

For extraction, solvents (like methanol, ethanol, sterile water, hexane, chloroform, ethyl acetate, benzene, diethyl ether, and acetone) are to be tried initially, on trial and error basis. According to the yield, solvent can be selected for final use of extraction. Plant extracts can be prepared using different extraction methods depending on the nature of the plant material and yield.

2.4.1.1 COLD METHOD

In this method, take about 20 g powder of the plant material (weight varies depending on the plant) and add 100 ml of AR grade methanol to it. Mix well and keep on a shaker at 300 rpm for 15hatroom temperature. Filter the extract using Whatman grade 1 filter paper. The procedure should be repeated for three times or until clear supernatant solvent is formed. Collect the supernatant and evaporate to dryness. Weigh the residue and use for further analysis. Calculate the percentage yield. This method can be employed for different plant materials using different or appropriate solvents [160, 199].

2.4.1.2 BIOASSAY-GUIDED FRACTIONATION

Sequential extraction of plant material is to be done using different solvents. Solvents are used in the order of its increasing polarity. Approximately 20 g of the powder of the plant material is first extracted with least polar solvent (100 ml) and then keep the mixture on a shaker at 250 rpm for 16h. Allow to settle. Then collect the supernatant, filter, and evaporate to dryness. Weigh the residue and store for further use. Then mix the leftover residue with solvent according to the polarity and the extraction has to be repeated with different solvents. The different fractions are then evaporated and then dissolved in a suitable solvent like DMSO or water. Prepare a stock solution with required concentration like 1 mg/ml. The extract can be stored under refrigeration (–20°C) and can be used for further analysis. It is filtered and used for testing the biological activities [185]. Polarity chart for bioassay-guided fractionation is as follows:

n- Hexane → Chloroform → Benzene → Diethyl ether → Ethyl acetate → Acetone
Ethanol → methanol → Water →

2.4.1.3   SOXHLET METHOD

In the Soxhlet method, high-temperature extraction is employed. In this method, about 10gof the powdered plant material is taken and wrapped in a good quality filter paper, and put in a Soxhlet. The solvent (100 ml) is taken in the round bottom flask of the apparatus and keeps on a heating mantle to boil for 10–20 min and solvent is collected in the bottom flask till the solvent in the soxhlet become clear. Collect the extract and evaporate to dryness. Weigh the residue and store. Mix the residue, which remained after removing the supernatant initially, with solvent according to the polarity. Repeat the extraction with next solvent. Collect different fractions by evaporation and dissolve in suitable solvents such as DMSO or water. Usually, a stock of 1% concentration can be prepared. The extracts should be kept under refrigeration (–20°C) till its use and analysis. Filter sterilization may also be done before the use.

2.5   BIOCHEMICAL TESTS FOR DETECTION OF VARIOUS SECONDARY METABOLITES

2.5.1   PHENOLIC COMPOUNDS

Phenolic compounds present in an extract can be detected using Folin's–Ciocalteu (FC) reagent. Take 1ml of the extract (1 mg/ml) in a test tube and add 5ml of FC reagent to it. Add 4 ml of 1.5% sodium carbonate, mix well and then keep for 30minat about 25°C for incubation. Development of blue color is confirmation of the phenolic compounds. Intensity of color is proportional to the quantity of phenolic compounds present in it [79]. The color is measured as optical density (OD) using a colorimeter, and the same will be graded as:

- + for low intensities,
- ++ for medium intensities, and
- +++ for high intensities.

2.5.2 FLAVONOIDS

2.5.2.1 FERRIC CHLORIDE METHOD

Ferric chloride method can be used to determine the presence of flavonoid compounds in an extract. In this method, take 1 ml of the extract in a test tube and add few drops of ferric chloride (0.1% made distilled water). Mix and keep at room temperature for10minutes. The presence of flavonoid compound can be noted by the development of green color. Note the OD using a colorimeter [79, 194].

2.5.2.2 DETECTION OF FLAVONOID COMPOUNDS USING ALKALINE TEST

Take1ml of the extract in a test tube and add 0.2 ml of 0.1 N NaOH. Presence of flavonoid compound is indicated by the increase in yellow color intensity. Add few drops of dilute HCl. The solution will become colorless confirming the presence of flavonoids [79, 195].

2.5.3 ALKALOIDS

Add 1 ml of 2 mM HCl to5 ml of extract (0.2%) in a test tube and then add 1 ml of Nessler's reagent. Mix well and wait for few minutes. Formation of orange precipitate is an indication of alkaloids present [196].

2.5.4   TERPENOIDS

Add few drops of concentrated H_2SO_4 to 1 ml of extract (0.2%) in test tube. Keep at 28°C for 30 min. If terpenoids are present, lower layer of the above mixture will turn to yellow [149].

2.5.5   SAPONINS

Few drops of Na_2CO_3 are added to 5ml of extract in a test tube and mix well. Presence of saponins will be indicated by the froth formation [203].

2.5.6   GLYCOSIDES

Two ml of extract is taken in a clean, dry test tube and add few drops of 0.1% of $FeCl_3$. Then add few drops of conc. H_2SO_4 to it. Keep it for few minutes. If glycosides are present, a reddish lower layer and bluish green upper layer will be formed [155]. Record the intensity of the color as: + for low level, ++ medium level, +++ high level, and (–) for complete absence.

2.5.7   BIOCHEMICAL ESTIMATION OF PHENOLIC CONTENT

Folin-Ciocalteu (FC) method can be used to estimate the phenolic content in an extract [184]. **Chemicals** used are: FC reagent, Gallic acid (SRL), Sodium carbonate (Merck); **Stock solutions** are: 1 mg/ml of Gallic acid stock in methanol, FC reagent with 1:2 dilution, and 1.5% sodium carbonate in water.

Take 1 ml of extract (1 mg/ml) in a test tube. Serial dilutions of 0.0 ml, 0.1 ml, 0.2 ml, 0.4 ml, 0.6 ml, 0.8 ml and 1.0 ml of the Gallic acid (standard 1 mg/ml) are prepared in series of test tubes. Makeup to 1 ml with double distilled water. Add 5 ml of FC reagent (1:2 diluted with distilled water) and 4 ml of sodium carbonate solution (1.5% in distilled water) to each test tube. Test tubes are then kept for incubation for 30min at 28°C. Measure the OD at 760nm with a UV-VIS spectrophotometer. Plot a graph with Gallic acid concentration on X-axis and OD on Y-axis. The quantity of phenolic compound present in extract can be calculated from the graph and expressed as Gallic acid equivalents.

$$C = (c \times V)/m \tag{1}$$

where: C = Total phenolic (mg/ml), c = concentration of Gallic acid (mg/ml), V = volume of the extract in ml, and m = weight of the extract

This estimation is to be done in triplicate for any extract using Gallic acid as standard. Average ± SD value of total phenolic is then determined.

2.5.8   BIOCHEMICAL ESTIMATION OF TOTAL FLAVONOIDS

Total flavonoid content in a sample can be estimated using the method as described earlier with some modifications [108, 184]. Stock solutions are: Quercetin standard–0.1 mg/ml prepared in methanol, 0.15% of $NaNO_2$, and 1.5% NaOH in water.

Prepare serial dilutions of standard by taking 0.0 ml, 0.2 ml, 0.4 ml, 0.6 ml, 0.8 ml, 1.0 ml of the quercetin standard and 0.5 ml of the extract in 7 separate test tubes. Make up the total volume to 2.0 ml with distilled water in each test tube and add 2.0 ml of 0.15% $NaNO_2$ in each test tube. Keep the test tubes at 30°C for 5min. Add 2 ml of 5% NaOH to each test tube, mix well and keep for 15min at room temperature. Take OD at 510 nm. Use distilled water as blank. The experiment is conducted in triplicate. Calculate Mean ± SD value. Plot a graph between quercetin concentration on X-axis and OD on Y-axis. A concentration of the total flavonoids can be calculated from the graph and expressed as quercetin equivalents.

2.5.9   CHROMATOGRAPHY

Particles or compounds in a mixture can be separated or isolated by chromatographic methods. Generally, in chromatography, particles are passed through a stationary phase along with a mobile phase. The holding or bonding of different particles with stationary phase varies depending on the nature of the particle size, affinity, chemical bonding, pH, etc., and will be separated from each other. Paper chromatography, high-pressure liquid chromatography, gas chromatography, thin layer chromatography, etc. work on this principle and are widely used for separation of particles. The components in a mixture can be separated on the basis of their partition coefficient in two immiscible phases.

2.5.9.1   PAPER CHROMATOGRAPHY

Separation is based on the liquid-liquid partition of amino acids between the water molecules that are adsorbed on the cellulose layer of the paper and the mobile organic phase.

2.5.9.2   THIN LAYER CHROMATOGRAPHY

Prepare silica gel plates by mixing silica gel with distilled water in the ratio 1:2. Make a uniform slurry and fill in the filling chamber. Place it at one end of the plates and drag it over the plates carefully to get a uniform thickness. Dry it. Pre-heat the plate at 100°C to get activated. Load the samples. Dry it and keep it in the equilibrated TLC tank. Allow to run till the solvent front reaches the top. Take out. Dry it. Spray the detecting agent. Dry and allow it to develop the color at 105°C for 20 min.

2.5.9.3   REVERSE PHASE HIGH-PRESSURE LIQUID CHROMATOGRAPHY: RP-HPLC SYSTEM

For example, HPLC unit includes Rheo dyne injector, dual pump, SPD photodiode array, and 6.12 SP5 integration software. The given chromatographic conditions are: (1) **Column:** Lichrosper RP 18 e 5µm, (2) **Detector used:** SPD PDA, (3) **Flow rate used:** 1ml/min, (4) **Injection volume:** 20 µl. The mobile system and wavelength can vary according to the nature of the sample. For example, the following mobile phases can vary according to sample:

- *Quercetin*–Acetonitrile: methanol (90:10);
- *Curcumin*–Methanol: water (75:25); and
- *Alstonia scholaris* (methanol extract)–Water: Acetonitrile (90:10).

The HPLC chromatogram or HPLC fingerprints are taken at different wavelengths and shown as overlaid chromatogram (Figure 2.4). PDA detector is used with this model. The peaks and area under each peak are to be analyzed from the recorded data [91, 120, 194].

HPTLC (High-performance thin layer chromatography), FPLC (Fast protein liquid chromatography) TOF (Time of flight), etc. are also different versions of chromatography. These work on the principle of chromatography.

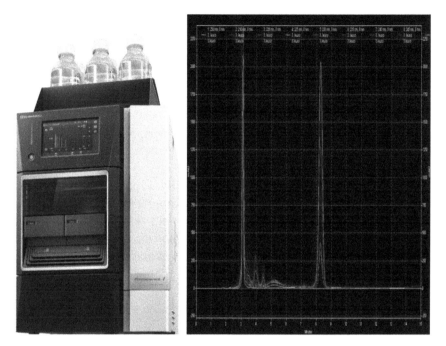

FIGURE 2.4 **(See color insert.)** HPLC instrument (Courtesy Shimadzu Co. Japan): Left: Overlaid chromatogram of plant extract *A. scholaris* (right).

2.5.10 HIGH-PERFORMANCE THIN LAYER CHROMATOGRAPHY (HPTLC)

High-performance thin layer chromatography (HPTLC) is an automated version of thin layer chromatography is widely used in the analytical method.

2.5.11 FOURIER TRANSFORM INFRA-RED SPECTROSCOPY (FTIR)

This is an automated version of infrared spectroscopy with higher resolution and sensitivity.

2.5.12 SCANNING ELECTRON MICROSCOPY (SEM)

Samples can also be identified by SEM. Here, electrons from an electron beam interact with atoms of the sample resulting in signals, depending on the composition of the sample. This produces the image. It has a resolution of 1 nm and a magnification of 100–500,000 times can be obtained. Conductive materials (such as osmium, rubidium, gold, silver, etc.) are commonly used as coating material in SEM.

2.5.13 LIQUID CHROMATOGRAPHY-MASS SPECTROSCOPY (LC-MS ANALYSIS)

LC-MS (Figure 2.5) can be used to identify a particular compound. The particles are separated by liquid chromatography. The particles are then subjected for ionization in an electric field. The particles are then separated by the e/m ratio, and this is used for the identification of the compound by suitable software. Generally, used mass ionizers are: single quadruple, triple quadruple, ion trap, time of flight (TOF) and Quadruple time of flight (Q-TOF).

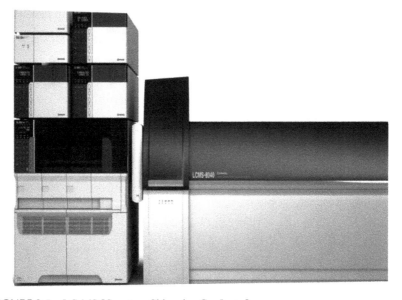

FIGURE 2.5   LC-MS [Courtesy Shimadzu Co. Japan].

2.5.13.1   LC-MS ANALYSIS

For Example, LC-MS system with following specifications can be used:

- Column: C–18.
- Probe: APCI (Atmospheric Pressure Chemical Ionization).
- Mode: Positive (which gives M+1 value); Negative (which shows M–1 value).

2.5.14   OTHER INSTRUMENTS IN PLANT EXTRACT RESEARCH

2.5.14.1   SPECTROPHOTOMETER

Spectrophotometer (Figure 2.6) is used for the measurement of OD. This is required for analysis, i.e., qualitative, and quantitative determination of food samples or extracts. Different models of spectrophotometer are available. Measurement is specific to a particular compound, and its absorption maximum or λ_{max} and its OD will be directed proportional to its concentration.

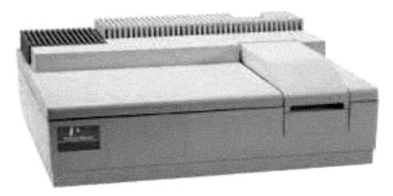

FIGURE 2.6   **(See color insert.)** UV-VISIBLE spectrophotometer [Courtesy Perkin-Elmmer].

2.5.14.2   LYOPHILIZER

Lyophilizer (Figure 2.7) is used for the preparation of dried/powdered samples using the principle of freeze-drying. In freeze -drying, samples are frozen at –80°C and kept in a lyophilizer. The low pressure created in the surrounding will allow the solid water to sublimate directly to gaseous phase at triple point. It works at very low temperature of –55°C to –110°C.

FIGURE 2.7 **(See color insert.)** Freeze dryer [http://www.chem.iastate.edu/faculty/ Wenyu_Huang/labtour; http://www.labogene.com/ScanVac].

2.5.14.3 CENTRIFUGE

Different types of centrifuges are used in food sample analysis. The speed varies from 1000 to 100,000 rpm.

2.5.14.4 CO_2 INCUBATOR

This is used for the incubation of cell lines in biological assays (Figure 2.8).

2.5.15 FLOW CYTOMETRY

It is employed for quantitative measurement of specific cell components. The separation of particle in flow cytometry is based on the particle size, scattering, dye used, etc. Fluorescent labeled antibodies, protein (gene expression), fluorogenic probes, etc. are used in flow cytometer for cell signaling, cell cycle analysis, cell identification based on gene expression, calcium efflux, enzyme activity, stem cell identification, etc. Particles are made into droplets and get charged by passing through an electric field. These are then scattered according to charge and recorded (Figure 2.9).

FIGURE 2.8 **(See color insert.)** CO_2 incubator [Courtesy Eppendorf India Ltd.].

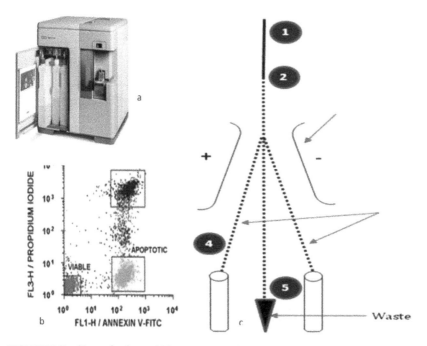

FIGURE 2.9 **(See color insert.)** Flow cytometer [Courtesy Bio-RAD]; **Legend: (a)** Flow cytometer; (b) Flow chart of flow cytometer steps: (1) Particle pass through laser and produce images, (2) Particles are partitioned to droplets, (3) Pass through an electric field, (4) Charged particles are pulled through each side. (5) Uncharged particles are passed through as waste); and (c) Report of flow cytometer analysis of viable, apoptotic, and necrotic cell separation.

2.5.16   REAL-TIME PCR

RT-PCR can be used for identification of samples and for specific gene expression and cell signal studies. It is based on binding of fluorescent dye on DNA and its measurement. FAM. SYBER green, VIC, HEX, ROX. TEXAS red. CY5, QUASAR, etc., are some of the calibrated fluorophores generally used in Real Time PCR. The intensity, at which it is above baseline, is called threshold and the respective cycle is called Cycle of Quantification, Cq (Figures 2.10a and 2.10b).

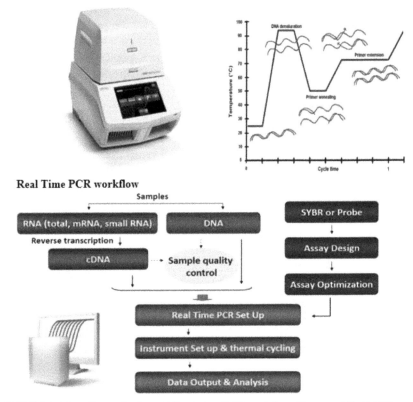

Real Time PCR workflow

FIGURE 2.10a   (See color insert.) Real-time PCR workflow [Courtesy Bio-RAD].

In Figure 2.10b, the quantification of specific gene expression in treated and untreated samples (Example IL beta Normal, treated for 1 h and 2 h); Ct values of GAPDH target diluted 10-fold confirm the ability to distinguish high and low copy targets when multiplexing.

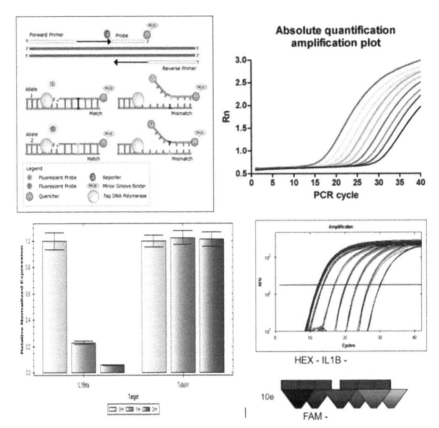

FIGURE 2.10b   **(See color insert.)** Real-time PCR [Courtesy Bio-RAD].

2.5.17   QUANTIFICATION OF DIFFERENT BIOACTIVE COMPOUNDS PRESENT IN THE EXTRACTS

Quantification of different compounds present in an extract can be done using pure compounds as standards. Specific or same HPLC conditions such as wavelength, solvent, flow rate, etc. are to be given for both sample; and standard. HPLC profiles thus obtained are to be compared for analysis and identification of compounds. Nature of column, wavelength, solvent system, flow rate, etc. will be changed depending on the compound [91, 194]. The method is illustrated below using some plant extracts and phytochemical standards.

The methanol extract of *Embelia ribes* is subjected for the quantitative analysis of quercetin and kaempferol at a wavelength of 254 nm. Pure quercetin and kaempferol are used as standards for comparison.

***Embelia ribes*:** RP-HPLC of the methanol extract of *Embelia ribes* (Figures 2.11–2.13; and Table 2.5) was done at 254 nm using quercetin and kaempferol as standard (100 µg/ml). The retention time and area are given by each compound in the HPLC chromatogram were noted, and the quantity of the quercetin and kaempferol in the sample was calculated using the formula:

$$\text{Concentration of the compound in the extract} = \frac{(\text{Area of sample/Area of standard})}{\text{Concentration of the standard}} \times \quad (2)$$

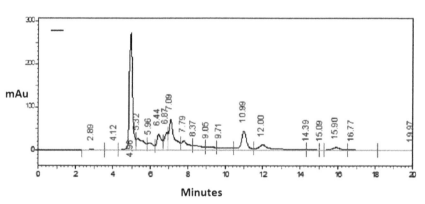

FIGURE 2.11   HPLC chromatogram of *Embelia ribes*. On X-axis time is in minutes, on Y-axis absorbance is in mAu.

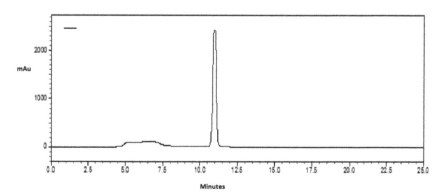

FIGURE 2.12   HPLC chromatogram of kaempferol.

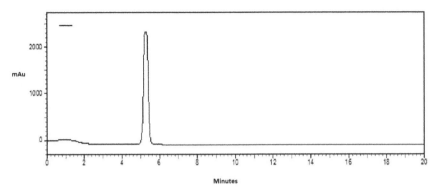

FIGURE 2.13 HPLC chromatogram of quercetin.

TABLE 2.5 Data Analysis for HPLC Chromatogram (Figure 2.11) of *Embelia Ribes*

			1: 254 nm, 8 nm				
Pk #	Retention Time	Area	Area %	Height	Height%	Start Time	Stop Time
1	2.89	42,157	0.39	1,982	0.34	2.34	3.57
2	4.12	24,448	0.22	1,074	0.18	3.57	4.31
3	**4.96**	**3,247,303**	29.85	273,330	46.40	4.31	5.24
4	5.32	741,325	6.81	26,664	4.53	5.24	5.81
5	5.96	373,551	3.43	16,857	2.86	5.81	6.25
6	6.44	644,730	5.93	37,547	6.37	6.25	6.68
7	6.87	530,798	4.88	40,793	6.92	6.68	6.94
8	7.09	1,450,464	13.33	72,386	12.29	6.94	7.64
9	7.79	572,218	5.26	21,681	3.68	7.64	8.26
10	8.37	395,357	3.63	11,801	2.00	8.26	8.94
11	9.05	263,541	2.42	7,823	1.33	8.94	9.56
12	9.71	302,836	2.78	6,487	1.10	9.56	10.44
13	**10.99**	**975,535**	8.97	45,235	7.68	10.44	11.52
14	12.00	748,607	6.88	12,773	2.17	11.52	14.33
15	14.39	65,409	0.60	1,803	0.31	14.33	15.02
16	15.09	19,766	0.18	1,327	0.23	15.02	15.27
17	15.90	254,514	2.34	6,984	1.19	15.27	16.49
18	16.77	95,098	0.87	1,715	0.29	16.49	18.12
19	19.97	129,622	1.19	812	0.14	18.12	22.72
Totals		10,878,989	100.00	589,127	100.00		

HPLC chromatogram of *Embelia ribes* shows two major peaks with retention time of 4.96 min and area of 3247303 mAu (milli Arbitrary units).

2.5.17.1   RP-HPLC PROFILING AND QUANTIFICATION OF PLANT EXTRACTS

2.5.17.2   QUANTIFICATION OF QUERCETIN AND KAEMPFEROL IN EMBELIA RIBES

- Concentration of Sample: **25 mg/ml** (w.r.t. dry extract weight);
- Concentration of Std Quercetin: **100 ug/ml** (1 mg in 10ml);
- Concentration of Std Kaempferol: **100 ug/ml** (1 mg in 10ml);
- Retention time for Quercetin: **4.90;**
- Retention Time for Kaempferol: **10.90;**
- Percentage purity of injected Quercetin: **98%;**
- Percentage purity of injected Kaempferol: **99%;**
- Area given by Standard Quercetin in the standard profile: **54,625,463;**
- Area given by Standard Kaempferol in standard profile: **74,887,452;**
- Area given by Quercetin in sample profile: **3,247,303** mAu;
- Area given by Kaempferol in sample profile: **975,535** mAu.

Results:
The percentage of Kaempferol in the sample: 0.98%(w/w)
The percentage of Quercetin in the sample: 0. 025%(w/w)

2.5.18   GAS CHROMATOGRAPHY-MASS SPECTROSCOPY (GC-MS)

GC-MS is also used in analytical methods making use of the principals of gas chromatography and mass spectrum. In the gas chromatogram, there is a capillary column. The particles are separated according to their affinity to stationary phase. The molecules retained in the column can be eluted out at different retention times. The separated particles are then subjected for ionization and are detected according to their e/m ratio.

2.6 BIOLOGICAL ASSAYS FOR TESTING DIFFERENT BIOLOGICAL ACTIVITIES

2.6.1 ANTIMICROBIAL ACTIVITY

National Committee for Clinical Laboratory Standard (NCCLS), European Committee for Antimicrobial Susceptibility Testing (EUCAST), and British Society for Antimicrobial Chemotherapy (BSAC) are some of approved international agencies who have designed the protocols for testing antimicrobial susceptibility [61]. Disc diffusion method and broth dilution methods are internationally accepted methods and are explained as follows [29, 81, 143]:

Media: Nutrient agar is used for the microbial cultures in general. Muller and Hinton agar (Hi-media, Mumbai) can be used for MRSA.

Nutrient agar: Peptone–5 g, Yeast extract–2 g, Beef extract–1 g, Glucose–10 g, NaCl–5 g, Agar–2% (all chemicals–Merck) and 100ml of distilled water. Nutrient agar is prepared by mixing these components. Adjust pH to 7.2 and sterilize at 121°C, 15 lbs for 20min. Pour to sterile Petri dishes of 90 mm diameter size and allow to solidify. These plates can be used for disc diffusion test to check antimicrobial activity of the extracts.

Nutrient broth: Peptone–5 g, Glucose–10 g, NaCl–5 g, and Yeast extract–2 g; Nutrient broth is prepared by mixing the components and making the volume up to 100 ml with distilled water. Take 10 ml broth in a 50 or 100 ml conical flask with side tube and sterilize at 121°C, 15 lbs for 20 min. This broth can be used for antimicrobial assay.

Muller and Hinton agar: MH agar (Hi media)–Prepare a 2% solution.

Potato dextrose agar: Boil and smash 20 g of potato after peeling of the skin. Add 100 ml of distilled water and filter through a muslin cloth. Then measure the volume of the filtrate. Add 2% dextrose to it and makes up to 100 ml with distilled water. Add Agar (Merck) and adjust pH to 5.5. Autoclave the media at 121°C, 15 lbs for 20 min and pour 20 ml each to sterile plates, allow to solidify and keep in a laminar flow or store in a sterile place.

2.6.1.1 EXTRACT PREPARATION

Dissolve the extracts in sterile DMSO or water at concentration of 100 mg/ml. This is used as stock. Stock is prepared in sterile water, if extracts are water soluble. The extracts thus prepared are either added to sterile medium for MIC (Minimum Inhibitory Concentration) determination or applied to sterile Whatman no. 1 disc of 5mm diameter, to carry out Bauer-Kirby test.

2.6.1.2   MICROBIAL CULTURES

Microbial strains can be purchased from different microbial collections such as: IMTECH, Chandigarh, India (MTCC) or ATCC (American Type Culture Collections). The following standard microbial strains can be used for antimicrobial assay in general:

Bacterial strains: *Escherichia coli* MTCC 41, *Pseudomonas aeruginosa* MTCC424, *Proteus vulgaris* MTCC426 (equivalent to ATCC 6380), *Klebsiella pneumonia* MTCC 3384, *Staphylococcus aureus* MTCC87.

Clinical strains: *Proteus vulgaris, Staphylococcus aureus, Klebsiella, Enterobacter, Acenetobacter, Proteus mirabilis, Shigella,* and *Salmonella typii* [196].

Fungal strains: *Candida albicans* MTCC183 (equivalent to ATCC 2091) and *Aspergillus niger* MTCC 281.

MDR strains: MRSA

2.6.1.3   INOCULUM PREPARATION

Revive the required culture in 2 ml freshly prepared and sterilized nutrient broth. Sub-culture to fresh medium and incubate at 36°C. The growth curve is drawn by recording OD at 600nm against time. When culture is at log phase (i.e., OD 0.5×10^5 to 0.8 or 6×10^5 CFU/ml), it can use for inoculation in nutrient broth either in test tubes for MIC determination or in plates for doing spread plate method to determine the zone of inhibition. Determine the MIC or Diameter of Zone of Inhibition.

2.6.1.4   DETERMINATION OF MINIMUM INHIBITORY CONCENTRATION (MIC)

Nutrient broth can be used to determine the MICs any extract or compound against a given microbial strain. Sterilize with 10 ml medium in a 50 ml conical flask. A test solution or any extract (stock diluted to 10 mg/ml in DMSO or in a suitable solvent and sterile) is then added to 10 ml of sterile medium kept in a series of conical flasks to get a serial dilution of 10000, 1000, 100, 10, 0 µg/ml. Inoculate each flask with a loop full of fresh culture at log phase (1×10^8 CFU/ml). Incubate at 36°C for 24 h. Monitor the OD of the culture at 600 nm with each dilution of the drug or extract. Blank is a tube with medium and extract. The lowest concentration, at which the growth of an organism is completely inhibited, is recorded as MIC of that extract by comparing with the OD of the blank. MIC can then be confirmed by inoculating to sterile nutrient agar plate. For obtaining most precise MIC value of an extract, the experiment has to be repeated making double dilutions, in between those adjacent dilutions, that showed positive and negative growth in the primary screening. The experiment is to be carried out in triplicate and determine average value of MIC. Analyze the result statistically using SPSS software. Repeat the experiment for each strain and extracts. The values can be compared between gram-positive and negative bacteria also [28, 137].

2.6.1.5   DETERMINATION OF ZONE OF INHIBITION

The diameter of zone of inhibition of each extract can be determined using sterile agar plates. Extracts at concentration of 10 mg/ml are prepared as stock and dilute to get a required concentration of 100 µg/ml, 10 µg/ml and 0.1 µg/ml. The 10 µl to 200 µl of the extract is applied on the sterile disc to get a serial concentration like 0.1 mg/ml, 1.0 mg/ml, 5.0 mg/ml, 10.0 mg/ml, 50.0 mg/ml per disc. Prepare similar discs for each extract. Dry and place it on the agar plates and incubate at 34°C overnight. The 3 or 4 sample discs can be tested on a single agar plate by placing the discs at equi-distance on the plate. Use DMSO as negative control, and gentamycin/ampicillin/penicillin (Hi-media) can be used as positive control. Measure the diameter of the inhibition zones for the extracts used and compare with that of the standard antibiotic disc used (Control). Table 2.6 shows the antimicrobial activity of extract of *Embelia ribes* against standard strains: Dilution test for MIC.

2.6.1.6   SYNERGISTIC EFFECT

The cumulative effect of each of the extracts along with antibiotic or different extracts is known as synergistic effect. It is illustrated with an example in the following subsections.

TABLE 2.6   Antimicrobial Activity of Extract *Embelia Ribes* Against Standard Strains: Dilution Test for MIC

Plant	Esche-richia coli (µg/ml)	Pseu-domonas aeruginosa (µg/ml)	K. pneu-monia (µg/ml)	Proteus mirabilis (µg/ml)	Staphylo-coccus aureus (µg/ml)	Candida albicans (µg/ml)	Asper-gillus niger (µg/ml)
Embelia ribes	250	250	280	312	187.50	300	500

2.6.1.6.1   Synergistic Effect of Methanol Extract of Alstonia Scholaris with Antibiotics

Synergistic effect of an extract with any other drug or compound can be tested using a microdilution method on a given microorganism. For example, methanol extract of *Alstonia scholaris* with penicillin on *Staphylococcus aureus* and the effect of this extract on *E. coli* growth are tested along with gentamicin.

In this method, antibiotic at standard concentrations (15 µg/ml for gentamicin and 20 µg for penicillin, respectively) is added along with 0, 10, 50, 100 and 200 µg/ml of the extracts to the medium taken in a series of conical flask. It is then inoculated with respective cultures. The growth is monitored at an interval of 3 h by checking the OD at 615 nm for 24 h. The rate and inhibition of growth are calculated. The concentration of extract and antibiotic combination, which the growth of a microorganism is inhibited completely, is noted. A flask having medium alone, without inoculums is taken as blank. The experiment is repeated thrice. Average ± SD is calculated.

The procedure can also be done on a 96-well plate. Add 200 µl of the sterile nutrient broth to each well. In the first series, i.e., A1, antibiotic solution is added in such way to get a concentration like 1000 µg/ml of antibiotic in well (200 µl of standard antibiotic solution with concentration 500 µg/ml). Add sample in the well B1 to have final extract concentration 500 µg/ml. In the well C1 both antibiotic and sample are added. Initial concentration of antibiotic and samples taken are, one corresponding MICs and one above the MIC values. Wells are then serially diluted, i.e., 100 µl solution from first well is transferred to second and the same volume from second to third and so on. Inoculate all wells with 5µl of culture with 0.1 OD to get a final concentration of 5×10^5 cfu/ml at log phase, mix well, cover, and incubate at 36°C for 18h. An example of FIC is shown in Table 2.7. Observe the growth and note MICs and Fractional inhibitory concentration (FIC) is then calculated [103], using formula [135]:

$$FIC = \left[\frac{MIC \text{ of drug in combination}}{MIC \text{ of drug alone}} \right] + \left[\frac{MIC \text{ of extract in combination}}{MIC \text{ of extract alone}} \right] \quad (3)$$

TABLE 2.7 FIC of Antibiotics Combination with Extracts Against *Staphylococcus Aureus* and *Escherichia Coli*

Micro-organism tested	Extract in combination	Antibiotic used	MIC antibiotic (µg/ml)	MIC antibiotic in combination (µg/ml)	MIC extract (µg/ml)	MIC extract in combination (µg/ml)	FIC
S. aureus	*A. scholaris*	Penicillin	20	4.50	150	35	0.46
E. coli	*E. ribes*	Gentamicin	15	4.00	250	50	0.49

2.6.2 STUDY OF MECHANISM OF ACTION OF THE EXTRACT ON MRSA AND E. COLI

2.6.2.1 PROTEIN PROFILING

The extract of known MIC value may be selected for studying action mechanism in *E. coli* and MRSA.

Method: Nutrient broth is prepared and sterilized. Extract at different sub-inhibitory concentrations (0, 10, 20, 50 and 100 µg/ml) is added serially to a series of test flasks containing 20 ml of sterile medium. All flasks are then inoculated with a loop full of culture at log phase, i.e., at an OD of 0.8 at 600nm, and are incubated at 36°C in an incubator. Collect the culture at log phase and centrifuge at 5000 rpm for 20 min. The pellet of the bacteria such collected is washed with sterile tris-buffer at pH 6.8. This can be lyophilized and used for protein isolation. The pellet is re-suspended in 5ml of lysis buffer (Tris pH 6.8, SDS 1%) and mix well. It is then centrifuged with speed 10,000 rpm for 20 min and re-suspended in tris buffer till further use. The 50 µl of the sample is mixed with loading buffer (Tris, sucrose, β-mercaptoethanol, SDS, bromophenol blue) and loaded on SDS-PAGE with untreated culture as control.

Protocol for SDS PAGE: The electrophoresis apparatus is set. Pre-heated 1% agarose in TE buffer is poured onto glass after proper cleaning. All chemicals can be purchased from Merck.

Preparation of the gel (Laemmli): The polyacrylamide gel is prepared using monomeric acrylamide and N,N'-methylene bisacrylamide. The TEMED (Tetramethylethyldiame) and ammonium persulphate are used as catalyst and initiator, respectively for the polymerization. At pH 7–10, the persulphate has maximum activity. SDS can denature the protein by breaking the disulfide bonds and impart negative charge to the proteins. Protein will move and can be separated according to their size in an electric field [180].

Ingredients are mixed as in the order given above and poured into glass plate assembly without forming any air bubbles. The gel is carefully overlaid with isopropanol to ensure a flat surface and to exclude air.

The ingredients are mixed as before and poured onto top of resolving gel. Insert a comb and allowed to settle. The comb is removed after solidifying the gel. The gel assembly is then placed in the buffer tank, and the tank is filled with the electrophoresis buffer. The samples are loaded in the well on the top of the gel. It is run at 70V, 100A till tracking dye has reached the bottom of the gel. Resolving gels are shown in Table 2.8 at gel concentration of 12.4% in 0.25M Tris-HCl, pH 8.8. Stacking gels are shown in Table 2.9 at gel concentration of 4.5% in 0.125M Tris-HCl, pH 6.8.

TABLE 2.8 Resolving Gels: Gel Concentration–12.4% in 0.20 M Tris-HCl, pH 8.8

Reagent	Volume, ml (to make 30 ml)	Volume, ml (to make10 ml)
40% Acryl amide stock*	9.40	3.10
Water (distilled)	12.30	3.80
10% SDS	0.30	0.10
Ammonium per sulphate, 1% (polymerizing agent)	0.50	0.50
1 M Tris-HCl pH 8.8	7.50	2.50
TEMED (added last)	20 µl	20 µl

* = 19:1 to 38:1, w:w ratio of acryl amide to N, N'-methylene bis–acryl amide.

TABLE 2.9   Stacking Gels: Gel Concentration–4.5% in 0.125 M Tris-HCl, pH 6.8

Reagent	Volume, ml (to make 15 ml)	Volume, ml (to make 10ml)
40% Acrylamide Stock*	1.70	1.10
Water	1.90	1.20
1M Tris-HCl pH 6.8	1.90	1.25
10% SDS	0.15	0.10
Ammonium per sulfate 1%	0.50	0.50
TEMED (stir quickly)	20 µl	20 µl

Buffer for Electrophoresis: Prepare buffer with composition of 196 mM glycine, 0.1% SDS, 50 mM Tris-HCl with pH 8.3is made by diluting a 10 x stock solution. The tank is filled with this buffer.

Staining of Gels: Materials required: Coomassie Brilliant Blue (CBB) or Page-Blue 83; –0.2% CBB is added to a mixture of water: methanol: acetic acid in the ratio 45:45:10. Gel is immersed in the staining solution taken in a plastic box or glass tray. Box is sealed and left for 20h on a shaker at 28°C or at 37°C for3h with agitation. The gel is de-stained with a mixture of water: methanol: acetic acid in the ratio 65:25:10 with agitation. The gel is observed using gel doc (alpha imager) with protein filters.

Silver staining of the gel: Ionic silver is converted to metallic silver to provide metallic silver images and can be easily observed. The following method is used for silver staining:

a. Materials used:
- Protein gel, Silver nitrate.
- **Solution 1**: 0.10gof silver nitrate in 5 ml distilled water.
- **Solution 2**: Mixture of 20 ml of NaOH (0.36%) and 5ml of Ammonium hydroxide (0.1%).
- **Solution 3**: Solutions 1 and 2 were mixed well and made up to 100 ml with distilled water.
- **Developer solution** is prepared by dissolving 3g of sodium carbonate in 80ml of water. Then 1 ml of 0.1% sodium thiosulfate and 1ml of 37% formaldehyde are also added and volume is made up to up to 100 ml using distilled water.
- **Stopper**: 5% acetic acid.

b. Procedure: Protein gel is washed with distilled water with gentle shake and is soaked in silver nitrate solution (solution 3), for 10 min along with shaking. Then the gel is washed with water and placed in developing tray and developer is poured into it and the same is shaken gently for some time. Dark brown bands are appeared. The reaction is stopped by adding 5% acetic acid, and the gel is observed using gel doc.

2.6.3   REVERSE TRANSCRIPTASE–POLYMERIZED CHAIN REACTION (RT-PCR)

RT PCR can be utilized for studying protein expression in MRSA and also for evaluating the effect of different compounds or extract on cell signaling related protein expressions, related to drug resistance. This is explained here using *A. scholaris* extract on MRSA. It is carried out by treating MRSA with the extract at subinhibitory concentration, keeping untreated MRSA as control. The action of the extract on MRSA growth is detected by the analysis of specific mRNA pattern.

The genes targeted are the gene responsible for membrane synthesis and β-lactam resistant gene in MRSA. Primers are designed using public domain (NCBI) and are selected using primer 3. The materials use dare: MRSA, β-mercaptoethanol, sodium acetate–2M, Chloroform, Phenol (All from M/s. Merck), MH agar (Hi-media), DEPC (Merck) RNAase, c DNA KIT (Genei, Bangalore). Primers used are shown in Table 2.10 and can be purchased from M/s. Genei, Bangalore.

TABLE 2.10   Primers Used

Target gene	Sequence of sense primer	Sequence of antisense primer
mecA	3'-AAATCGATGGTAAAGGTTGGC–5'	5'-AGTTCTGCAGTACCGGATTTGC–3'
MprF	3'-GTATCGGGAGTTATCTGG–5'	5'-TCAACCTACGTGCTCTAC–3'
msrR	3'-GGTGATAGTCTTCGGCTTG–5'	5'-GGAGGTTGCTTTTGGTGTA–3'

2.6.3.1   ISOLATION OF TOTAL RNA FROM BACTERIAL CELLS

Procedure: In this method, total RNA can be extracted from MRSA cells in the logarithmic phase of growth and purify using Guanidium

Isothiocyanate [129]. Total RNA is isolated from treated MRSA, untreated MRSA and from MSSA. Complementary DNA (cDNA)is then prepared according to the manufacturer's instructions by Pharmacia cDNA kit. Then Reverse transcription PCR is carried out. Cells are grown to mid-logarithmic phase in MH (pH 6.6) at 36°C. The 750 µl culture is centrifuged, and the pellet is re-suspended in 600 µl lysis buffer (freshly supplemented with 0.7% β-ME) and is then mixed well by vortexing.

The 60 µl of 2M Na-acetate (pH–4.0) is added and mixed again by vortexing. An equal volume of hot phenol (68°C) saturated with DEPC water (pH 4.0) is then added and vortexed vigorously for 5 min. The mixture is incubated at 68°C for 10 min and allowed to cool. Then 120 µl of chloroform is added and vortexed vigorously for 15 min with intermittent incubation on ice. The mixture is centrifuged. About 150 µl of the aqueous phase is collected and transferred to a fresh microcentrifuge tube, and an equal volume of isopropanol is added to it. The solutions are then mixed well and incubated at –20°C for 1–2 h. RNA is precipitated by centrifugation at a speed 13,000 rpm for 20 min. Pellet obtained is dissolved in 500 µl of lysis solution. RNA is then re-precipitated by adding an equal volume alcohol, kept at –20°C for 1–2 h. After centrifugation, the pellet is washed in 80% ethanol, dried at room temperature and dissolved in 10 µl of DEPC treated water. Quantify RNA by measuring the absorbance at 260nm. Note: Extensive precautions should be taken against RNase contamination for all experiments with RNA.

2.6.3.2 SEMI-QUANTITATIVE RT-PCR

Reverse transcription of the isolated RNA is performed to synthesize the first strand of cDNA with reverse primer, and then amplification of cDNA is done using specific primer sets as follows (Figure 2.14):

Procedure

- Isolated RNA samples are subjected to DNAse treatment to make them free from any contaminating DNA.
- For cDNA synthesis, take a total of 200ng of RNA, incubate at 70°C with specific antisense primer for 10 min, annealing in a thermal cycler.
- Add 5X buffer and the Superscript RT at 42°C. Reverse transcription is carried out in a thermal cycler.

- Following cDNA synthesis, amplification of specific genes responsible for cell wall synthesis (mecA genes) is done using specific primer sets.
- Amplification is done for 40 cycles (each cycle with 94°C for 30 s, 50°C for 30 s and 72°C for 30 s, and finally seven minutes extension at 72°C).
- DNAse treated RNA that had not been reverse transcribed, is used as a negative control.
- Remove aliquots (on completing 25, 30 and 34 cycles for each PCR product) are electrophoresed; and the gels are analyzed with a Gel Doc system.
- 16S rRNA is a housekeeping gene and is constitutively expressed and is used as a positive control.
- Each set of experiments is done in triplicate.
- Similarly, amplification for msrR and mpF gene is also done using specific primers as shown in Table 2.10.
- Samples are then observed after electrophoresis on agarose gel using a gel documentation system.

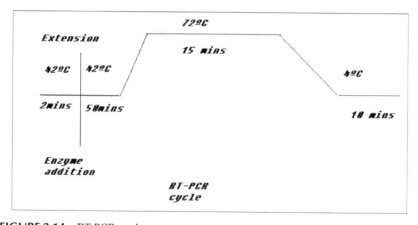

FIGURE 2.14   RT-PCR cycle program.

2.6.3.3   EXPERIMENT SET UP FOR DNASE

DNAse treatment of isolated RNA samples has to be done before cDNA preparation to remove any DNA contamination with RNA, so that reverse transcriptase could only reverse transcribe the mRNA to prepare the complementary DNA. The procedure generally follows the given set up:

- Autoclaved water: As required to make the total volume 10 µl.
- RNAse Inhibitor: 0.5µl.
- RNA: Desired volume in µl to have total 1 µg.
- DNAse Enzyme: 1 µl.

Note: Total 10 µl of reaction set up.

This reaction mixture is kept at 37°C for 30 min in a water bath. The reaction is stopped by adding 1 µl of EDTA to each microfuge tube to chelate Mg^{2+} ions. Finally, the heat inactivation of DNAse enzyme is done at 70°C in a water bath for 10 m.

2.6.4   DETERMINATION OF CYTOTOXICITY

Cytotoxicity test using Mice Spleen cells can be done by Tryphan blue exclusion method [140].

Principle: The dead and damaged cells stain blue inside on treating with trypan blue and can be distinguished from viable cells. This is due to the toxic effect of the drug, which makes pores on the membrane and trypan blue can enter inside the cells.

Materials required: PBS (Phosphate buffered saline)–NaCl 4g, KCl 0.1g, Na_2HPO_4 0.72g, KH_2PO_4 0.10g are dissolved in 500 ml of double distilled water, gentamicin 50 µg/ml, trypan blue 1.0% and sterile distilled water. Chemicals for this test can be purchased from M/s. SRL, Mumbai–India.

Preparation of cell lines: Collect spleen cells from experimental mice and culture in medium, RPM1 1640 (Roswell Park Memorial Institute). Dissect mice, collect spleen, crush in PBS and centrifuged at 1000 rpm. Collect the pellet and wash with PBS. Final pellet was diluted with phosphate buffer. Concentration is a dusted to 10^6cells/ml. It is then incubated in RPM1 medium at 36°C for 18 h in CO_2 incubator. Cytotoxicity of plant extracts or any compound can be detected using these spleen cells cultured in RPMI 1640 medium. The cell number at $10–^3$ dilution is to be counted using a hemocytometer.

Prepare different dilutions (10^{-1}, 10^{-2} and 10^{-3}) of extract using sterile water (10 mg/ml). Add different concentration of test compound (like 0.0, 5.0, 10.0, 50.0, 100, 200, 500 and 1000 µg/ml) with the cell lines having concentration 10^6cells/ml per well. Make up to 1.0 ml with RPMI/PBS and incubate for 4 h at 37°C. Add 1 ml, 1% trypan blue (in distilled water) to each tube and mix well. Place one drop in a hemocytometer and count the number of dead cells (10x).

% toxicity = [(Number of dead cells/Total number of cells)] x 100. Plot a graph with concentration of extract on X-axis and% toxicity on Y-axis and calculate IC_{50} value of each extract.

2.6.5   DETERMINATION OF ANTIOXIDANT ACTIVITY

2.6.5.1   DIPHENYL DI PICRYL PHENYL HYDRAZYL (DPPH) METHOD

Antioxidant activity of an extract can be determined using DPPH method [113, 205].

Principle: Antioxidant compounds can react with DPPH and reduce it to hydrazine. The color or OD of the DPPH solution can decrease as it getting reduced by an antioxidant compound. Therefore, a decrease in absorbance of reaction mixture is an indication of antioxidant activity of the compound added. The difference in absorbance can be calculated by measuring the OD using a spectrophotometer.

Materials required: DPPH (SRL)–0.1mM in methanol, extract concentrations of 10, 25, 50, 100, and 200 µg/ml in methanol, ascorbic acid standard (100 µg/ml) and distilled water. Chemicals used were of AR grade.

Procedure: Take500 µl of the sample in a small test tube. Prepare0.01Mm. DPPH solution using methanol. Add 1.0ml of DPPH to the extract and mix well. Keep at room temperature for 20 min. Measure OD at 517nm using a UV-VIS spectrophotometer. Ascorbic acid can be taken as positive control. Keep a negative control, i.e., DPPH without the extract. Calculate the percentage activity using the formula:

$$\% \text{ activity} = \{1\text{-}([A_{sample}/A_{control}]\} \times 100 \qquad (4)$$

Experiment is repeated thrice. Mean value ± SD is calculated. Determine IC_{50} (concentration for 50% inhibition of growth) for the compound or extract using linear regression formula by plotting graph viz. concentration on X-axis and percentage of inhibition on Y-axis.

2.6.6 METHOD TO DETERMINE THE PROLIFERATION OR INHIBITION OF ADIPOGENESIS

Adipocytes Staining: Oil Red O Staining

Adipocytes cells are washed three times using PBS. Fixation of cells is done by keeping in 10% formalin for 1 h at room temperature. After fixation, cells are washed once with PBS, stained with freshly diluted Oil Red O solution (3 parts of 0.6% Oil Red O in isopropanol and 2 parts of water). Cells are then washed twice with distilled water and visualized under a microscope. For quantitative analysis, Oil Red O stain is dissolved in isopropanol and OD is read at 520 nm by ELISA plate reader.

2.6.7 DETERMINATION OF ANTI-INFLAMMATORY EFFECT

2.6.7.1 CYCLOOXYGENASE AND 5-LIPOOXYGENASE INHIBITION METHOD

Peripheral lymphocyte from human blood can be cultured in RPMI medium. Add 20% heat-inactivated fetal calf serum, agglutinin, and antibiotic such as gentamycin or streptomycin to 20mlofmedium. Filter through 0.22 μm cellulose acetate membrane filters. Incubate the culture at 36°C after adding fresh plasma (1×10^6 cells/ml) aseptically. Add 1 μl of lipopolysaccharide for activation. After 24 h of incubation, add aqueous extract at requisite concentration such as 10 μg/ml, 100 μg/ml, 1000 μg/ml and incubate for 24 to 36 h. Ibuprofen at 100 μg/ml can be used as standard. Pellet the sample by centrifugation (5000 rpm 15 min) after the incubation. Add lysis buffer and again centrifuged. Suspend the pellet in small amount of supernatant and use for anti-inflammatory assay.

Assay of Cyclooxygenase: Add Tris-HCl buffer (pH 7.2) and glutathione (0.1 mg)to the prepared lymphocyte suspension (5 ml). Add arachidonic acid and incubate for 30min at 36°C. Stop the reaction by adding 0.2 ml 10%

TCA in 1N HCl and 0.2 ml TBA. Centrifuge at 3000 rpm for 5 min. Read OD at 632 nm and compare reading of the sample with that of standard.

2.6.7.2   INHIBITION OF ALBUMIN DENATURATION METHOD

Add extract at different concentrations such as 10 μg/ml, 100 μg/ml, 1000 μg/ml to the 1% aqueous bovine album solution. Incubate the samples at 36°C for 20 min and then heat at 57°C for 20 min. Cool the samples to room temperature. Measure the turbidity at 660 nm. IC_{50} value is calculated by plotting the graph between percentage inhibition and concentration of the extract. The percentage inhibition is calculated as:

$$\text{Percentage inhibition} = \left[\frac{(\text{Absorbance of control-Absorbance of extract})}{\text{absorbance of control}} \right] \times 100 \quad (5)$$

2.6.7.3   MEMBRANE STABILIZATION TEST

In this method, RBC suspension is prepared. Collect 10 ml of fresh human blood in heparinized tubes or vials. Centrifuge at 300 rpm for 10 min. Wash pellets with equal volume of PBS. Repeat the washing for three times. Prepare a 10% suspension in normal saline or PBS. Take 1 ml of the suspension and add 1ml of test solution. Aspirin can be used as standard. Incubate at 56°C for 30minand cool the tubes. Centrifuge at 3000 rpm for5min. Collect the supernatant and read the absorbance at 560 nm. The experiments are done in triplicate, and percentage inhibition is calculated using Eq. (5).

2.6.7.4   PROTEASE INHIBITION METHOD

Make the reaction and mixture by adding Tris-HCl–1 ml (pH 7.2), 0.06 mg of trypsin and 1 ml of test sample at different concentrations. Incubate at 36°C for 7 min and add 1 ml of 0.8% casein. Again incubate at 36°C for 20 min and terminate the reaction by adding 2 ml of 70% perchloric acid ($HClO_4$). Centrifuge at 2000 rpm for 10 min. Collect supernatant and measure OD at 210 nm using buffer as blank. Experiments are repeated in triplicate. The% inhibition and IC_{50} value are calculated as described in this section.

2.6.8   WESTERN BLOT ANALYSIS FOR DETERMINATION OF ADIPOCYTE DIFFERENTIATION BY SPECIFIC ADIPONECTIN AND LEPTIN

The 3T3 L–1 cells are cultured in DMEM for 24 to 48 h in a CO_2 incubator at 38°C with different concentrations and also without the extract of interest. Centrifuge at 1000 rpm and collect the pellet, lyse with lysis buffer (20mM Tris with pH of 7.3, 2 Mm EDTA, 1% tritron, 0.1% SDS, 1 mM PMSF, 10 µg/ml leupeptin) and separate protein. Estimate the protein by Lawry's method. Perform SDS-PAGE as explained above using 12% polyacrylamide gel. Transfer the gel to PVDF membrane. Block with 5% milk powder in PBST (0.05% TWEEN 20 in PBS, pH 7.2) for 1–2 h. Incubate with primary antibody at 4°C overnight and then with secondary antibody for 1 h. Detect the band with densitometer (ECL–Amersham Pharmacia Biotech) and calculate the relative expression of the protein using suitable software. Actin can be used as standard.

2.6.8.1   REAL-TIME PCR ANALYSIS FOR GENE EXPRESSION LEVEL STUDY

Culture 3T3-L1 cells in a 96-well plate in medium with different concentrations of the extract. One well is kept without extract as negative standard. After 24–36 h of incubation, collect the culture. Isolate RNA with TRIZOL-reagent. Prepare cDNA as per manufacturer's instructions. This cDNA can then be used for Real-Time PCR analysis for PPRAγ gene expression using reaction mix with syber green, primers (Table 2.11), dNTPs. Taq, buffer, etc. Actin is used as positive control. Real-time PCR gives a quantitative measurement of gene expression.

TABLE 2.11   The Primers Used for Adipose Gene Expression Studies

Gene-Targeted	Primers
VEGF	Forward 3'- GAAAGGCTTCAGTGTGG–5'
	Reverse 5'- CAGGAATGGGTTTGTCG–3'
PPRAγ	Forward 3'–TCACAATGCCATCAGGT- 5,'
	Reverse 5- GCGGGAAGGACTTTATGTA–3'
Actin	Forward 3'-CCTCTATGCCAACACAGT–5
	Reverse 5'-AGCCACCAATCCACACAG- 3'

2.6.9   BIO-ASSAY FOR ANTICANCER ACTIVITIES OF PLANT EXTRACTS

2.6.9.1   METHYL THIAZOLYL DIPHENYL-TETRAZOLIUM BROMIDE (MTT) ASSAY FOR THE ANALYSIS OF CELL PROLIFERATION

Principle: Methyl thiazolyl diphenyl-tetrazolium bromide (MTT) can form a yellow colored solution on dissolving in water. MTT will be reduced, by metabolically active or live cells to form an insoluble product, formazan. This leads to a change in color of the solution from yellow to blue or pink. The OD is monitored using a spectrophotometer [140].

Materials required: DLA cell lines, RPM1–1640 medium (Hi-media), DMSO, MTT stock–55 µg/ml (SRL), plant extract, CO_2 incubator, and 96-well plates.

Procedure: DLA cells are purchased or subcultured. Collected cell lines are to be washed using PBS or culture medium and transfer to fresh RPM1 medium for reviving. Take 500 µl the cell lines in RPM1 medium and load the wells with a concentration of 5000 cells/well. Then incubate the plates at 37°C and 5% CO_2 for 24 h in a CO_2 incubator. 200 µl extract is added serially to get a concentration of 0, 10, 20, 50, 100, 200, 500 and 1000 µg/ml in a series of well. Then 20 µl of MTT (5 µg/ml) is added to it. Cells are then incubated for 48 h at 37°C and at 5% CO_2 in a CO_2 incubator. Then 100 µl DMSO is added and incubated at 37°C for 30 min. OD of culture is measured at 545 nm and blank at 630nm. A negative control is well with MTT and without extract. The experiment is carried out in triplicate, and average% inhibition is determined and used for calculating IC_{50}. Regression analysis used for calculating IC_{50} value. The% of dead cells is calculated by the formula:

$$\% \text{ inhibition} = 100 - \left[\frac{\text{OD of the well with drug}}{\text{OD of the control}}\right] \times 100 \qquad (6)$$

Example: Calculation of IC_{50} value for aqueous extract of *Embelia ribes*is (Figure 2.15).

The regression equation of concentration of the extract taken µg/ml and% Inhibition is:

$$Y = 45.170 + 14.766 \ X \qquad (7)$$

where: Y = concentration of the extract taken in µg/ml and X =% inhibition.

The value of IC_{50} = [−45.170 + 14.766*50] = 693.13. The values are then analyzed using **Chi Test** and p-value < 0.05.

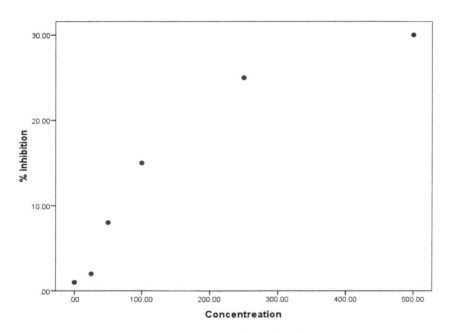

FIGURE 2.15 IC_{50} value for aqueous extract of *Embelia ribes.*

2.6.9.2 APOPTOSIS ASSAY FOR DNA LADDERING

Principle: Apoptosis is the natural programmed cell death, and occurs in normal cells and are controlled by genes. The enzyme caspases are activated during apoptosis. This results in DNA fragmentation and forming smaller DNA fragments of size 200 bp. The DNA fragments can be viewed on agarose gels. There is no programmed cell death in cancer cells, which results in abnormal growth of cells. Inducing apoptosis is a solution to prevent the growth of abnormal cells. Therefore, a compound, which can induce apoptosis, can be used as an anticancer compound [10].

   Materials required: Cell lines (Different cell lines are available), EDTA−10 mM, protease (0.2 mg/ml), RNAase−1 mg/ml(10 μl) (Genei, Bangalore), buffer (Lysis)−50 mM tris HCl (pH−8.0) (SRL), 40% sucrose, buffer (loading)−10 mM EDTA, and a pinch of Bromophenol blue (Merck).

Procedure: The cell lines (5 x 10^5 cells/ml) in fresh sterile RPM1* with different concentrations of the extract are taken and kept for incubation. One well is kept as control, i.e., without extract. It is then incubated for 24 h and centrifuged at 400 rpm for 5 min. The pellet is collected. The 20 µl lysis buffer is added to the pellet. It is kept at 50°C for 1h. The sample is then loaded with 10 µl of the loading buffer on a 1.5% agarose gel and electrophoresis is done with 70v for 3 h. The gel is then observed under gel documentation system. The experiment is done with different extracts and repeated thrice.

Note: Medium depending on the nature of cell lines is selected.

2.7   SUMMARY

MAPK, JAK/STAT 3. PPRγ, etc. are some of the cell signaling pathways. Phytochemicals such as quercetin, capsaicin, lupeol, etc. act on such cellular pathways and have beneficiary effects in controlling diseases like diabetes, obesity, allergy, and liver sclerosis, etc. Tea, turmeric, papaya, avocado, and ginger are consumed regularly. The quantity of active components can be estimated using biochemical tests. The HPLC, HPTLC, FPLC, LC-MS, GC-MS, etc. are different tools for characterization of the phytochemicals. Some extracts or compounds present in different foods affect synergically in preventing cellular damages or proliferation. Sometimes these act as molecular triggers in cell cycle pathways and sometimes as inhibitors of irregular cell signal pathways and prevent abnormal developments. A modern molecular method like real-time PCR, next-generation sequencing, flow cytometry, etc., can be used to study the cellular pathways.

KEYWORDS

- adiponectin
- adipose cells
- alkaloids
- catechins
- cell signaling

- flow cytometry
- MAPK-mitogen-activated protein kinase
- nutrigenomics
- phytochemicals
- quercetin
- resveratrol
- terpenoids

REFERENCES

1. Zheng, W., & Shiow, Y. W., (*2001*). Antioxidant activity and phenolic compounds in selected herbs. *Journal of Agriculture and Food Chemistry, 49*(11), 5165–5170.
2. Adwan, G., & Mohammad, M., (2008). Synergistic effects of plant extracts and antibiotics on *Staphylococcus aureus* strains isolated from clinical specimens. *Middle-East Journal of Scientific Research, 3*(3), 134–139.
3. Adwan, M. G., Abu-Shanab, A. B., & Adwan, M. K., (2008). *In vitro* activity of certain drugs in combination with plant extracts against *Staphylococcus aureus* infections. *Pak. Journal of Medicine, 24*, 541–544.
4. Aggarwal, B. B., Aggarwal, S. R., Shishodia, S., & Takada, Y., (2004). Role of resveratrol in prevention and therapy of cancer: Preclinical and clinical studies. *Anticancer Research, 24*(5), 2783–2840.
5. Adhami, V. M., Malik, A., Zaman, N., Sarfaraz, S., Siddiqui, I. A., Syed, D. N., et al., (2007). Combined inhibitory effects of green tea polyphenols and selective cyclooxygenase-2 inhibitors on the growth of human prostate cancer cells both *in vitro* and *in vivo*. *Clinical Cancer Research, 13*(2), 1611–1619.
6. Agnes, M. R., Muriel, C., Cristian, D., Rajendra, G. M., John, M. P., & Stephen, O. D., (2002). Cancer chemopreventive and antioxidant activities of pterostilbene, a naturally occurring analogue of resveratrol. *J. Agric. Food Chemistry, 50*(12), 3453–3457.
7. Ahamed, B., (2001). Antimicrobial and phytochemical studies on 45 Indian medicinal plants against multi-drug resistant human pathogens, *Journal of Ethnopharmacology, 74*, 113–123.
8. Ahmad, N., Feyes, D. K., Niemen, A. L., Agarwal, R., & Mukhtar, H., (1997). Green tea constituent epigallocatechin–3-gallate and induction of apoptosis and cell cycle arrest in human carcinoma cells. *J. Natl. Cancer Inst., 89*, 1881–1886.
9. Ajaya, K. R., Sridevi, K., Kumar, V. N., Nanduri, S., & Rajagopal, S., (2004). Anticancer and immunostimulatory compounds from Andrographis paniculata. *Journal of Ethnopharmacology, 92*(2 & 3), 291–295.
10. Ajith, T., Avu, V., & Riji, T., (2008). Antitumor and apoptosis promoting properties of atorvastatin, an inhibitor of HMG-CoA reductase, against Dalton's Lymphoma

Ascites tumor in mice. *Journal of Experimental and Theoretical Oncology, 7*(4), 291–298.

11. Akiko, S., Yoshiyuki, M., Akira, T., & Noriyuki, N., (2009). Versatile synthesis of epicatechin series procyanidin oligomers, and their antioxidant and DNA polymerase inhibitory activity. *Tetrahedron, 65*(36), 7422–7428.

12. Ali, H., Houghton, P. J., & Amala, S., (2006). Alpha-amylase inhibitory activity of some Malaysian plants used to treat diabetes, with particular reference to *Phyllanthus amarus*. *Journal of Ethnopharmacology, 107*, 449–455.

13. Amelia, P., Almeida, H., Adriana, F., Daniela, A. M., Silva, M., & Beatriz, A. G., (2006). Antibacterial activity of coffee extracts and selected coffee chemical compounds against enterobacteria. *J. Agric. Food Chem., 54*(23), 8738–8743.

14. Andrea, G. A., Renata, S. M., & Adriana, F., (2010). Species, roasting degree and decaffeination influence the antibacterial activity of coffee against *Streptococcus mutans*. *Food Chemistry, 118*(3), 782–788.

15. Androutsopoulos, V., Randolph, R. J., John, F. H., Somchaiya, I., Fernandes, E., Lima, J., Costa, P., & Bahia, M. F., (2008). Walnut (*Juglans regia*) leaf extracts as strong scavengers of pro-oxidant reactive species. *Food Chemistry, 106*, 1014–1020.

16. Anna, P., Paulina, D., Magdalena, B., & Krystyna, P., (2012). Polyphenolic content and comparative antioxidant capacity of flavored black teas. *International Journal of Food Sciences and Nutrition, 63*(6), 742-748.

17. Apati, P., Agnes, K., Peter, J. H., Glyn, B. S., & Geoffrey, K., (2006). *In-vitro* effect of flavonoids from *Solidago canadensis* extract on glutathione S-transferase. *Journal of Pharmacy and Pharmacology, 58*(2), 251–256.

18. Aqil, F., & Ahmad, I., (2007). Antibacterial properties of traditionally used Indian medicinal plants. *Methods Find Exp. Clinical Pharmacology, 29*(2), 79–92.

19. Archer, A. W., (1988). Determination of safrole and myristicin in nutmeg and mace by high-performance liquid chromatography. *Journal of Chromatography, 438*(5), 117–121.

20. Arima, H., Ashida, H., & Genichi, D., (2002). Rutin-enhanced antibacterial activities of flavonoids against *Bacillus cereus* and *Salmonella enteritidis*. *Bioscience, Biotechnology, and Biochemistry, 66*(5), 1009–1014.

21. Ashok, D. B., Vaidya, P., & Thomas, P. A., (2007). Current status of herbal drugs in India: An overview. *J. Clin. Biochem. Nutrition, 41*(1), 1–11.

22. Atef, D. A., & Erodou, O. T., (2003). Antimicrobial activities of various medicinal and commercial plant extracts. *Turkish Journal of Biology, 27*, 157–162.

23. Ayo, R. G., Amupitan, J. O., & Zhao, Y., (2007). Cytotoxicity and antimicrobial studies of 1, 6, 8- trihydroxy–3-methyl-anthraquinone (emodin) isolated from the leaves of *Cassia nigricans* Vahl. *African Journal of Biotechnology, 6*(11), 1276–1279.

24. Babu, B. H., Shylesh, B. S., & Padikkala, L., (2001). Antioxidant and hepatoprotective effect of *Acanthus ilicifolius*. *Fitoterapia, 72*, 272–277.

25. Bae, H. C., Lee, J. Y., & Myoung, S. N., (2005). Effect of red ginseng extract on growth of *Lactobacillus sp., Escherichia coli* and *Listeria monocytogenes* in pH controlled medium. *Korean Journal for Food Science of Animal Resources, 25*(2), 257–264.

26. Balasubramaian, S., & Echet, R. L., (2007). Curcum suppresses AP1 transcription factor-dependent differentiation and activates apoptosis in human epidermal keratinocytes. *Journal of Biology and Chemistry, 282*, 6707–6715.

27. Barber, N. J., Zhang, X., Zhu, Z., Pramanik, R., Barber, J. A., Martin, F. L., Morris, J. D. H., & Muir, G. H., (2006). Lycopene inhibits DNA synthesis in primary prostate epithelial cells *in vitro* and its administration is associated with a reduced prostate-specific antigen velocity in a phase II clinical study. *Prostate Cancer and Prostatic Diseases, 9*, 407–413.

28. Baris, O., Gulluce, M., Sahin, F., Ozer, H., Kilic, H., Ozkan, H., et al., (2006). Biological activities of the essential oil and methanol extract of *Achillea biebersteinii* Afan. (*Asteraceae*). *Turk, J. Biol., 30*, 65–73.

29. Bauer, A. W., Perry, D. M., & Kirby, W. M., (1959). Single disc antibiotic sensitivity testing of Staphylococci. *Arch. Int. Med., 104*, 208–216.

30. Biglow, R. L., & Cardelli, J. A., (2006). The green tea catechins, epigallocatechin and epicatechin inhibit HGF/Met signaling in immortalized and tumorigenic breast epithelial cells. *Oncogenes, 25*(6), 1922–1930.

31. Birdsall, T., & Kelly, G., (1997). Berberine: Therapeutic potential of an alkaloid in several medicinal plants. *Alternative Medicine, 2*, 94–103.

32. Bischoff, S., (2008). Quercetin: Potentials in the prevention and therapy of disease. *Current Opinion in Clinical Nutrition & Metabolic Care, 11*(6), 733–740.

33. Borris, R. P., (1996). Natural products research: Perspectives from a major pharmaceutical company. *Journal of Ethnopharmacology, 51*, 29–38.

34. Bozin, B., Neda, M., Isidora, S., & Emilija, J., (2007). Antimicrobial and antioxidant properties of rosemary and sage (*Rosmarinus officinalis L.* and *Salvia officinalis L., Lamiaceae*) essential oils. *Journal of Agriculture and Food Chemistry, 55*(19), 7879–7885.

35. Candice, C. N. P., Anna, K. J., Dulcie, A. M., & Evan, S., (2001). Cyclooxygenase inhibiting and anti-bacterial activities of South African Erythrina species, *Journal of Ethnopharmacology, 74*(30), 231–237.

36. Carbera, C., Artacho, R., & Gimenz, R., (2006). Beneficial effects of green tea-review. *Journal of Am. Coll. Nutr., 25*, 79–99.

37. Chan, E. W. C., Lim, Y. Y., Ling, S. K., Tan, S. P., Lim, K. K., & Khoo, M. G. H., (2009). Caffeoylquinic acids from leaves of *Etlingera* species (Zingiberaceae). *LWT–Food Science and Technology, 42*(5), 1026–1030.

38. Chandrasekaran, C. V., Gupta, A., & Agarwal, A., (2010). Effect of an extract of *Andrographis* paniculata leaves on inflammatory and allergic mediators *in vitro*. *Journal of Ethnopharmacology, 129*(2), 203–207.

39. Charles, G., Jean, L., Lebrun, M., Dufour, P., & Pichette, A., (2006). Glycosidation of lupine-type triterpenoids as potent *in vitro* cytotoxic agents. *Bioorganic & Medicinal Chemistry, 14*(19), 6713–6725.

40. Chen, Y., Yu, Q. J., Li, X., Luo, Y., & Liu, H., (2007). Extraction and HPLC characterization of chlorogenic acid from tobacco residuals. *Sep. Sci. Technol., 42*, 3481–3492.

41. Cheng, S. S., Liu, J. L., Chang, H., & Shang, T. C., (2008). Antifungal activity of cinnamaldehyde and eugenol congeners against wood-rot fungi. *Bioresource Technology–Exploring Horizons in Biotechnology: A Global Venture, 99*(11), 5145–5149.

42. Chukwujekwu, J. C., Coombes, P. H., Mulholland, D. A., & Van Staden, J., (2003). Emodin–an antibacterial anthraquinone from the roots of *Cassia occidentalis*. *South African Journal of Botany, 72*(2), 295–297.

43. Cordell, G., (1981). *Introduction to Alkaloids: A Biogenetic Approach* (pp. 23–28). Wiley and Sons, New York.
44. Cowan, M. M., (1999). Plant products as antimicrobial agents, Clin. Microbiol. Rev., *12*(4), 564–582.
45. Cragg, G. M., & Newman, D. J., (2005). Plants as a source of anti-cancer agents. *J. Ethnopharmacol., 100*, 72–79.
46. Cseke, L. J., Kirakosyan, A., Kaufman, P. B., Warber, L., Duke, J. A., & Brelmann, H. L., (2009). *Phytochemicals, Chemical Components from Plants: Natural Products from Plants* (pp. 3–40). CRC Press, Taylor and Francis Group, Boca Raton, FL.
47. Cushnie, T. P., & Lamb, A. J., (2005). Review: Antimicrobial activity of flavonoids. *International Journal of Antimicrobial Agents, 26*, 343–356.
48. Damien, L., Soizic, P., Dennis, K., & Florence, B., (2011). Antiplasmodial and cytotoxic activities of medicinal plants traditionally used in the village of Kohima, Uganda. *Journal of Ethanopharmacology, 133*(2), 850–855.
49. Daneshtalab, M., (2008). Discovery of chlorogenic acid-based peptidomimetics as novel class of antifungals–A success story in rational drug design. *J. Pharm. Pharmaceut. Sci.,* (www.cspsCanada.org), *11*(2), 44–55.
50. Daniela, A. P., Rejane, B., Vanessa, C., & Batista, D. C. F., (2011). Chlorogenic acids from Tithoniadiversifolia demonstrate better anti-inflammatory effect than indomethacin and its sesquiterpene lactones. *Journal of Ethanopharmacology, 136*(2), 355–362.
51. De León, M. R., & López, L., (2010). Antibacterial properties of zeylasterone, a triterpenoid isolated from *Maytenus blepharodes*, against *Staphylococcus aureus. Microbiological Research, 165*(8), 617–626.
52. Djeddi, S., Anastasia, K., Sokovic, M., Catherine, K., & Helen, S., (2007). Mor Sesquiterpene lactones from *Centaurea pullata* and their antimicrobial activity. *Journal Natural Products, 70*(11), 1796–1799.
53. Dong, K. R., Dong-Wook, H., Hyun, S., & Suong-Hyu, H., (2005). Prevention of reactive oxygen species-induced oxidative stress in human microvascular endothelial cells by green tea polyphenol. *Toxicology Letters, 155*(2), 269–275.
54. Dorman, H. J. D., & Deans, S. G., (2008). Antimicrobial agents from plants: Antibacterial activity of plant volatile oils. *Journal of Applied Microbiology, 88*(2), 308–316.
55. Earnshaw, W. C., (1995). Nuclear changes in apoptosis. *Current Biology, 7*, 337–343.
56. Ejaz, A., Muhammad, I., Abdul, M., & Muhammad, A., (2006). Antioxidant activity with flavonoidal constituents from *Aerva persica. Archives of Pharmaceutical Research, 29*(5), 467–470.
57. Eloff, J. N., & Katerere, D. R., (2008). The biological activity and chemistry of the Southern African *Combretaceae. Journal of Ethanopharmacology, 119*(3), 686–699.
58. Elzawely, A. A., Xuan, T. D., & Tawata, S., (2005). Antioxidant and antibacterial activities of *Rumex japonicus Houtt*: aerial parts. *Biology Pharm. Bulletin, 28*, 2225–2230.
59. Emeline, H., Bertan, S., Geneviève, B., & Didier, S., (2009). Quassinoid constituents of *Quassia amara L.* leaf herbal tea: Impact on its antimalarial activity and cytotoxicity. *Journal of Ethanopharmacology, 126*(1), 114–118.
60. Erasto, P., Bojase-Moleta, G., & Majinda, R. R. T., (2004). Antimicrobial and antioxidant flavonoids from the roots wood of *Bolusanthus spesiosus. Phytochemistry, 65*, 875–880.

61. EUCAST, (2003). Discussion document–determination of minimum inhibitory concentrations (MICs) of antibacterial agents by broth dilution. *Clin. Microbiol. Infect.*, *9*,(8), 1–7.

62. Eugenia, P., Luís, V., & Carlos, C., (2009). Antifungal activity of the clove essential oil from *Syzygium aromaticum* on *Candida*, *Aspergillus* and dermatophyte species. *Journal of Medical Microbiology*, *58*(11), 1454–1462.

63. Faizi, S., Najma, R., & Shaheena, A., (2003). Evaluation of the antimicrobial property of *Polyalthia longifolia* var. pendula: Isolation of a lactone as the active antibacterial agent from the ethanol extract of the stem, *Phytotherapy Research*, *17*(10), 1127–1238.

64. Frank, R., Stermitz, L., Tawara, J. N., & Lauren, A., (2000). Synergy in a medicinal plant: Antimicrobial action of berberine potentiated by 5*-methoxyhydnocarpin, a multidrug pump inhibitor. *Proc. Natl. Acad. Sci. USA*, *97*(4), 1433–1437.

65. Freitas, A. M., Almeida, M. T. R., Andrighetti-Fröhner, C. R., Cardozo, F. T. G. C., & Barardi, C. R. M., (2009). Antiviral activity-guided fractionation from *Araucaria angustifolia* leaves extract. *Journal of Ethnopharmacology*, *126*(3), 512–517.

66. Garcia, C. V. M., Irene, C., Collado, P. S., & Javier, G., (2007). The anti-inflammatory flavones quercetin and kaempferol cause inhibition of inducible nitric oxide synthase, cyclooxygenase–2 and reactive C-protein, and down-regulation of the nuclear factor kappa B pathway in Chang liver. *European Journal of Pharmacology*, *557*(2 & 3), 221–229.

67. Geetha, T., & Varalakshmi, P., (2001). Anti-inflammatory activity of lupeol in rats. *Journal of Ethanopharmacology, 76,* 77–80.

68. Gelareh, M., Zahra, E., Karamatollah, R., & Mohammad, H., (2009). Identification and quantification of phenolic compounds and their effects on antioxidant activity in pomegranate juices of eight Iranian cultivars. *Food Chemistry*, *115*(4), 1274–1278.

69. George, A., Pesewu, R. R. C., & David, P. H., (2008). Antibacterial activity of plants used in traditional medicines of Ghana with particular reference to MRSA. *Journal of Ethnopharmacology, 116*(1), 102–111.

70. Getie, M., Gebre-Mariam, T., Rietz, R., Hohne, C., Huschka, C., Schmidtke, M., Abate, A., & Neubert, R. H., (2003). Evaluation of the anti-microbial and anti-inflammatory activities of the medicinal plants *Dodonaea viscose, Rumex nervosus, Rumex abyssinicus. Fitoterapia, 74*, 139–143.

71. Gibbons, S., (1992). Plants as a source of bacterial resistance modulators and anti-infective agents. *Phytochemistry Reviews*, *4*(1), 63–78.

72. Green, R. J., (2004). *Antioxidant Activity of Peanut Plant Tissues* (p. 156). Master Thesis, North Carolina State University, Raleigh USA.

73. Grohs, P., Marie-Domique, K., & Laurent, G., (2003). *In vitro* bactericidal activities of linezolid in combination with vancomycin, gentamicin, ciprofloxacin, fusidic acid, and rifampin against *Staphylococcus aureus. Antimicrobial Agents and Chemotherapy, 47*(1), pp. 418–420.

74. Gulcin, I., Oktay, M., Kufrevioglu, I., & Aslan, A., (2002). Determination of antioxidant activity of lichen Cetraria islandica (L) Ach. *J. Ethnopharmacol., 79*, 325–329.

75. Gupta, H. P., Singh, R., Srivastava, O. P., Khana, J. M., Mathur, I. S., & Gupta, S. K., (1981). Antifungal property of some substituted naphthoquinones, anthraquinone and imidazolquinones. *Indian Journal of Microbiology, 21*, 57–59.

76. Gupta, R., Gabrielson, B., & Ferugson, S. M., (2005). Nature's medicines: Traditional knowledge and intellectual property management–case studies from the National Institute of Health. *Current Drug Discovery Technology, 2,* 203.

77. Halliwell, B., Aeschbach, R., Loliger, J., & Aruoma, O. I., (1995). The characterization of antioxidants. *Food Chem. Toxic., 33,* 601–617.

78. Handa, S. S., (2006). An overview of extraction techniques for medicinal and aromatic plants. *South East Asia Regional Workshop on Extraction Technologies for Medicinal and Aromatic Plants.* Sukhdev Swami Handa ... Development Division,. Central Institute of Medicinal and Aromatic Plants, P. O. CIMAP, Lucknow, India; pages 21.

79. Harborne, J. B., (1989). Methods in plant biochemistry. In: *Plant Phenolics* (pp. 79–100). Academic Press, London, UK.

80. Hata, K., Kazuyuki, H., & Takahashi, S., (2002). Differentiation and apoptosis-inducing activities by pentacyclic triterpenes on a mouse melanoma cell line. *J. Nat. Prod., 65*(5), pp. 645–648.

81. Heatley, N. G., (1944). Method for the assay of penicillin. *Biochem. J., 38,* 61–65.

82. Heim, K. E., Tagliaferro, A. R., & Bobilya, D. J., (2002). Flavonoids antioxidants: Chemistry, metabolism and structure-activity relationships. *J. Nutr. Biochem., 13,* 572–584.

83. Heinonen, M., (2007). Antioxidant activity and antimicrobial effect of berry phenolics-a Finnish perspective. *Mol. Nutr. Food Research, 51*(6), 684–691.

84. Hifur, R., (2011). Beneficial effects of lupeol. *Life Sciences, 88*(7 & 8), 285–293.

85. Hong, J. H., Ahn, K. H., Bae, E., Jeon, S. S., & Choi, H. Y., (2006). The effects of curcum on the invasiveness of prostate cancer *in vitro* and *in vivo. Prostate Cancer and Prostatic Diseases, 9,* 147–152.

86. Honigsbaum, M., (2002). *The Fever Trail in Search of Cure for Malaria* (p. 326). Farrar stratus Giroux, New York.

87. Hougee, S., Faber, J., & Smit, H. F., (2005). Selective COX–2 inhibition by *Pterocarpus marsupium* extract is characterized by pterostilbene and its activity in healthy human volunteers. *Planta. Medica., 71*(5), 387–392.

88. Hymavathi, A., Babu, K. S., Naidu, P. G. M., Krishna, S. R., Diwan, P. V., & Rao, J. M., (2009). Bioactivity-guided isolation of cytotoxic constituents from stem-bark of Premnaomentosa. *Bioorganic & Medicinal Chemistry Letters, 19*(19), 5727–5731.

89. Hyun-Joo, J., & Song, Y., (2009). Anti-inflammatory, anti-angiogenic and anti-nociceptive activities of an ethanol extract of *Salvia plebeia*n. *Journal of Ethnopharmacology, 126*(2), 355–360.

90. Ivanova, D., Gerova, T., & Chervenkov, T., (2005). Polyphenols and antioxidant capacity of Bulgarian medicinal plants. *Journal of Ethanopharmacology, 96*(1 & 2), 145–150.

91. Jana, S., Hana, B., Hana, P., Pavel, M., & Eva, T., (2007). HPLC quantification of seven quaternary benzo[c]phenanthridine alkaloids in six species of the family *Papaveraceae. Journal of Pharmaceutical and Biomedical Analysis, 44,* 283–287.

92. Jayshree, D. P., & Shreyas, A. B., (2009). Antimicrobial screening a phytochemical analysis of the resin part of *Acacia catech. Pharmaceutical Biology, 47*(1), 34–37.

93. Jean, D. W., Marie-Christine, L., & Zacharias, T., (2007). α-Glucosidase inhibitory constituents from stem bark of *Termalia superba* (Combretaceae). *Phytochemistry, 68*(15), 2096–2100.

94. Jeong, H. J., Hyun, W., & Hyun, P., (2010). Anti-inflammatory and anti-arthritic activity of total flavonoids of the roots of Sophora *flavescens. Journal of Ethanopharmacology, 127*(3), 589–595.

95. Jimoh, F. O., Adedapo, A. A., Aliero, A. A., & Afolayan, A. J., (2008). Polyphenolic contents and biological activities of *Rumex ecklonianus. Pharmaceutical Biology, 46*, 5, 333–340.

96. Jian-Yu, P., Chen, S., Mei-Hua, Y., Jun, W., Jari, S., & Kun, Z., (2009). Update on lignans: Natural products and synthesis. *Natural Product Reports, 26*, 1251–1292.

97. Jin, T. T., (2005). Genistein, EGCG, and capsaicin inhibit adipocyte differentiation process via activating AMP-activated protein kinase. *Biochemical and Biophysical Research Communications, 238*, 694–699.

98. John, G., (2010). *Economic Botany and Ethnobotany* (p. 243). International Publishing Academy, New Delhi.

99. Joshi, C. R., (1986). Metal chelates of juglone and their antimicrobial activity. *Indian Journal of Pharmaceutical Science, 48*(5), 101–104.

100. Jun, L., Eng-Hui, C., & Arne, H., (2004). Targeting thioredoxin reductase is a basis for cancer therapy by arsenic trioxide, *PNAS, 30*, 12288–12293.

101. Kahkonen, M. P., Hopia, H. J., Vuorela, J. P., Rauha, K., & Pihlaja, T. S. K., (1999). Antioxidant activity of plant extracts containing phenolic compounds. *Journal of Agricultural and Food Chemistry, 47*(6), pp. 3954–3962.

102. Kalsi, J. S., Cellek, A., Muneer, P., Kell, D., & Mhas, S., (2002). Current oral treatments for erectile dysfunction. *Expert Opinion in Pharmacotherapy, 3*, 1613–1629.

103. Kamicker, J., & Michael, T., (2008). Bacterial efflux pump inhibitors. In: Totowa, N. J., (ed.), *Methods in Molecular Medicine* (pp. 112–118). Humana Press Inc.

104. Kaneda, Y., (1991). *In vitro* effects of berberine sulphate on the growth and structure of *Entamoeba histolytica, Giardia lamblia*, and *Trichomonas vaginalis. Annals of Tropical Medicine and Parasitology, 85*(4), 417–425.

105. Kao, Y. H., Hiipakka, R. A., & Liao, S., (2000). Modulation of obesity by a green tea catechin. *Am. J. Clin. Nutr., 72*, 1232–1234.

106. Kaur, C., & Kapoor, H. C., (2002). Antioxidant activity and total phenolic content of some Asian vegetables. *International Journal of Food Science and Technology, 37*, 153–161.

107. Kelkel, M., Marc, S., Mario, D., & Marc, D., (2011). Antioxidant and anti-proliferative properties of lycopene. *Free Radical Research, 45*(8), 925–940.

108. Kessler, M., Ubendu, G. C., & Jung, L., (2003). Antoprooxidant activity of rutin and quercetin derivatives. *J. Pharm. Pharmacol., 55*(2), 131–142.

109. Khan, A., Haque, E., Rahman, M., & Nessa, F., (2008). Bioactivity of roots of *Laportea crenulata. Pharmaceutical Biology, 46*(10), 695–699.

110. Khan, N., Afaq, F., & Mukhtar, H., (2007). Apoptosis by dietary factors: The suicide solution for delaying cancer growth. *Carcinogenesis, 28*(7), 233–239.

111. Kianbakht, S., & Jahaniani, F., (2003). Evaluation of antibacterial activity of *Tribulus terrestris* L. growing in Iran. *Iran. J. Pharmacol. Ther., 2*, 22–24.

112. Kim, L., (2001). In search of natural substrates and inhibitors of MDR Pumps. *Journal of Molecular Microbiology and Biotechnology, 3*(2), 247–254.

113. Koleva, I., Teris, A., Van Beek, P. H., Linssen, A., & Lyuba, N. E., (2001). Screening of plant extracts for antioxidant activity: Comparative study on three testing methods. *Phytochemical Analysis, 13*(1), 1–17.

114. 114. Kong, L. D., Zhang, Y., Pan, X., Tan, R. X., & Cheng, C. H., (2000). Inhibition of xanthine oxidase by liquiritigenin and isoliquiritigenin isolated from *Sinofranchetia chinensis*. *Cell and Molecular Life Sciences, 57*, 500–505.

115. 115. Kostova, S., (2006). Synthetic and natural coumarins as antioxidants. *Reviews in Medicinal Chemistry, 6*(4), 365–374.

116. Kuzma, L., Elzbieta, B., & Halina, W., (2009). Methyl jasmonate effect on diterpenoid accumulation in *Salvia sclarea* hairy root culture in shake flasks and sprinkle bioreactor. *Enzyme and Microbial Technology, 44*(6 & 7), 406–410.

117. Lansky, E. P., & Robert, A. N., (2007). *Punica granatum* (pomegranate) and its potential for prevention and treatment of inflammation and cancer. *Journal of Ethnopharmacology,* 109(2), 177–206.

118. Lee, J., Young-Ju, J., & Mi-Hee, Y., (2009). Antimicrobial effect and resistant regulation of *Glycyrrhiza uralensis* on methicillin-resistant *Staphylococcus aureus*. *Natural Product Research, 23*(2), 101–111.

119. Lee, Y. L., Ok-Hwa, K., Jang-Gi, C., You-Chang, O., & Dong-Yeul, K., (2010). Synergistic effect of emodin in combination with ampicillin or oxacillin against methicillin-resistant *Staphylococcus aureus*. *Pharmaceutical Biology, 48*(11), 1285–1290.

120. Li, L., & M-Hui, L., (2010). Distribution of seven polyphenols in several medicinal plants of Boraginaceae in *China*. *Journal of Medicinal Plants Research, 4*(12), 1216–1221.

121. Linassier, C., Pierre, M., Jean-Bernard, P., & Pierre, J., (1990). Mechanisms of action in NIH–3T3 cells of genistein, an inhibitor of EGF receptor tyrosine kinase activity. *Biochemical Pharmacology, 39*(1), 187–193.

122. Liz, R., Teresinha, N., & Frode, S., (2009). Evaluation of antimicrobial and antiplatelet aggregation effects of *Solidago chilensis Meyen*. *International Journal of Green Pharmacy, 3*(1), 35–39.

123. López, P., Sánches, C., Batlle, R., & Nerín, C., (2005). Solid and vaporphase antimicrobial activities of six essential oils: Susceptibility of selected foodborne bacterial and fungal strains. *J. Agric. Food Chem., 53*, 6939–6946.

124. Lopez, T., De Esparza, R., Ruiz, R., & Meckes, M., (2007). Antibacterial activity of *Piqueria trinervia*, a Mexican medicinal plant used to treat diarrhea. *Pharmaceutical Biology, 45*(6), 446–452.

125. Magid, A., Laurence, V. N., & Geoffroy, B., (2008). Tyrosinase inhibitors and sesquiterpene diglycosides from *Guioa villosa*. *Planta. Med., 74*(1), 55–60.

126. Maheshwari, R. K., Singh, A. K., Jaya, G., & Rikhab, C. S., (2006). Multiple biological activities of curcum: A short review. *Life Sciences, 78*(18), 2081–2087.

127. Mao, L., Zhang, Z., Donald, L. H., Wang, H., & Ruiwen, Z., (2007). Curcum, a dietary component, has anticancer, chemosensitization, and radiosensitization effects by down-regulating the *MDM2* oncogene through the PI3K/mTOR/ETS2 pathway. *Cancer Research, 67*(3), 19–28.

128. Maria, J. G., & Alexander, A. N., (2006). Genes involved in intrinsic antibiotic resistance of *Acenetobacter bayleyi*. *Antimicrobial Agents and Chemotherapy, 50*(11), 3562–3567.

129. Mariana, N. S., Norfarrah, M. A., Yusoff, F. M., & Arshad, A., (2009). Selective *in vitro* activity of marine extract on genes encoding membrane synthesis of methicillin resistance *Staphylococcus aureus*. *Biotechnology, 8*(1), 180–183.

130. Marimuthu, S., Adluri, R., & Sudheer, K. M., (2007). Ferulic acid: Therapeutic potential through its antioxidant property. *Journal of Clinical Biochemistry and Nutrition, 40*(2), 92–100.

131. Marja, P. K., & Anu, I. H., (1999). Activity of plant extract containing phenolic compounds. *Journal of Agricultural and Food Chemistry, 40*(5), 3954–3962.

132. Marta, C. T. D., Glyn, M. F., Adilson, S., Camila, D., Martinello, F., Soares, M. M., et al., (2005). Anti-candida activity of Brazilian medicinal plants. *Journal of Ethnopharmacology, 97,* 305–311.

133. Martinello, F., Soares, M. M., Santos, A. C., Sugohara, A., Garcia, S. B., & Curti, C., (2012). Anti-inflammatory and antinociceptive activities of a hydroethanolic extract of *Tamarindus indica* leaves. *Sci. Pharm., 80,* 685–700.

134. Mathabe, M., Ahmed, A., Hussein, V., & Nikolova, E., (2008). Antibacterial activities and cytotoxicity of terpenoids isolated from *Spirostachys Africana*. *Journal of Ethnopharmacology, 116,* 194–197.

135. Mazumdar, K., Dutta, N. K., Kumar, K. A., & Dastidar, S. G., (2005). *In vitro* and *in vivo* synergism between tetracycline and the cardiovascular agent oxyfedrine HCl against common bacterial strains. *Biol. Pharm. Bull., 28,* 713–717.

136. McDonald, S., Prenzler, P. D., Autolovich, M., & Robards, K., (2001). Phenolic content and antioxidant activity of olive extracts. *Food Chemistry, 73,* 73–84.

137. Mendoza, M. T., (1998). What's new in antimicrobial susceptibility testing? *Microbiology and Infectious Diseases, 27*(3), 113–115.

138. Miliauskas, G., & Venskutonis, T. A., (2004). Screening of radical scavenging activity of some medicinal and aromatic plant extracts. *Food Chemistry, 85*(20), 231–237.

139. Monika, S., Monika, Ł., Paweł, D., & Edward, K., (2011). The antimicrobial activity of thyme essential oil against multidrug resistant clinical bacterial strains, *Microbial Drug Resistance, 18*(2), 137–148.

140. Mosmann, T., (1983). Rapid colorimetric assay for cellular growth and survival: Application to proliferation and cytotoxicity assays. *Journal of Immunology Methods, 65,* 55–63.

141. Nair, R., Vaghasiya, Y., & Chanda, S., (2007). Antibacterial activity of *Eucalyptus citriodora* Hk. oil on few clinically important bacteria. *African Journal of Biotechnology, 7*(1), 25–26.

142. National Committee for Clinical Laboratory Standards (NCCLS), (1997). *Methods for Antimicrobial Susceptibly Tests for Bacteria that Grow Aerobically* (p. 332). Approved Standard M7-A5, NCCLS, Wayne, PA, USA.

143. National Committee for Clinical Laboratory Standards, (2000). *Methods for Dilution Antimicrobial Susceptibly Tests for Bacteria that Grow Aerobically* (p. 340). Approved Standard M7-A5, NCCLS, Wayne, PA, USA.

144. Naval, M. V., Gomez, M. P., Carretero, M. E., & Villar, A. M., (2007). Neuroprotective effect of a ginseng (*Panax ginseng*) root extract on astrocytes primary culture. *Journal of Ethnopharmacology, 112*(2), 262–270.

145. Ni, Y., (1995). Therapeutic effect of berberine on 60 patients with non-insulin dependent diabetes mellus and experimental research. *Chinese Journal of Integrated Traditional and Western Medicine, 1*(2), 91–95.

146. Nidhi, R., Vaghasiya, Y., & Chanda, S., (2007). Antibacterial activity of *Eucalpytus citriodora* Hk. oil on few clinically important bacteria. *African Journal of Biotechnology*, *7*(1), 25–26.

147. Nirmala, A., Eliza, J., Rajalakshmi, M., Priya, E., & Daisy, P., (2008). Effect of hexane extract of *Cassia fistula* barks on blood glucose and lipid profile in streptozotocin diabetic rats. *International Journal of Pharmacology*, *4*(5), 292–296.

148. Nostro, A., Germanpo, V. D., Angelo, A., & Marino, M. A., (2000). Extraction methods and bioautography for evaluation of medicinal plant antimicrobial activity. *Letters in Applied Microbiology*, *30*(5), 379–384.

149. Odebiyi, O. O., (1979). Antimicrobial alkaloids from a Nigerian chewing stick (*Fagara zanthoxyloides*). *Planta. Medica.*, *36*, 204–207.

150. Oliveira, I., Sousa, A., Ferreira, I., Bento, A., Estevinho, L., & Pereira, J. A., (2008). Total phenols, antioxidant potential and antimicrobial activity of walnut (*Juglans regia* L.) green husks. *Food Chemistry and Toxicology*, *46*, 2326–2331.

151. Olukemi, O. A., Ilori, O., Sofidiya, M., Aniunoh, O., & Lawal, B., (2005). Antioxidant activity of Nigerian dietary spices. *Journal of Environment and Agriculture Food Chemistry*, *4*(6), 1086–1093.

152. Omer, E., (2006). Antibacterial and antifungal activity of ethanolic extracts from eleven spice plants. *Biologia*, *61*(3), 275–278.

153. Ono, K. H., Nakane, M., & Fukishima, J. C., (1997). Inhibition of reverse transcriptase activity by a flavonoid compound, 5, 6, 7-trihydroxyflavone. *Biochem. Biophys. Res. Commun.*, *3*, 982–987.

154. Parekh, J., Karathia, N., & Chanda, S., (2006). Screening of some traditionally used medicinal plants for potential antibacterial activity. *Indian Journal of Pharmaceutical Science*, *68*(6), 832–834.

155. Parekh, J., Karathia, N., & Chanda, S., (2006). Antibacterial activity of *Bauhinia variegata*. *Journal of Biomedical Research*, *9*, 53–56.

156. Park, Y. H., & Han, D. W. H., (2003). Protective effects of green tea polyphenol against reactive oxygen species-induced oxidative stress in cultured rat calvarial osteoblast. *Cell Biology and Toxicology*, *19*(5), 325–337.

157. Pezzuto, M. J., (2008). Resveratrol as an inhibitor of carcinogenesis. *Pharmaceutical Biology*, *46*(7 & 8), 443–573.

158. Pietta, P. G., (2000). Flavonoids as antioxidants. *Journal of Natural Products*, *63*(7), 1035–1042.

159. Poole, K., (2000). Efflux-mediated resistance to fluoroquinolones in gram-negative bacteria. *Antimicrobial Agents and Chemotherapy*, *44*, 2233–2241.

160. Pourmorad, S., Hosseinimehr, S. J., & Shahabimajd, N., (2006). Antioxidant activity, phenol and flavonoid contents of some selected Iranian medicinal plants. *African Journal of Biotechnology*, *5*(11), 1142–1145.

161. Pranjpe, P., (2001). *Indian Medicinal Plants Forgotten Healers: A Guide to Ayurvedic Herbal Medicine with Identity, Habitat, Botany, Photochemistry, Ayurvedic Properties, Formulations & Clinical Usage* (p. 438). Chaukhamba Sanskrit Pratishthan, New Delhi.

162. Rabbani, C., (1986). The effect of vinca alkaloid anticancer drug, vinorelbine, on chromatin and histone proteins in solution. *European Journal of Pharmacology*, *613*(1–3), 34–38.

163. Ramos, F. A., Yoshihisa, T., Miki, S., Yousuke, K., Tsuchiya, K., & Takeuchi, M., (2006). Antibacterial and antioxidant activities of quercetin oxidation products from yellow onion (*Allium cepa*) skin. *Journal of Agriculture and Food Chemistry, 54*(10), 3551–3557.

164. Randolph, R. J. A., Androutsopoulos, V., Somchaiya, S., & Gerry, A. P., (2008). Phytoestrogens as natural prodrugs in cancer prevention: Dietary flavonoids. *Phytochemistry Reviews, 8*(2), 375–386.

165. Rao, K., Ch, B., Narasu, L. M., & Giri, A., (2010). Antibacterial activity of *Alpinia galanga* (L) wild crude extracts. *Applied Biochemistry and Biotechnology, 162*(3), 871–884.

166. Rastogi, N., Domadia, P., Shetty, S., & Dasgupta, D., (2008). Screening of natural phenolic compounds for potential to inhibit bacterial cell division. *Indian Journal of Experimental Biology, 46*(11), 783–787.

167. Raven, P. H., (1998). Medicinal plants and global sustainability. In: *Medicinal Plants: A Global Heritage–Proceedings of the International Conference on Medicinal Plants for Survival* (pp. 14–18). International Development Research Center, New Delhi.

168. Ravindranath, M. H., Vaishali, R., & Songeun, M., (2009). Differential growth suppression of human melanoma cells bytea (*Camellia sinensis*) epicatechins (ECG, EGC and EGCG). *eCAM, 6*(4), 523–530.

169. Rayalam, S., Della-Fera, M. A., & Baile, C. A., (2008). 19 Phytochemicals and regulation of the adipocyte life cycle. *Journal of Nutritional Biochemistry*, 717–726.

170. Rice, E. C., Nicholas, J., Milleran, G. A., Ronald, R. C., & David, P. H., (2008). Antibacterial activity of plants used in traditional medicines of Ghana with particular reference to MRSA. *Journal of Ethnopharmacology, 116*(1), 102–111.

171. Robinson, M. M., & Zang, X., (2011). *The World Medicine Situation 2011: Traditional Medicine: Global Situation, Issues and Challenges*. WHO Press, Geneva, http://www.who.int/medicines/areas/policy/world_medicines_situation/WMS_ch18_wTraditionalMed.pdf (accessed on 31 March 2018).

172. Rogerio, A. P., Kanashiro, A., Fontanari, C., Da Silva, E. V. G., & Lucisano, V. Y. M., (2007). Anti-inflammatory activity of quercetin and isoquercitrin in experimental murine allergic asthma. *Inflammation Research, 56*(6), 402–408.

173. Rossi, A., Ligresti, A., Longo, R., Russo, A., Borrelli, F., & Sautebin, L., (2002). The inhibitory effect of propolis and caffeic acid phenethyl ester on cyclooxygenase activity in J774 macrophages, *9*(6), 530–535. http://www.ncbi.nlm.nih.gov/pubmed/12403162.

174. Sacchetini, J. C., & Poulter, C. D., (1997). Creating isoprenoid diversity. *Science, 277*, 1788–1789.

175. Saha, M. R., Ashraful, A., Akter, R., & Jahangir, R., (2008). *In vitro* free radical scavenging activity of *Ixora coccinea* L. *Bangladesh Journal of Pharmacology, 3*(10), 90–96.

176. Saleem, M., Kweon, M. H., Yun, J. M., Syed, D. N., Adhami, V. M., & Mukhtar, H., (2005). A novel dietary triterpene Lupeol induces Fas-mediated apoptotic death of androgen-sensitive prostate cancer cells and inhibits tumor growth in a xenograft model. *Cancer Research, 65*(23), 11203–11213.

177. Saleem, M., (2005). Lupeol, a fruit and vegetable-based triterpene, induces apoptotic death of human pancreatic adenocarcinoma cells via inhibition of RAS signaling pathway. *Carcinogenesis, 26*(11), 1956–1964.

178. Sanchez-Moreno, C., Jimenez-Escrig, A., & Saura-Calixto, F., (2006). Study of low-density lipoprotein oxidizabily indexes to measure the antioxidant activity of dietary polyphenols. *Nutr. Res.*, *24*(5), 324–328.

179. Sevil, E., Bircan, G. M., & Murat, K., (2009). DNA damage protecting activity and *in vitro* antioxidant potential of the methanol extract of *Cyclotrichium niveum*. *Pharmaceutical Biology*, *47*(3), 219–2297.

180. Shai, L. J., McGaw, l., Aderogba, M. K., & Eloff, J. N., (2008). Four pentacyclic triterpenoids with antifungal and antibacterial activity from *Curtisia dentata* (Burm.f) C.A. Sm. leaves. *Journal of Ethanopharmacology*, *119*(2), 238–244.

181. Shukla, Y., (2009). Induction of apoptosis by lupeol in human epidermoid carcinoma A431 cells through regulation of mitochondrial, Akt/PKB and NFkappaB signaling pathways. *Cancer Biol. Ther.*, *8*(17), 1632–1639.

182. Silver, L. L., (2007). Multitargeting monotherapautic antibacterials. *Nature Reviews on Drug Discovery*, *6*, 41–55.

183. Singh, P., & Goyal, G. K., (2008). Dietary lycopene: Its properties and anticarcinogenic effects. *Comprehensive Reviews in Food Science and Food Safety*, *7*(3), 255–270.

184. Singleton, V. L., Orthofer, R., & Lamuela-Raventos, R. M., (1999). Analysis of total phenols and other oxidation substrates and antioxidants by means of Folin-Ciocalteu reagent. *Methods Enzymol.*, *299*, 152–178.

185. Sokmen, A., Brian, M. J., & Erturk, M., (1999). The *in vitro* antibacterial activity of Turkish medicinal plants. *Journal of Ethnopharmacology*, *67*(1), 79–86.

186. Sreeja, S., (2009). An *in vitro* study on antiproliferative and antiestrogenic effects of *Boerhaavia diffusa* L. extracts. *Journal of Ethanopharmacology*, *126*(2), 221–225.

187. Srinivasagam, R. S., Arumugam, K., & Manickam, R., (2007). The neuroprotective effect of *Withania somnifera* root extract in MPTP-intoxicated mice: An analysis of behavioral and biochemical variables. *Cellular & Molecular Biology Letters*, *12*(4), 473–481.

188. Stermitz, F. R., Peter, L., Tawara, J. N., & Lauren, A., (1999). Synergy in a medicinal plant: Antimicrobial action of berberine potentiated by 5*-methoxyhydnocarpin, a multidrug pump inhibitor. *Proc. Natl. Acad. Sci. USA*, *2*(8), 124–128.

189. Stiborova, M. J., Sejba, L., Aimova, D., Poljkova, J., Wisener, J., & Frei, E., (2004). The anticancer drug elipticine forms covalent DNA adducts, mediated by human cytochrome P450, through metabolism to 13-hydroxyellipticine and ellipticine N2 oxide. *Cancer Research*, *64*, 8374–8380.

190. Tada, M., Hiroe, Y., Kiyohara, S., & Suzuki, S., (1998). Noematicidal and antimicrobial constituents from *Allium grayi* Regel and *Alliumfistulosum* L. var. *caespitosum*. *Agricultural Biology and Chemistry*, *52*(7), 2383–2385.

191. Tanira, M. O., Wasfi, I. A., Al-Homsi, M., & Bashir, A. K., (1996). Toxicological effects of *Teucrium stocksianum* after acute and chronic administration in rats. *J. Pharm. Pharmacol.*, *48*, 1098–1102.

192. Taylor, C. E., & Greenough, W. B., (1989). Control of diarrheal diseases. *Annual Review of Public Health*, *10*, 221–224.

193. Tepe, B., (2005). Antimicrobial and antioxidant activities of the essential oil and various extracts of Salvia *tomentosa* Miller (Lamiaceae). *Food Chemistry*, *90*(3), 333–340.

194. Teramachi, F., Koyano, T., Kowithayakorn, T., Hayashi, M., Komiyama, K., & Ishibashi, M. J., (2005). Collagenase inhibitory quinic acid esters from *Ipomoea pes-caprae*. *Journal of Natural Products, 68*(5), 794–796.

195. Tereschuk, M., Marta, V. Q. R., Guillermo, R. C., & Lidia, R. A., (*1997*). Antimicrobial activity of flavonoids from leaves of *Tagetes muta*. *Journal of Ethanopharmacology, 56(3),* 227–232.

196. Thara, K. M., & Zuhara, K. F., (2016). Biochemical analysis and antiproliferation action of *Centratherum Anthelmticum* (L) Kuntze seed extract on cancer cell lines like Dalton's *Lymphoma Ascites* (DLA) and *Ehrlich Ascites Carcinoma* (EAC). *Indo American Journal of Pharmaceutical Research, 6*(1), 3941–3948.

197. Thara, K. M., & Fathimath, Z., (2013). Biochemical, HPLC, LC-MS analysis and biological activities of methanol extract of *Alstonia scholaris*. *Inter. J. Phytotherap., 3*(2), 61–74.

198. Thara, K. M., Zuhara, K. F., & Raji, T. K., (2013). Inhibitory and synergistic effect of *Pterocarpus marsupeum* extracts on clinical and standard strains of microorganisms and its biochemical analysis. *Indo American Journal of Pharmaceutical Research, 3*(7), 5583–5593.

199. Thisoda, P., Nuchanart, R., Nanthanit, P., Worasuttayangkurn, L., Ruchirawat, S., & Jutamaad, S., (2000). Inhibitory effect of *Andrographis paniculata* extract and its active diterpenoids on platelet aggregation. *European Journal of Pharmacology, 553*(1–3), 39–45.

200. Toshio, F., Marumo, A., Kiyoshi, K., Toshihisa, K., Sumio, T., & Taro, N., (2002). Antimicrobial activity of licorice flavonoids against methicillin-resistant *Staphylococcus aureus*. *Fitoterapia, 73*(6), 536–539.

201. Turner, A., (1970). Terpenoids and steroids. *Annual Reports on the Progress of Chemistry, Section B, Organic Chemistry, 66*, 389–441.

202. Viljoen, A., Van-Vuuren, S., Ernst, E., Klepser, M., Demirci, B., Basser, H., & Van, W. B. E., (2003). *Osmitopsis asteriscoides* (Asteraceae)–the antimicrobial and essential oil composition of a Cape-Dutch remedy. *J. Ethnopharmacol., 88*(2 & 3), 137–143.

203. Vogel, A. I., (1958). *Textbook of Practical Organic Chemistry* (pp. 90–92). Longman, London.

204. Wagner, K. H., & Elmadfa, I., (2003). Biological relevance of terpenoids: Overview focusing on mono-, di- and tetraterpenes. *Ann. Nutr. Metabolism., 47*(3 & 4), 95–106.

205. Wamtinga, R. S., (2006). Phenolic content and antioxidant activity of six Acanthaceae from *Burkina Faso. Journal of Biological Sciences, 6*, 249–252.

206. Wang, J., Chen, J., & Lu, Y., (2011). Determination of phenolic compounds from different fractions of *Solidago Canadensis*. *Chinese Journal of Information on Traditional Chinese Medicine, 10*, 25–34.

207. Wang, Z., Bertram, J. C., Jeffrey, C. B., Kassisà, S., Melanie, H. C., Peter, R., et al., (1998). Structural basis of inhibitor selectivity in MAP kinases. *Structure, 6*(9), 1117–1128.

208. Wang, W., Ke, T., Hui-Ling, W., & Wei-Dong, H., (2010). Distribution of resveratrol and stilbene synthase in young grape plants (Vitis *vinifera* L. cv. Cabernet Sauvignon) and the effect of UV-C on its accumulation. *Plant Physiology and Biochemistry, 48*(2 & 3), 142–152.

209. Weinberg, B., & Bealer, B., (2002). The world of caffeine. In: *The Science and Culture of World's Popular Drug* (pp. 58–68). Routledge, New York.
210. Whitsett, J., Timothy, G., & Lamartiniere, C. A., (2006). Genistein and resveratrol: Mammary cancer chemoprevention and mechanisms of action in the rat. *Expert Review of Anticancer Therapy, 6*(12), 1699–1706.
211. Williams, C. A., & Grayer, R. J., (2004). Anthocyanins and other flavonoids. *Nat. Prod. Rep., 21*, 539–573.
212. Wu, D., Kong, Y., Cong, H., Jing, C., Lihong, H., & Xu, S., (2008). D-Alanine: Ligase as a new target for the flavonoids quercetin and apigenin. *Antimicrobial Agents and Chemotherapy, 32*(5), 421–426.
213. Xiaowei, S., Mark, Y., Sangster, P., & Doris, H. D., (2010). *In vitro* effects of pomegranate juice and pomegranate polyphenols on foodborne viral surrogates. *Foodborne Pathogens and Disease, 7*(12), 1473–1479.
214. Xu, H., & Lee, S. F., (2001). Activity of plant flavonoids against antibiotic-resistant bacteria. *Phytotherapy Research, 15*, 39–43.
215. Yan, P., (2011). *In vitro* modulatory effects of *Andrographis paniculata, Centella asiatica* and *Orthosiphon stameus* on cytochrome P450 2C19 (CYP2C19). *Journal of Ethnopharmacology, 133*(2), 881–887.
216. Yean-Yean, S., & Philip, J. B., (2006). Quantification of Gallic acid and ellagic acid from longan (*Dimocarpus longan* Lour.) seed and mango (*Mangifera indica* L.) kernel and their effects on antioxidant activity. *Food Chemistry, 97*(3), 524–530.
217. Young-Hee, L., (2007). *In vitro* activity of kaempferol isolated from the *impatiens balsama* in combination with erythromycin or clindamycin against *Propionibacterium acnes. The Journal of Microbiology, 47*(5), 473–477.
218. Yu, W. B., Jian-Guang, L., Wang, J. S., Zhang, D. M., & Kong, Y., (2011). Pentasaccharide resin glycosides from *Ipomoea pes-caprae. Journal of Natural Products, 74*(4), 620–628.
219. ZebSaddiqe, I. N., & Alya, M., (2010). Review of the antibacterial activity of *Hypericum perforatum* L. J. Ethnopharmacol., *131*(3), 511–521.
220. https://en.wikipedia.org/wiki/Phenols (accessed on November 20, 2016).
221. http//en.m.wikipedia.org/wiki/atropine, (accessed on November 20, 2016).
222. www.google.co.in/camptothecin (accessed on November 20, 2016).
223. www.ncbi.nih.gov/terminalia (accessed on 20 November 2016).
224. www.drugs.com/npc/ginseng (accessed on 20 November 2016).
225. www.himalayawellness.com/himpyrin (accessed on 20 November 2016).
226. https://en.wikipedia.org/wiki/Digitalis (accessed on 20 November 2016).
227. www.google.co.in/camptothecin (accessed on 20 November 2016).
228. www.wikipedia.org//wiki/qunin (accessed on 20 November 2016).
229. https://www.ncbi.nlm.nih.gov › NCB/resiprine (accessed on 20 November 2016).
230. www.wikipedia.org/wiki/tamiflu (accessed on 20 November 2016).
231. www.wikipedia.org/wiki/taxol (accessed on 20 November 2016).
232. www.wikipedia.org/wiki/vincristine (accessed on 20 November 2016).

PART II
Plant-Based Drugs

CHAPTER 3

ROLE OF HERBAL DRUGS IN MANAGEMENT OF LIFE STYLE DISEASES

V. SREELAKSHMI and ANNIE ABRAHAM*

*Corresponding author. E-mail: annieab2001@gmail.com

ABSTRACT

Medicinal plants are rich sources of therapeutic agents for the prevention of diseases and ailments. These may emerge as good substitutes or better alternatives for synthetic chemicals based drugs or may even replace them. This chapter investigates the potential of herbal medicines in health and disease management with special reference to lifestyle diseases.

3.1 INTRODUCTION

The discovery of the relationship between oxidant species and degenerative diseases has revolutionized the field of medical research and guarantees a new age of disease management. Oxygen is a necessary element in our daily life and cells utilize oxygen for energy generation. Oxidant species are considered as the metabolic by-products of cellular respiration. These free radical by-products are generally classified as: reactive oxygen species (ROS) and reactive nitrogen species (RNS). They have both good and bad effects on the human body. The intricate balance between its dual roles forms the basic aspect of normal life [97].

This chapter investigates: a. Role of free radicals and oxidative stress in biological systems; and b. Potential of herbal medicines in health and disease management of lifestyle diseases.

3.2   FREE RADICALS

Free radicals are chemical entities in which their outer orbital has an unpaired electron [109]. They are highly reactive and are strong oxidants due to the contribution of an electron or reductants due to acceptance of an electron from other molecules [14]. The major free radicals are [34] listed below:

- Dinitrogen trioxide (N_2O_3);
- Hydrogen peroxide (H_2O_2);
- Hydroxyl radical (OH•);
- Hypochlorous acid (hocl);
- Lipid peroxide (LOOH);
- Lipid peroxyl radical (LOO•),
- Nitric oxide radical (NO•);
- Nitrogen dioxide (NO_2);
- Nitrous acid (HNO_2);
- Ozone (O_3);
- Peroxynitrite ($ONOO^-$);
- Peroxyl radical (ROO•);
- Singlet oxygen ($^1O_2$); and
- Superoxide anion radical ($O_2•^-$).

Following are three important sources of oxidant species [103]:

a. **Endogenous sources** are the normal cellular metabolic pathways in a biological system such as: mitochondrial electron transport chain and cytochrome p450 mediated xenobiotic removal.
b. **Exogenous sources** include cigarette smoke, alcohol, industrial wastes, chlorinated compounds, ionizing radiations, microbial infections, etc.
c. **Pathological sources** include inflammatory responses, mental stress, some drugs, etc.

In the cellular energy production process, molecular oxygen is reduced to series of oxidant species as follows [98]:

$$O_2 + e^- + H^+ \longrightarrow HO_2•$$
$$HO_2• \longrightarrow H^+ + O_2•^-$$
$$O_2•^- + 2H^+ + e^- \longrightarrow H_2O_2$$

$$H_2O_2 + e^- \longrightarrow HO^- + OH\bullet$$
$$OH\bullet + H^+ + e^- \longrightarrow H_2O \qquad (1)$$

Superoxide is the precursor of most ROS, and it is used for killing microbes by immune cells. ROS can also be produced by the myeloperoxidase-halide-H_2O_2 system. Myeloperoxidase (MPO) is a heme enzyme that is present in granules of activated neutrophils, macrophages, and monocytes. MPO synthesize the potent oxidant, hypochlorous acid from hydrogen peroxide and chloride ion [81].

$$H_2O_2 + H^+ + Cl^- \longrightarrow HOCl + H_2O \qquad (2)$$

Divalent metal ions (copper or iron) catalyze the generation of OH• by Fenton reaction. Iron is oxidized by H_2O_2 and forms OH•.

$$H_2O_2 + Fe^{2+} \longrightarrow OH^- + Fe^{3+}$$
$$O_2\bullet^- + H_2O_2 \longrightarrow OH\bullet + OH^- + O_2 \qquad (3)$$

The enzyme nitric oxide synthase (NOS) produces RNS from amino acid arginine. NO reacts with $O_2\bullet^-$ to produce $ONOO^-$.

$$L\text{-Arginine} + O_2 + NADPH \longrightarrow NO\bullet + citrulline$$
$$NO\bullet + O_2\bullet^- \longrightarrow ONOO^- \qquad (4)$$

ROS and RNS have both favorable and harmful potential depending on the environment [35, 64]. The favorable effects include cellular signaling, defense against infectious agents, etc. However, the unstable configuration of free radicals makes them high energy entities. The excess energy is liberated through their reactions with any molecules it interacts. According to the free radical theory of aging, the aging of organisms is because of the accumulation of free radicals in the cells [44]. Among the various ROS, OH• is the extremely reactive species and interacts with all the biomolecules. The long-term effects of oxidative stress are inflicted in DNA modifications [2].

Oxidation of proteins results in the unfolding and misfolding of functional protein with the loss of three-dimensional structures of proteins. Attack of oxidants on protein thiol groups results in oxidative damage along with carbonylation, thus leading to the formation of advanced glycation end (AGE) products [23, 126]. The RNS, $ONOO^-$ is one of the major oxidants associated with protein alteration. $ONOO^-$ interact with metalloproteins hemoglobin, myoglobin, and cytochrome C. It oxidizes the metal ions and leaves them nonfunctional. It also nitrates proteins. All of these reactions affect protein architecture and thus alter the enzymatic activities and cell signaling cascades [88].

Lipids are most vulnerable biomolecules that undergo free radical attack through a process called lipid peroxidation. Free radicals act directly on the polyunsaturated fatty acids (PUFA) because of the presence of multiple double bonds and the reactive hydrogen atoms in their methylene bridges ($-CH_2-$). In the process of peroxidation, free radicals take electrons from the lipids in cell membranes; and it advances by a chain reaction mechanism [28, 40, 53].

Lipid peroxidation initiated, when a radical removes a hydrogen from PUFA and forms lipid radical (L•). L• undergo some chemical rearrangement and forms a conjugated diene radical. In the presence of oxygen, the conjugated radical forms lipid peroxyl radicals (LOO•), the carriers of the chain reactions, as shown below:

$$LH + R\bullet \longrightarrow L\bullet + RH$$
$$L\bullet + O_2 \longrightarrow LOO\bullet \tag{4}$$

LOO• capture a hydrogen atom from adjacent PUFA and forms lipid hydroperoxide (LOOH) and a second lipid radical (L•). L• can continue the same reactions and produce additional lipid hydroperoxides and propagated the process of lipid peroxidation. LOOH are able to break down to aldehydes. They are able to diffuse to different parts of the body-spread peroxidation. The aldehyde, 4-hydroxynonenal binds to proteins and makes them non-functional [23].

$$LOO\bullet + LH \longrightarrow LOOH + L\bullet$$
$$LOO\bullet + LH \longrightarrow LOO\bullet$$
$$LOOH \longrightarrow LO\bullet + LOO\bullet + aldehydes \tag{5}$$

All these events are implicated in tissue injury and free radical-mediated alteration of biomolecules [29, 42, 99, 136].

3.3   ANTIOXIDANTS

An antioxidant is a molecule capable of preventing the oxidation of other molecules; contain monohydroxy/polyhydroxy phenol and has lower activation energy for hydrogen donation. The presence of an electron-donating group especially a hydroxyl group on the *o-* or *p*-positions makes a compound polar. These hydrogen atoms deactivate the free radicals and break off the chain reaction. This accomplishes the antioxidant ability of a molecule. Cells are equipped with various antioxidants to alleviate ROS/RNS damage and for cellular repair [82]. According to Gutteridge and Halliwell [73], antioxidants are classified as:

a. *Primary antioxidants:* These are concerned with the inhibition of oxidants generation.
b. *Secondary antioxidants:* These scavenge oxidant species.
c. *Tertiary antioxidants:* These repair the oxidized molecules.

Antioxidants are also classified into natural and synthetic based on their source. Natural antioxidants are synthesized through metabolism in the human body and are also supplemented to the body from other natural sources. Synthetic antioxidants are synthesized in the laboratory and supplied for boosting the antioxidant potential of the body. Butylated hydroxyl anisole, butylated hydroxy toluene, and propyl gallate are examples of synthetic antioxidants. Natural antioxidants can be classified further into enzymatic or non-enzymatic antioxidants. Enzymatic antioxidants are exclusively produced in the human body. Superoxide dismutase (SOD), catalase, glutathione peroxidase (GPx), glutathione reductase (GR), glutathione S transferase (GST), etc. are examples of enzymatic antioxidants. Non-enzymatic antioxidants are reduced glutathione (GSH), vitamins, carotenoids, etc.

3.3.1   ENZYMATIC ANTIOXIDANTS

3.3.1.1   SUPEROXIDE DISMUTASE (SOD)

SOD is an oxidoreductase that converts hydrogen peroxide into water and molecular oxygen [30].

$$2O_2^{\bullet-} + 2H^+ \longrightarrow H_2O_2 + O_2 \qquad (6)$$

SOD is also known to compete NO for superoxide anion, which inactivates NO to form $ONOO^-$ [30]. It is an important primary antioxidant that affords protection in nearly all the living cells exposed to oxygen. There are three isoforms of SOD found in mammals. Each form is a product of distinct genes and having different sub-cellular localization with the same function.

- **SOD1** is located in the cytoplasm and organelles of cells. It is a dimeric protein (32,000 kDa) with copper and zinc in each monomer.
- **SOD2** is located in the mitochondria of cells. It is a tetrameric protein (40,000 kDa) with a single manganese atom in each subunit.
- **SOD3** is a secretory protein with copper and zinc as co-factors and expressed in fibroblasts and endothelial cells.

3.3.1.2   CATALASE

Catalase is a primary antioxidant located in peroxisomes, and it detoxifies hydrogen peroxide [15].

$$2H_2O_2 \longrightarrow 2H_2O + O_2 \tag{7}$$

Structurally catalase is a tetramer, and each polypeptide chain contains a porphyrin heme (iron) group and a molecule of NADPH. Heme group permits the enzyme to counter with hydrogen peroxide [11]. Catalase is the enzyme with the highest turnover number among all the enzymes. One molecule of catalase can decompose 40 million hydrogen peroxide molecules per second. It thus counteracts and balances the continual production of hydrogen peroxide in the biological system. Catalase also oxidizes toxins such as formaldehyde and nitrite using use hydrogen peroxide [37].

3.3.1.3   GLUTATHIONE SYSTEM

The glutathione system of antioxidants includes: GSH, GPx, GR, and GST. The GPx is a group of selenium-dependent primary antioxidant enzymes that function in the decomposition of hydrogen peroxide. Different isoforms of GPx are encoded by different genes, and more than eight forms have been identified in humans. The isozymes vary in cellular location and substrate specificity. GPx1 is the most abundant isoform present in the mammalian cytoplasm. The most preferred substrate of GPx1 is hydrogen peroxide lipid, and hydroperoxides are a substrate for GPx4. The GPx2 is an intestinal and extracellular enzyme, and GPx3 is extracellular [79]. The reaction catalyzed by GPx is as follows, where GSSG represents glutathione disulfide.

$$2GSH + H_2O_2 \rightarrow GSSG + 2H_2O$$

$$2GSH + LOOH \rightarrow GSSG + LOH + 2H_2O \tag{8}$$

Hydrogen peroxide generated is neutralized by catalase. For the continuous functioning of GPx, glutathione disulfide must be converted back to the reduced form. The ratio of GSSG/GSH in the cell is a major factor in the redox maintenance of the cell. It is essential to maintain high levels of the GSH and a low level of GSSG. This equilibrium is retained by GR, which catalyzes the reduction of GSSG to GSH [19].

$$GSSG + NADPH + H^+ \rightarrow 2GSH + NADP^+ \tag{9}$$

The GR is a secondary antioxidant enzyme with tissue distribution similar to GPx. It is a flavor-enzyme and NADPH necessary for its action is obtained from the hexose monophosphate pathway.

GST represents a family of isozymes capable of catalyzing the conjugation of xenobiotic compounds to glutathione and render them less or nontoxic [84]. Lipid peroxidation end products (such as 4-hydroxynonenal) is a predominant substrate of GST, and thus it is a secondary antioxidant enzyme [114].

3.3.2 METABOLIC ANTIOXIDANTS

3.3.2.1 REDUCED GLUTATHIONE

GSH (L-γ-glutamyl-L-cysteinylglycine) is the major non-protein thiol found in cells [56]. The major antioxidant functions of GSH are [67, 123]:

a. Functions as a cofactor for various enzymes, such as multiple peroxidases, and GST, etc.
b. Interacts directly with ROS/RNS and electrophiles.
c. Maintaining the sulfhydryl groups of proteins in reduced form.
d. Functions in DNA synthesis as cofactor of ribonucleotide reductase.
e. Regeneration of other antioxidants like vitamin C and vitamin E.
f. Maintaining the cellular cysteine reserve.

3.3.2.2 VITAMINS

Vitamin C or ascorbic acid is a major water-soluble, antioxidant molecule in the biological system [5]. Ascorbic acid is oxidized to a reactive electrophile, dehydroascorbic acid by contributing an electron to the lipid radical and terminates the lipid peroxidation chain reaction. Dehydroascorbic acid reacts with nucleophiles on proteins, resulting in non-enzymatic modifications of proteins. Dehydroascorbic acid is reduced back to ascorbic acid with the addition of two electrons by GSH. It promotes the regeneration of vitamin E (α tocopherol) from α Tocopheroxyll radical produced during ROS scavenging.

Vitamin E or tocopherols are fat soluble, chain-breaking antioxidants that are present in all cellular membranes and α tocopherol is the most active and abundant form in the biological system. It protects the integrity of lipid membranes. Tocopherols are able to donate its phenolic hydrogen atom to lipid peroxy radicals and form Tocopherol Radical, and it is reduced back by ascorbic acid and retinol [12].

3.3.2.3 CAROTENOIDS

Carotenoids are a group of natural pigments that are found in plants and microorganisms. They are present in liver, egg yolk, milk, butter, spinach, carrots, tomato, and grains [133]. These are lipophilic compounds and have antioxidant functions in lipid phases at low oxygen partial pressures. Carotenoids can be classified as: carotenoid hydrocarbons or carotenes and oxygenated carotenoids or xanthophylls. The antioxidant activity of carotenoids is by its system of conjugated double bonds for the delocalization of unpaired electron [77]. Besides being a precursor to vitamin A, β carotene has potent antioxidant properties as it traps singlet oxygen without degradation and protects against a free radical attack. Ascorbic acid and tocopherols are the antioxidant partners of carotenoids. Carotenoids can afford protection against peroxidative damage of lipids at high concentrations.

3.3.2.4 POLYPHENOLS

Polyphenols or polyhydroxy phenols constitute a group of plant secondary metabolites that represent an integral part of the human diet [108].

Polyphenols are potent antioxidants that are able to reduce the generation of ROS/RNS, scavenge them and thus preventing damage to biomolecules. More than 8,000 polyphenolic compounds have been identified in various plant species.

3.4 OXIDATIVE STRESS

Under steady-state physiological conditions, the oxidative molecules produced from external sources and normal are scavenged by various antioxidative defense mechanisms described earlier in this chapter. Oxidative stress is the disparity in the production of oxidant species and body's ability to detoxify these oxidants.

When the level of ROS is up-regulated, this affects the normal cellular metabolism and the destructive nature of oxidants affects almost all organs and is reflected in degenerative disorders. The pathology initiates from the disturbance in either reactive species formation, their elimination or in both, simultaneously [66]. The effects of oxidative stress on the human body have been investigated by researchers.

3.5 SELECTED LIFE STYLE DISEASES

Lifestyle diseases or diseases of civilization or diseases of longevity are diseases associated with mode of life. Oxidative stress is well known to be involved in the pathogenesis of lifestyle-related diseases. Some of the notable lifestyle-linked diseases caused by oxidative stress are described in this section.

3.5.1 DIABETES MELLITUS

It is a group of metabolic diseases associated with high blood sugar levels for a long period [134]. Growing scientific data on both experimental and clinical trials suggest the connecting link between oxidative stress and diabetes pathology.

Pancreas functions in both exocrine and endocrine ways. The process of digestion is aided through its exocrine function by producing enzymes. Blood sugar level regulation is through the endocrine function by producing hormones. Islets of Langerhans are irregularly shaped patches of endocrine tissues distributed throughout the pancreas. There are five different types of cells present in the pancreatic islets with diversified functions [26], as follows:

a.   Alpha cells (produce glucagon).
b.   Beta cells (produce insulin and amylin).
c.   Delta cells (produce somatostatin).
d.   Epsilon cells (produce ghrelin).
e.   Gamma cells (produce pancreatic polypeptides).

Glucagon and insulin produced by pancreatic cells are responsible for homeostasis of glucose in the body. When the concentration of glucose in the bloodstream is low, the pancreas releases glucagon. Glucagon promotes the process of glycogenolysis and causes the liver to convert stored glycogen into glucose for the immediate energy need of the body. On the other hand, a high blood glucose level stimulates the release of insulin. It allows glucose to be used by insulin-dependent tissues and promotes glycogen synthesis for storage. Glucagon and insulin are linked in a feedback system that maintains the blood glucose levels at a steady level.

Diabetes mellitus is a condition characterized either because of the reduction in insulin production or inability of cells to respond properly to the insulin produced [18, 118].

3.5.1.1   DIABETIC PATHOLOGY AND COMPLICATIONS

Metabolic abnormalities of diabetes are closely associated with oxidative stress. Intracellular carbohydrate metabolism is altered as a result of hypo-glycemic conditions. With unrelenting hyperglycemia, unbalanced level of glucose is transported to the cells. It increases the glucose instability through glycolysis and tricarboxylic acid cycle. This directs the electron transport chain in the mitochondria and excessive superoxide generation over and above the detoxifying potential of antioxidant enzymes [140]. This alters the optimal balance between oxidant generation and degrada-tion and favors oxidative stress. Further, higher glucose level results in its auto-oxidation. Glucose is oxidized in a transition-metal dependent reaction to keto-aldehydes and superoxide anion radicals.

Formation of AGEs-products is another way for oxidant generation in hyperglycemic condition through the reaction of sugar with protein.

Interaction of receptor for advanced glycation end products (RAGE) with its ligand AGE is activation of pro-inflammatory pathways. This again promotes the free radical formation and activates nuclear transcription factor-κB (NF-κB). NF-κB enhances production of nitric oxide. ROS/RNS mediated pathways ultimately result in the destruction of pancreatic beta cell damage [24, 72].

The complications of free radical and oxidative stress induced diabetes mellitus are: Cardiovascular diseases (CVDs), neuropathy, nephropathy, retinopathy, stroke, hearing impairment, and skin problems.

3.5.2 CARDIOVASCULAR DISEASES (CVDS)

CVD is a group of diseases of heart and blood vessels. The morbidity and mortality rate of CVD is expected to grow to 23.6 million by 2030 [74]. The conditions of CVDs are: Aortic aneurysms, cardiomyopathy, carditis, congenital heart disease, coronary artery disease (CAD: such as angina and myocardial infarction), heart arrhythmia, hypertensive heart disease, peripheral artery disease, rheumatic heart disease, stroke, valvular heart disease, and venous thrombosis.

In most CVDs, a symptomatic atherosclerosis commences from early days and progresses with age. Atherosclerosis or arteriosclerotic vascular disease (ASVD) is the hardening of the arteries, where deposits of fat-laden macrophages (foam cells) along with cholesteryl esters, free cholesterol, calcium, cell wastes collectively called plaque buildup inside the arteries [52]. These plaques diminish the elasticity of the artery walls but blood flow is not affected because of the enlargement of the artery muscular artery wall enlarges at the site of plaques, this make ASVD asymptomatic for decades. As long as these plaques remain thick and stable, CVD tend to be asymptomatic. Over time, the cap begins to thin and weaken, making the plaque more vulnerable to rupture. The condition is usually asymptomatic, and signs happen only when the blood flow is severely impeded by the narrowing of arteries. This is a catastrophic event resulting in ischemia, a condition characterized by constraint in blood supply to tissues, causing the lack of oxygen and glucose needed for cellular metabolism. Ischemia further leads to a sudden heart attack and stroke.

Marked narrowing of coronary arteries and cardiac ischemia produces cardiac arrest with symptoms such as: Arrhythmias, breathlessness, chest pain (angina), dizziness, nausea, and palpitations.

Carotid arteries transport blood to the brain and neck. Stroke is a condition of lack of adequate blood supply to brain tissues because of narrowing or closure of carotid arteries. Ischemia in carotid arteries is associated with the symptoms such as [119]: Blurred vision difficulty in speaking, difficulty in walking or standing up straight, losing consciousness, and weakness.

3.5.2.1   RISK FACTORS OF ATHEROSCLEROSIS

- Cigarette smoking;
- Diabetes;
- Dyslipidemia;
- Foods high in saturated and trans-fats;
- Hypertension;
- Lack of physical activity; and
- Obesity.

3.5.2.2   MECHANISM OF ATHEROSCLEROSIS

Lipoproteins are particles that contain triacylglycerol, cholesterol, phospholipids, and amphipathic proteins called apolipoproteins. Low-density lipoproteins (LDL) deliver cholesterol to cells by receptor-mediated endocytosis. Atherosclerosis is initiated by inflammatory processes commenced by LDL accumulation in artery walls [62]. Endothelial cells line the inner surface of blood vessels. Stress imposed by oxidants modifies LDL cholesterol and instigate atherogenic plaque formation [90]. The injured endothelia draw white blood cells and macrophages and initiate chronic inflammatory pathways. Oxidized LDL accumulates inside the cytoplasm of macrophages, chemokines liberated by these cells stimulate smooth muscle proliferation in the arterial wall and impedes blood flow [57].

3.5.3   CATARACT

Cataract is a serious visual impairment that accounts for the major cause of blindness globally [92]. It is the clouding or opacity of the natural intraocular crystalline lens with the loss of lens transparency.

The lens is a unique organ because of its composition and properties [131]. The main constituents of the lens are water (65%) and proteins (34%). The extraordinarily high protein content makes the lens an unusual organelle, and this enables the lens to have a refractive index considerably greater than its fluid environment [111]. The transparency of the crystalline lens depends on its vascularity, absence of light-scattering organelles within the mature lens fibers, a standard organization of its cells, packing of the lens fibers, protein assembly, extensive cytoskeleton for maintaining the precise shape, ion balance by membrane pumps, etc. [6]. Any modification in the normal architecture of eye lens is associated with the change in the clarity of the lens or opacification and eventually forms the cataract.

3.5.3.1 MECHANISM OF CATARACTOGENESIS

The lens is a specialized tissue designed to focus light onto the retina and photo-oxidation of the lenticular structure is a long-term effect of this process. The lens is usually equipped with an effective system of antioxidants to handle the oxidative attack and a decline in the antioxidant potential is linked with cataractogenesis [121].

Protein turnover in the lens is exceptionally slow, and most of the proteins produced during developmental stages are preserved in the whole lifetime [43]. Oxidative damage of lens proteins is considered as the key factor of cataract formation. Oxidative stress modifies lens proteins and alteration of structural and functional proteins in the lens as a result of oxidation, transamidation, carbamoylation, phosphorylation, proteolysis, etc. Accumulation and precipitation of modified proteins are linked with lenticular opacities; and the major lens proteins reported to be modified in cataract are: crystallins, tubulin, vimentin, connexins, cadherins, membrane ATPases, etc.

Maintenance of calcium level is one of the criteria for lens transparency, and this is achieved by membrane Ca^{2+} ATPase. The activity of Ca^{2+} ATPase is prone to oxidants, and sulfhydryl oxidation and the resulting calcium accumulation in the lens is lethal to the survival of epithelial cells [71]. Another membrane pump assaulted by oxidants is Na^+K^+ ATPase. It maintains the cellular balance of sodium and potassium. A lower intracellular potassium concentration promotes the conversion of inactive pro-caspases to active caspases and alteration of $Na^+ K^+$ ATPase is associated with cataract formation [45, 46].

The crystalline lens is purely epithelial structure originally derived from the surface ectoderm. Any factor that affects the normal physiology of lens epithelia will result in vision impairment. Oxidative damage to lens epithelial cells appears as the mechanism of non-congenital cataract pathology [120]. Oxidative stress appears to be a signal for lenticular apoptosis through the activation of calpains and caspases [60, 61]. Lens proteins are predominant targets of these enzymes, and it further results in the accumulation of damaged proteins.

3.6   HERBAL MEDICINES

According to the World Health Organization (WHO), about 80% of the population in developing countries utilizes traditional medicinal plants as the primary health care. Modern medicine identifies herbalism as alternative medicine, and many of the pharmaceuticals currently prescribed (such as opium, aspirin, digitalis, quinine, etc.) have a history of use as herbal preparations. 25% of modern drugs have been derived from plants. Today much attention is drawn to the traditional system as a resource for drug development and contributes largely to the commercial drug preparations manufactured today. Various plants have been reported in the modern pharmacopeia against degenerative conditions as a cost-effective way of medication.

3.7   ROLE OF HERBAL MEDICINES TO PREVENT LIFE STYLE DISEASES

Functional foods are natural products that enhance physiological functions and prevent the abnormal processes behind various diseases. Medicinal plants have an enormous range of bioactivities and are used in the management of chronic infectious diseases since the beginning of civilization. Most of the lifestyle diseases are basically preventable, and medicinal plants as a complementary treatment play a significant role in the management of these chronic conditions.

Oral hypoglycemic agents and insulin administrations are major treatment modalities effective in controlling hyperglycemia [105]. A large number of studies have assessed the impact of various medicinal plants on the diabetic pathology and complications (Table 3.1).

TABLE 3.1 Antidiabetic Potential of Medicinal Plants

Medicinal plant	Effect on metabolism	Reference
Achyranthesaspera	Prevents hyperglycemia	[59]
Allium cepa	Prevents hyperglycemia and increases HDL	[13]
Allium sativum	Reduces glucose, cholesterol, triglycerides, urea, uric acid and creatinine	[25]
Aloe barbadensis	Prevents hyperglycemia	[115]
Andrographispaniculata	Normalizes glucose, cholesterol, and triglyceride level	[83]
Asparagus racemosus	Normalizes glucose, cholesterol, and triglyceride level	[135]
Azadirachtaindica	Normalizes insulin level and glycogen content	[8]
Biophytum sensitivum	Prevents hyperglycemia	[100]
Bombax ceiba	Hypoglycemic and hypolipidemic	[9]
Brassica juncea	Normalizes insulin level	[130]
Centella asiatica	Reduces LDL, Hypoglycemic	[54]
Cocciniaindica	Normalizes lipid profile	[70]
Cynodondactylon	Normalizes lipid profile and prevents hemoglobin glycation	[50]
Eclipta alba	Inhibits alpha-glucosidase and aldose reductase	[48]
Eucalyptus globules	Hypoglycemic	[95]
Ficusbengalenesis	Reduces lipid peroxidation	[32]
Ficusracemosa	Hypoglycemic	[38]
Gymnema sylvestre	Reduces glucose, urea, uric acid, and creatinine levels	[113]
Heliotropiumindicum	Hypoglycemic	[58]
Hemidesmusindicus	Restoration of activities of glycolytic enzymes	[33]
Hibiscus rosasinesis	Hypoglycemic	[138]
Lantana camara	Hypoglycemic and prevents hemoglobin glycation	[137]
Mangiferaindica	Hypolipidemic and prevents hemoglobin glycation	[36]
Momordicacharantia	Hypoglycemic	[96]
Momordicacymbalaria	Normalizes insulin level and increases HDL	[55]
Morus alba	Normalizes lipid profile and prevents hemoglobin glycation	[76]
Mucunapruriens	Hypoglycemic	[68]

TABLE 3.1   *(Continued)*

Medicinal plant	Effect on metabolism	Reference
Murrayakoeingii	Reduces the level of glucose, creatinine and glycated hemoglobin	[3]
Nelumbo nucifera	Normalizes insulin level and prevents hemoglobin glycation	[69]
Ocimum sanctum	Hypoglycemic	[106]
Phyllanthus emblica	Hypoglycemic	[125]
Phyllanthusniruri	Reduces hemoglobin glycation and inhibits alpha-glucosidase	[87]
Picrorrhizakurroa	Hypoglycemic	[47]
Pterocarpus marsupium	Hypolipidemic and normalizes insulin level	[75]
Punicagranatum	Increases HDL and reduces sugar level	[101]
Salacia oblonga	Normalizes lipid profile and insulin level	[7]
Salacia reticulate	Inhibits aldose reductase and reduces pancreatic peroxide generation	[142]
Tinosporacordifolia	Normalizes insulin level	[94]
Trigonellafoenumgraecum	Hypoglycemic	[78]
Vinca rosea	Improves glucose tolerance and regenerates pancreatic beta cells	[1]

Research studies suggest a possible role of numerous indigenous plants in alleviating CVD pathology. *Allium sativum* or garlic has been endowed with medicinal values in addition to its culinary uses. Garlic has various effects on the cardiovascular system such as: hypoglycemic, hypolipidemic, etc. [102]. *Commiphora Mukul* has been used in Ayurveda for centuries for heart problems. *Commiphora Mukul* extract or Guggul Gum reduces blood pressure and obesity and has hypolipidemic activity by blocking cholesterol biosynthesis [65].

Ginkgo biloba leaf extract has been found to possess cardioprotective activity by its flavonoids and terpenoids [89]. Another plant, traditionally used by Chinese systems of medicine, *Lingusticumwallichii* is also validated for hypotensive and circulatory stimulatory activities. Tetramethylpyrazine, the active compound in *Lingusticumwallichii*, has vasodilatory potential. The roots of *Rauwolfia serpentine* resources of reserpine and are hypotensive in nature [107]. *Stephaniatetrandra* is also hypotensive, and the alkaloid, tetrandrine present in *Stephaniatetrandra* has been reported for having a cardioprotective effect [85, 128].

The plants *Aesculushippocastanum* [22], *Ananascomosus* [114], *Curcuma longa* [27], *Emblicaofficinalis* Gaertn [86], *Garcinia indica* Linn. [93], *Limoniumwrightii* [141], *Ocimum sanctum* [124], *Psidiumguajava* [142], *Rosmarinusofficinalis* [132], *Terminalia arjuna* [91], *Vitisvinifera* [16], *Withaniasomnifera* [21] have been extensively studied for cardioprotective activity. Table 3.2 indicates health benefits of selected plants on cardiac systems.

TABLE 3.2 List of Medicinal Plants with Cardioprotective Activity

Medicinal plant	Effects on the cardiovascular system	Reference
Aesculushippocastanum	Vasodilation	[22]
Allium sativum	Hypotensive, hypolipidemic, fibrinolytic, etc.	[104]
Ananascomosus	Hypolipidemic, anti-inflammatory	[114]
Commiphora Mukul	Hypolipidemic, weight loss agent, anti-inflammatory	[65]
Curcuma longa	Reduces cardiomyopathy	[27]
*Emblicaofficinalis*Gaertn	Homeostasis of hemodynamic function	[86]
Garcinia indica	Reduces myocardial damage	[93]
Ginkgo biloba	Hypolipidemic, improves circulatory flow	[129]
Limoniumwrightii	Reduces myocardial ischemia	[141]
Lingusticumwallichii	Hypertensive, vasodilation	
Ocimum sanctum	Hypolipidemic	[124]
Psidiumguajava	Reduces myocardial ischemia	[142]
Rauwolfiaserpentina	Hypotensive	[107]
Rosmarinusofficinalis	Improves circulatory flow	[132]
Stephaniatetrandra	Hypertensive	[85, 128]
Terminalia arjuna	Hypolipidemic, anti-inflammatory	[91]
Vitisvinifera	Improves post-ischemic ventricular recovery	[16]
Withaniasomnifera	Anti-inflammatory	[21]

Cataract is an irreversible condition, and numerous studies indicate the method of preventive protection as an effective way of cataract management. The oxidative stress is regarded as the major factor responsible for cataractogenesis. Many plants and phenolic compounds have shown anticataractogenic potential by preventive protection. Table 3.3 indicates plants, which have the potential to prevent cataract formation.

TABLE 3.3 Anticataractogenic Potential of Medicinal Plants

Medicinal plant	Beneficial effects on anticataractogenic	Reference
Adhatodavasica	Inhibits aldose reductase in diabetic cataract	[31]
Allium cepa	Improves antioxidant status	[51]
Allium sativum	Prevents diabetic cataract	[104]
Angelica dahurica	Inhibits aldose reductase in diabetic cataract	[117]
Aralia elata	Prevents diabetic cataract	[143]
Azadirachtaindica	Inhibits aldose reductase in diabetic cataract	[43]
Brassica oleracea var. italic	Prevents lipid peroxidation and maintain membrane integrity	[139]
Cassia tora	Prevents lenticular apoptosis	[122]
Citrus aurantium	Reduces protein and lipid oxidation	[80]
Cochlospermumreligiosum	Reduces oxidative stress and prevents lipid peroxidation	[20]
Curcuma longa	Inhibits aldose reductase in diabetic cataract	[41]
Dregeavolubilis	Improves antioxidant status and reduces lipid peroxidation	[10]
Embelicaofficinalis	Inhibits aldose reductase in diabetic cataract	[127]
Emilia sonchifolia	Improves antioxidant status	[63]
Erigeron annuus	Reduces protein gycation	[49]
Ginkgo biloba	Antiapoptotic and cytoprotective	[129]
Moringaoleifera	Reduces oxidant attack and boosing antioxidant status	[112]
Origanumvulgare	Improves antioxidant status	[17]
Trigonellafoenumgraecum	Prevents lipid peroxidation and boosts antioxidant status	[39]
Vernoniacinerea	Improves antioxidant status	[4]
Vitexnegundo	Calcium homeostasis	[110]

3.8 FUTURE TRENDS

Due to the wide biological activities and low costs, there is a great demand for herbal medicine in developed and developing countries. Quality checked herbal medicine products are sure to cause beneficial therapeutic effects on the users, and it is an active area of research in the field of pharmaceutical biology.

3.9 SUMMARY

Oxidant-antioxidant imbalance plays a significant role in the pathology of various lifestyle diseases. Some medicinal plants possess a significant function in the prevention/management of these conditions by the presence of various bioactive phenolic compounds. Most of the medicinal plants studied are edible, and it is another aspect of herbalism called functional foods. Thus the incorporation of any functional plant food in the daily diet is a better endeavor to prevent the progression of such chronic disorders.

KEYWORDS

- antioxidants
- atherosclerosis
- carotenoids
- glutathione
- lipid peroxidation
- oxidative stress
- polyphenolic compounds
- reactive oxygen species
- superoxide dismutase

REFERENCES

1. Ahmed, M. F., Kazim, S. M., Ghori, S. S., Mehjabeen, S. S., Ahmed, S. R., Ali, S. M., & Ibrahim, M., (2010). Antidiabetic activity of *Vinca rosea* extracts in alloxan-induced diabetic rats. *International Journal of Endocrinology, 6,* e-Article ID 841090.
2. Al-Dalaen, S. M., & Al-Qtaitat, A. I., (2014). Oxidative stress versus antioxidants. *American Journal of Bioscience and Bioengineering, 2,* 60–71.
3. Arulselvan, P., Senthilkumar, G. P., Kumar, S. D., & Subramanian, S., (2006). Antidiabetic effect of *Murrayakoenigii* leaves on streptozotocin-induced diabetic rats. *Die. Pharmazie., 61,* 874–877.
4. Asha, R., Devi, V. G., & Abraham, A., (2016). Lupeol, a pentacyclic triterpenoid isolated from *Vernonia cinerea* attenuate selenite-induced cataract formation in Sprague Dawley rat. *Chemico Biological Interactions, 245,* 20–29.

5. Barros, A. I., Nunes, F. M., Gonçalves, B., Bennett, R. N., & Silva, A. P., (2011). Effect of cooking on total vitamin C contents and antioxidant activity of sweet chestnuts (*Castanea sativa* Mill.). *Food Chemistry, 128,* 165–172.

6. Bassnet, S., Shi, Y., & Vrensen, G. F., (2011). Biological glass: Structural determinants of eye lens transparency. *Philosophical Transactions of the Royal Society B: Biological Sciences, 366,* 1250–1264.

7. Bhat, B. M., Raghuveer, C. V., D'Souza, V., & Manjrekar, P. A., (2012). Antidiabetic and hypolipidemic effect of *Salacia oblonga* in streptozotocin-induced diabetic rats. *Journal of Clinical and Diagnostic Research, 6,* 1685–1687.

8. Bhat, M., Kothiwale, S. K., Tirmale, A. R., Bhargava, S. Y., & Joshi, B. N., (2011). Antidiabetic properties of *Azardiractaindica* and *Bougainvillea spectabilis*: *In vivo* studies in murine diabetes model. *Evidence-Based Complementary and Alternative Medicine, 9,* e-Article ID 561625.

9. Bhavsar, C., & Talele, G. S., (2013). Potential anti-diabetic activity of *Bombax ceiba*. *Bangladesh Journal of Pharmacology, 8,* 102–106.

10. Biju, P. G., Devi, V. G., Lija, Y., & Abraham, A., (2007). Protection against selenite cataract in rat lens by Drevogenin D, a triterpenoid aglycone from *Dregeavolubilis*. *Journal of Medicinal Food, 10,* 308–315.

11. Boon, E. M., Downs, A., & Marcey, D., (2017). Catalase: H_2O_2:H_2O_2 Oxidoreductase, http://biology.kenyon.edu/BMB/Chime/catalase/frames/cattx.htm. (accessed on August 31).

12. Burton, G. W., & Traber, M. G., (1990). Vitamin E antioxidant activity, biokinetics and bioavailability. *Annual Review of Nutrition, 10,* 357–382.

13. Campos, K. E., Diniz, Y. S., Cataneo, A. C., Faine, L. A., Alves, M. J., & Novelli, E. L., (2003). Hypoglycaemic and antioxidant effects of onion, *Allium cepa*: Dietary onion addition, antioxidant activity and hypoglycaemic effects on diabetic rats. *International Journal of Food Sciences and Nutrition, 54,* 241–246.

14. Cheeseman, K. H., & Slater, T. F., (1993). An introduction to free radical chemistry. *British Medical Bulletin, 49,* 481–493.

15. Chelikani, P., Fita, I., & Loewen, P. C., (2004). Diversity of structures and properties among catalases. *Cellular and Molecular Life Sciences, 61,* 192–208.

16. Cui, J., Cordis, G. A., Tosaki, A., Maulik, N., & Das, D. K., (2002). Reduction of myocardial ischemia-reperfusion injury with regular consumption of grapes. *Annals of the New York Academy of Sciences, 957,* 302–307.

17. Dailami, K. N., Azadbakht, M., Pharm, Z. R., & Lashgari, M., (2010). Prevention of selenite-induced cataractogenesis by *Origanum Vulgare* extract. *Pakistan Journal of Biological Sciences, 13,* 743–747.

18. Daneman, D., (2006). Type 1 diabetes. *Lancet, 367,* 847–858.

19. Deponte, M., (2013). Glutathione catalysis and the reaction mechanisms of glutathione-dependent enzymes. *Biochimica et Biophysica Acta, 1830,* 3217–3266.

20. Devi, V. G., Rooban, B. N., Sasikala, V., Sahasranamam, V., & Abraham, A., (2010). Isorhamnetin–3-glucoside alleviates oxidative stress and opacification in selenite cataract *in vitro*. *Toxicology In Vitro, 24,* 1662–1669.

21. Dhuley, J. N., (2000). Adaptogenic and cardioprotective action of *ashwagandha* in rats and frogs. *Journal of Ethnopharmacology, 70,* 57–63.

22. Diehm, C., Vollbrecht, D., Amendt, K., & Comberg, H. U., (1992). Medical edema protection-Clinical benefit in patients with chronic deep vein incompetence: A

placebo-controlled double-blind study. *VASA Zeitschrift fur Gefasskrankheiten, 21*, 1888–1892.

23. Doorn, J. A., & Petersen, D. R., (2003). Covalent adduction of nucleophilic amino acids by 4-hydroxynonenal and 4-oxononenal, *Chemico Biological Interactions, 143–144*, 93–100.

24. Drews, G., Krippeit-Drews, P., & Dufer, M., (2010). Oxidative stress and beta-cell dysfunction. *Pflugers Archiv- European Journal of Physiology, 460*, 703–718.

25. Eidi, A., Eidi, M., & Esmaeili, E., (2006). Antidiabetic effect of garlic (*Allium sativum* L.) in normal and streptozotocin-induced diabetic rats. *Phytomedicine, 13*, 624–629.

26. Elayat, A. A., El-Naggar, M. M., & Tahir, M., (1995). An immunocytochemical and morphometric study of the rat pancreatic islets. *Journal of Anatomy, 186*, 629–637.

27. El-Sayed, E. M., Abd El-azeem, A. S., Afify, A. A., Shabana, M. H., & Ahmed, H. H., (2011). Cardioprotective effects of *Curcuma longa* L. extracts against doxorubicin-induced cardiotoxicity in rats. *Journal of Medicinal Plants Research, 5*, 4049–4058.

28. Esterbauer, H., Dieber-Rotheneder, M., Waeg, G., Striegl, G., & Jürgens, G., (1990). Biochemical, structural and functional properties of oxidized low-density lipoprotein. *Chemical Research in Toxicology, 3*, 77–92.

29. Florence, T. M., (1995). The role of free radicals in disease. *Australian and New Zealand Journal of Ophthalmology, 23*, 3–7.

30. Fukai, T., & Ushio-Fukai, M., (2011). Superoxide dismutases: Role in redox signaling, vascular function, and diseases. *Antioxidants & Redox Signaling, 15*, 1583–1606.

31. Gacche, R. N., & Dhole, N. A., (2011). Aldose reductase inhibitory, anti-cataract and antioxidant potential of selected medicinal plants from the Marathwada region, India. *Natural Product Research, 25*, 760–763.

32. Gayathri, M., & Kannabira, K., (2008). Antidiabetic and ameliorative potential of *Ficusbengalensis* bark extract in streptozotocin-induced diabetic rats. *Indian Journal of Clinical Biochemistry, 23*, 394–400.

33. Gayathri, M., & Kannabiran, K., (2008). Hypoglycemic activity of *Hemidesmusindicus* R. Br. on streptozotocin-induced diabetic rats. *International Journal of Diabetes in Developing Countries, 28*, 6–10.

34. Genestra, M., (2007). Oxyl radicals, redox-sensitive signaling cascades, and antioxidants: Review. *Cell Signaling, 19*, 1807–1819.

35. Glade, M. J., (2003). The role of reactive oxygen species in health and disease. *Amherst Nutrition, 19*, 401–403.

36. Gondi, M., Basha, S. A., Bhaskar, J. J., Salimath, P. V., & Rao, U. J., (2015). Anti-diabetic effect of dietary mango (*Mangiferaindica* L.) peel in streptozotocin-induced diabetic rats. *Journal of the Science of Food and Agriculture, 95*, 991–999.

37. Goodsell, D. S., (2017). Catalase- Molecule of the Month, RCSB Protein Data Bank, http://mgl.scripps.edu/people/goodsell/illustration/pdb. (Accessed on August 31).

38. Gul-e-Rana, Karim, S., Khurhsid, R., Saeed-ul-Hassan, S., Tariq, I., Sultana, M., et al., (2013). Hypoglycemic activity of *Ficusracemosa* bark in combination with the oral hypoglycemic drug in diabetic human. *Acta Poloniae Pharmaceutica, 70*, 1045–1049.

39. Gupta, S. K., Kalaiselvan, V., Srivastava, S., Saxena, R., & Agrawal, S. S., (2010). *Trigonellafoenum-graecum* (fenugreek) protects against selenite-induced oxidative stress in experimental cataractogenesis. *Biological Trace Element Research, 136*, 258–268.

40. Gutteridge, J. M., (1995). Lipid peroxidation and antioxidants as biomarkers of tissue damage. *Clinical Chemistry, 41,* 1819–1828.

41. Halder, N., & Joshi, S., (2003). Lens aldose reductase inhibiting the potential of some indigenous plants. *Journal of Ethnopharmacology, 86,* 113–116.

42. Halliwell, B., (2001). Role of free radicals in the neurodegenerative diseases: Therapeutic implications for antioxidant treatment. *Drugs & Aging, 18,* 685–716.

43. Harding, J. J., & Crabbe, M. J. C., (1984). The lens: Development, proteins, metabolism, and cataract. In: Davson, H., (ed.), *The Eye* (Vol. 1B, pp. 207–492). Academic Press, Orlando, Florida.

44. Hekimi, S., Lapointe, J., & Wen, Y., (2011). Taking a good look at free radicals in the aging process. *Trends in Cell Biology, 21,* 569–576.

45. Hightower, K., & McCready, J., (1994). Selenite-induced damage to lens membranes. *Experimental Eye Research, 58,* 225–229.

46. Hughes, F. Jr., Bortner, C., Purdy, G., & Cidlowski, J., (1997). Intracellular K^+ suppresses the activation of apoptosis in lymphocytes. *The Journal of Biological Chemistry, 272,* 30567–30576.

47. Husain, G. M., Singh, P. N., & Kumar, V., (2009). Antidiabetic activity of standardized extract of *Picrorhizakurroa* in a rat model of NIDDM. *Drug Discoveries &Therapeutics, 3,* 88–92.

48. Jaiswal, N., Bhatia, V., Srivastava, S. P., Srivastava, A. K., & Tamrakar, A. K., (2012). Antidiabetic effect of *Eclipta alba* associated with the inhibition of alpha-glucosidase and aldose reductase. *Natural Products Research, 26,* 2363–2367.

49. Jang, D. S., Yoo, N. H., Kim, N. H., Lee, Y. M., Kim, C. S., Kim, J., et al., (2010). 3, 5-Di-O-caffeoyl-epi-quinic acid from the leaves and stems of *Erigeron annuus* inhibits protein glycation, aldose-reductase, and cataractogenesis. *Biological and Pharmaceutical Bulletin, 33,* 329–333.

50. Jarald, E. E., Joshi, S. B., & Jain, D. C., (2008). Antidiabetic activity of aqueous extract and non-polysaccharide fraction of *Cynodondactylon* Pers. *Indian Journal of Experimental Biology, 46,* 660–667.

51. Javadzadeh, A., Ghorbanihaghjo, A., Bonyadi, S., Rashidi, M. R., Mesgari, M., Rashtchizadeh, N., & Argani, H., (2009). Preventive effect of onion juice on selenite-induced experimental cataract. *Indian Journal of Ophthalmology, 57,* 185–189.

52. Jay, R., (2007). *Recognize Atherosclerosis Symptoms That Indicate Heart Disease, 2016.* https://universityhealthnews.com/daily/heart-health/recognize-atherosclerosis symptoms-that-indicate-heart-disease/ (accessed on August 31).

53. Jenkinson, A. M., Collins, A. R., Duthie, S. J., Wahle, K. W., & Duthie, G. G., (1999). The effect of increased intakes of polyunsaturated fatty acids and vitamin E on DNA damage in human lymphocytes. *Federation of American Societies for Experimental Biology Journal, 15,* 2138–2142.

54. Kabir, A. L., Samad, M. B., D'Costa, N. M., Akhter, F., Ahmed, A., & Hannan, J. M. A., (2014). Anti-hyperglycemic activity of *Centellaasiatica* is partly mediated by carbohydrase inhibition and glucose-fiber binding. *BioMed. Central Complementary and Alternative Medicine, 14,* 1–14.

55. Kameswararao, B., Kesavulu, M. M., & Apparao, C., (2003). Evaluation of antidiabetic effect of *Momordicacymbalaria* fruit in alloxan-diabetic rats. *Fitoterapia, 74,* 7–13.

56. Karoui, H., Hogg, N., Frejaville, C., Tordo, P., & Kalyanaraman, B., (1996). Characterization of sulfur-centered radical intermediates formed during the oxidation of thiols and sulfite by peroxynitrite-ESR-SPIN trapping and oxygen uptake studies. *The Journal of Biological Chemistry, 271,* 6000–6009.

57. Karp, G., (2010). *Cell and Molecular Biology: Concepts and Experiments* (6th edn., p. 307). John Wiley & Sons, Inc., New York.

58. Kujur, R. S., Singh, V., Ram, M., Yadava, H. N., Singh, K. K., Kumari, S., & Roy, B. K., (2010). Antidiabetic activity and phytochemical screening of crude extract of *Stevia rebaudiana* in alloxan-induced diabetic rats. *Pharmacognosy Research, 2,* 258–263.

59. Kumar, S. A., Gnananath, K., Gande, S., Goud, R. E., Rajesh, P., & Nagarjuna, S., (2011). Antidiabetic activity of ethanolic extract of *achyranthesaspera* leaves in streptozotocin-induced diabetic rats. *Journal of Pharmacy Research, 4,* 3124–3125.

60. Lee, J. C., Kim, H. R., Kim, J., & Jang, Y. S., (2002). Antioxidant property of an ethanol extract of the stem of *Opuntia ficus-indica* var. Saboten. *Journal of Agricultural and Food Chemistry, 50,* 6490–6496.

61. Lenzlinger, P. M., Saatman, K. E., Raghupathi, R., & Mcintosh, T. K., (2000). Overview of basic mechanisms underlying neuropathological consequences of head trauma. In: Newcomb, J. K., Miller, L. S., & Hayes, R. L., (eds.), *Head Trauma: Basic, Preclinical and Clinical Directions* (p. 230). Wiley-Liss, New York.

62. Li, X., Fang, P., Li, Y., Kuo, Y. M., Andrews, A. J., Nanayakkara, G., et al., (2016). Mitochondrial reactive oxygen species mediate lysophosphatidylcholine-induced endothelial cell activation. *Arteriosclerosis, Thrombosis and Vascular Biology, 36,* 1090–1100.

63. Lija, Y., Biju, P. G., Reeni, A., Cibin, T. R., Sahasranamam, V., & Abraham, A., (2006). Modulation of selenite cataract by the flavonoid fraction of *Emilia sonchifolia* in experimental animal models. *Phytotherapy Research, 20,* 1091–1095.

64. Lopaczynski, W., & Zeisel, S. H., (2001). Antioxidants, programmed cell death and cancer. *Nutrition Research, 21,* 295–307.

65. Lugg, L. T., (2003). Herbs in cardiovascular diseases. In: *Botanical Medicine in Modern Clinical Practice* (8th Annual Course, Lecture 3), Columbia University, New York.

66. Lushchak, V. I., (2011). Environmentally induced oxidative stress in aquatic animals. *Aquatic Toxicology, 101,* 13–30.

67. Lushchak, V. I., (2012). Glutathione homeostasis and functions: Potential targets for medical interventions. *Journal of Amino Acids,* p. 26, e-Article ID 736837.

68. Majekodunmi, S. O., Oyagbemi, A. A., Umukoro, S., & Odeku, O. A., (2011). Evaluation of the anti-diabetic properties of *Mucunapruriens* seed extract. *Asian Pacific Journal of Tropical Medicine, 4,* 632–636.

69. Mani, S. S., Subramanian, I. P., Pillai, S. S., & Muthusamy, K., (2010). Evaluation of hypoglycemic activity of inorganic constituents in *Nelumbo Nucifera* seeds on streptozotocin-induced diabetes in rats. *Biological Trace Element Research, 138,* 226–237.

70. Manjula, S., & Ragavan, B., (2007). Hypoglycemic and hypolipidemic effect of *coccinia indica* Wight and Arn in alloxan-induced diabetic rats. *Ancient Science of Life, 27,* 34–37.

71. Marcantonio, J. M., Duncan, G., & Rink, H., (1986). Calcium-induced opacification and loss of protein in the organ-cultured bovine lens. *Experimental Eye Research, 42*, 617–630.

72. Maritim, A. C., Sanders, R. A., & Watkins, J. B., (2003). Diabetes, oxidative stress, and antioxidants: A review. *Journal of Biochemical and Molecular Toxicology, 17*, 24–38.

73. Mehta, S. K., & Gowder, S. T. J., (2015). Members of antioxidant machinery and their functions. Chapter 4, In: Gowder, S. J. T., (ed.), *Basic Principles and Clinical Significance of Oxidative Stress* (pp. 59–85). InTech.

74. Mendis, S., Puska, P., & Norrving, B., (2011). *Global Atlas on Cardiovascular Disease Prevention and Control* (pp. 3–18). World Health Organization in collaboration with the World Heart Federation and the World Stroke Organization, Geneva.

75. Mishra, A., Srivastava, R., Srivastava, S. P., Gautam, S., Tamrakar, A. K., Maurya, R., & Srivastava, A. K., (2013). Antidiabetic activity of heartwood of *Pterocarpus marsupium* Roxb. and analysis of phytoconstituents. *Indian Journal of Experimental Biology, 51*, 363–374.

76. Mohammadi, J., & Naik, P. R., (2008). Evaluation of hypoglycemic effect of *Morus alba* in an animal model. *Indian Journal of Pharmacology, 40*, 15–18.

77. Mortensen, A., Skibsted, L. H., & Truscott, T. G., (2001). The interaction of dietary carotenoids with radical species. *Archives of Biochemistry and Biophysics, 385*, 13–19.

78. Mowla, A., Alauddin, M., Rahman, M. A., & Ahmed, K., (2009). Anti-hyperglycemic effect of *Trigonellafoenum-graecum* (fenugreek) seed extract in alloxan-induced diabetic rats and its use in diabetes mellitus: A brief qualitative phytochemical and acute toxicity test on the extract. *African Journal of Traditional, Complementary and Alternative Medicines, 6*, 255–261.

79. Muller, F. L., Lustgarten, M. S., Jang, Y., Richardson, A., & Van Remmen, H., (2007). Trends in oxidative aging theories. *Free Radical Biology & Medicine, 43*, 477–503.

80. Muthuswamy, U., Kuppusamy A., Vivekanandhan, L., Thirumalaisamy, S. A., & Varadharajan, S., (2011). Anticataract and antioxidant activities of *Citrus aurantium* L. peel extract against naphthalene induced cataractogenesis in rats. *Journal of Pharmacy Research, 4*, 680–682.

81. Nimse, S. B., & Palb, D., (2015). Free radicals, natural antioxidants and their reaction mechanisms. *RSC Advances, 5*, 27986–28006.

82. Noori, S., (2012). An overview of oxidative stress and antioxidant defensive system. *Open Access Scientific Reports, 1*, 1–9.

83. Nugroho, A. E., Andrie, M., Warditiani, N. K., Siswanto, E., Pramono, S., & Lukitaningsih, E., (2012). Antidiabetic and anti-hiperlipidemic effect of *Andrographispaniculata* (Burm. f.) Nees and andrographolide in high fructose fat-fed rats. *Indian Journal of Pharmacology, 44*, 377–381.

84. Oakley, A., (2011). Glutathione transferases: A structural perspective. *Drug Metabolism Reviews, 43*, 138–151.

85. Ody, P., (1993). *The Complete Medicinal Herbal* (p. 442). Dorling Kindersley, New York.

86. Ojha, S., Golechha, M., Kumari, S., & Arya, D. S., (2012). Protective effect of *Emblicaofficinalis* (amla) on isoproterenol-induced cardiotoxicity in rats. *Toxicology and Industrial Health, 28*, 399–411.

87. Okoli, C. O., Obidike, I. C., Ezike, A. C., Akah, P. A., & Salawu, O. A., (2011). Studies on the possible mechanisms of antidiabetic activity of extract of aerial parts of *Phyllanthusniruri*. *Pharmaceutical Biology, 49*, 248–255.

88. Pacher, P., Beckman, J. S., & Liaudet, L., (2007). Nitric oxide and peroxynitrite in health and disease. *Physiological Reviews, 87*, 315–424.

89. Panda, P. S., & Naik, S. R., (2009). Evaluation of cardioprotective activity of *Ginkgo biloba* and *Ocimum sanctum* in rodents. *Alternative Medicine Review, 14*, 161–172.

90. Parthasarathy, S., Raghavamenon, A., Garelnabi, M. O., & Santanam, N., (2010). Oxidized low-density lipoprotein. *Methods in Molecular Biology, 610*, 403–417.

91. Parveen, A., Babbar, R., Agarwal, S., Kotwani, A., & Fahim, M., (2011). Mechanistic clues in the cardioprotective effect of *Terminalia arjuna* bark extract in isoproterenol-induced chronic heart failure in rats. *Cardiovascular Toxicology, 11*, 48–57.

92. Pascolini, D., & Mariotti, S. P., (2010). Global estimates of visual impairment: 2010. *British Journal of Ophthalmology, 96*, 614–618.

93. Patel, K. J., Panchasara, A. K., Barvaliya, M. J., Purohit, B. M., Baxi, S. N., Vadgama, V. K., & Tripathi, C. B., (2015). Evaluation of cardioprotective effect of aqueous extract of *Garcinia indica* Linn. fruit rinds on isoprenaline-induced myocardial injury in *Wistar* albino rats. *Research in Pharmaceutical Sciences, 10*, 388–396.

94. Patel, M. B., & Mishra, S., (2011). Hypoglycemic activity of alkaloidal fraction of *Tinosporacordifolia*. *Phytomedicine, 18*, 1045–1052.

95. Patra, A., Jha, S., & Sahu, A. N., (2009). Antidiabetic activity of aqueous extract of *Eucalyptus citriodorahook* in alloxan-induced diabetic rats. *Pharmacognosy Magazine, 5*, 51–54.

96. Perumal, V., Khoo, W. C., Abdul-Hamid, A., Ismail, A., Saari, K., Murugesu, S., et al., (2015). Evaluation of antidiabetic properties of *Momordicacharantia* in streptozotocin-induced diabetic rats using metabolomics approach. *International Food Research Journal, 22*, 1298–1306.

97. Pham-Huy, L. A., He, H., & Pham-Huy, C., (2008). Free radicals, antioxidants in disease and health. *International Journal of Biomedical Sciences, 4*, 89–96.

98. Pisoschi, A. M., & Pop, A., (2015). The role of antioxidants in the chemistry of oxidative stress: A review. *European Journal of Medicinal Chemistry, 97*, 55–74.

99. Poli, G., Leonarduzzi, G., Biasi, F., & Chiarpotto, E., (2004). Oxidative stress and cell signaling. *Current Medicinal Chemistry, 11*, 1163–1182.

100. Puri, D., & Baral, N., (1998). Hypoglycemic effect of *Biophytum sensitivum* in the alloxan diabetic rabbits. *Indian Journal of Physiology and Pharmacology, 42*, 401–406.

101. Radhika, S., Smila, K. H., & Muthezhilan, R., (2011). Antidiabetic and hypolipidemic activity of *Punicagranatum* Linn on alloxan induced rats. *World Journal of Medical Sciences, 6*, 178–182.

102. Rahman, K., & Lowe, G. M., (2006). Garlic and cardiovascular disease: A critical review. *Journal of Nutrition, 136*, 736–7340.

103. Rahman, K., (2007). Studies on free radicals, antioxidants and co-factors. *Clinical Interventions in Aging, 2*, 219–236.

104. Raju, T. N., Kanth, V. R., & Lavanya, K., (2008). Effect of methanolic extract of *Allium sativum* (AS) in delaying cataract in STZ-induced diabetic rats. *Journal of Ocular Biology, Diseases and Informatics, 1*, 46–54.

105. Rang, H. P., & Dale, M. M., (1991). *The Endocrine System Pharmacology* (2nd edn., pp. 504–508). Longman Group Ltd, UK.

106. Rao, A. S., Vijay, Y., Deepthi, T., Lakshmi, S. C. H., Rani, V., Rani, S., et al., (2013). Anti-diabetic effect of ethanolic extract of leaves of *Ocimum sanctum* in alloxan-induced diabetes in rats. *International Journal of Basic & Clinical Pharmacology, 2,* 613–616.

107. Rauwolfia Alkaloids, (1996). In: Rockville, M. D., (ed.), *USP DI: Drug Information for the Health Care Professional* (16th edn., Vol. I). United States Pharmacopeial Convention.

108. Rice-Evans, C., (2001). Flavonoid antioxidants. *Current Medicinal Chemistry, 8,* 797–807.

109. Riley, P. A., (1994). Free radicals in biology: Oxidative stress and effects of ionizing radiation. *International Journal of Radiation Biology, 65,* 27–33.

110. Rooban, B. N., Lija, Y., Biju, P. G., Sasikala, V., Sahasranamam, V., & Abraham, A., (2009). *Vitexnigundo* attenuates calpain activation and cataractogenesis in selenite models. *Experimental Eye Research, 88,* 575–582.

111. Sakthivel, M., Elanchezhian, R., Thomas, P. A., & Geraldine, P., (2010). Alterations in lenticular proteins during ageing and selenite-induced cataractogenesis in Wistar rats. *Molecular Vision, 16,* 445–453.

112. Sasikala, V., Rooban, B. N., Priya, S. G., Sahasranamam, V., & Abraham, A., (2010). *Moringaoleifera* prevents selenite-induced cataractogenesis in rat pups. *Journal of Ocular Pharmacology and Therapeutics, 26,* 441–447.

113. Sathya, S., Kokilavani, R., & Gurusamy, K., (2008). Hypoglycemic effect of *Gymnema Sylvestre* (retz.,) R. Br leaf in normal and alloxan induced diabetic rats. *Ancient Science of Life, 28,* 12–14.

114. Saxena, P., & Panjwani, P., (2014). Cardioprotective potential of hydro-alcoholic fruit extract of *Ananascomosus* against isoproterenol-induced myocardial infarction in Wistar Albino rats. *Journal of Acute Disease, 3,* 228–234.

115. Sharma, S., Siddiqui, S., Ram, G., Chaudhary, M., & Sharma, G., (2013). Hypoglycemic and hepatoprotective effects of processed *Aloe vera* gel in a mice model of alloxan-induced diabetes mellitus. *Journal of Diabetes and Metabolic Disorders, 4,* 1–6.

116. Sharma, S., Yang, Y., Sharma, A., Awasthi, S., & Awasthi, Y. C., (2004). Antioxidant role of glutathione-S-transferases: Protection against oxidant toxicity and regulation of stress-mediated apoptosis. *Antioxidants & Redox Signaling, 6,* 289–300.

117. Shin, K. H., Lim, S. S., & Kim, D. K., (1998). Effect of byakangelicin, an aldose reductase inhibitor on galactosemic cataracts, the polyol contents and Na^+K^+ ATPase activity in sciatic nerves of streptozotocin-induced diabetic rats. *Phytomedicine, 5,* 121–127.

118. Shoback, (2011). Chapter 17. In: Gardner, D. G., & Dolores, (eds.), *Greenspan's Basic & Clinical Endocrinology* (pp. 112–118). McGraw-Hill Medical, New York.

119. Sims, N. R., & Muderman, H., (2010). Mitochondria, oxidative metabolism and cell death in stroke. *Biochimica et Biophysica Acta, 1802,* 80–91.

120. Spector, A., (1991). The lens and oxidative stress. In: Sies, H., (ed.), *Oxidative Stress, Oxidants and Antioxidants* (pp. 529–558), Academic Press, London.

121. Sreelakshmi, V., & Abraham, A., (2016). Age-related or senile cataract: Pathology, mechanism and management. *Austin Journal of Clinical Ophthalmology, 3,* 1–8.

122. Sreelakshmi, V., & Abraham, A., (2016). Polyphenols of *Cassia tora* leaves prevents lenticular apoptosis and modulates cataract pathology in Sprague-Dawley rat pups. *Biomedicine & Pharmacotherapy, 81*, 371–378.

123. Steenvoorden, D. P., & Van Henegouwen, G. M., (1997). The use of endogenous antioxidants to improve photoprotection. *Journal of Photochemistry and Photobiology B., 41*, 1–10.

124. Suanarunsawat, T., Boonnak, T., Na Ayutthaya, W. D., & Thirawarapan, S., (2010). Anti-hyperlipidemic and cardioprotective effects of *Ocimum sanctum* L. fixed oil in rats fed a high fat diet. *Journal of Basic and Clinical Physiology and Pharmacology, 21*, 387–400.

125. Sultana, Z., Jami, S. I., Ali, E., Begum, M., & Haque, M., (2014). Investigation of antidiabetic effect of ethanolic extract of *Phyllanthus Emblica* Linn. fruits in experimental animal models. *Pharmacology & Pharmacy, 5*, 11–18.

126. Sung, C. C., Hsu, Y. C., Chen, C. C., Lin, Y. F., & Wu, C. C., (2013). Oxidative stress and nucleic acid oxidation in patients with chronic kidney disease. *Oxidative Medicine and Cellular Longevity*, p. 15. e-article ID 301982.

127. Suryanarayana, P., Kumar, P. A., Saraswat, M., Petrash, J. M., & Reddy, G. B., (2004). Inhibition of aldose reductase by tannoid principles of *Emblicaofficinalis*: Implications for the prevention of sugar cataract. *Molecular Vision, 12*, 148–154.

128. Sutter, M. C., & Wang, Y. X., (1993). Recent cardiovascular drugs from Chinese medicinal plants. *Cardiovascular Research, 27*, 1891–1901.

129. Thiagarajan, G., Chandani, S., Rao, H. S., Samuni, A. M., Chandrasekaran, K., & Balasubramanian, D., (2002). Molecular and cellular assessment of *ginkgo biloba* extract as a possible ophthalmic drug. *Experimental Eye Research, 75*, 421–430.

130. Thirumalai, T., Therasa, S. V., Elumalai, E. K., & David, E., (2011). Hypoglycemic effect of *Brassica juncea* (seeds) on streptozotocin-induced diabetic male albino rat. *Asian Pacific Journal of Tropical Biomedicine, 1*, 323–325.

131. Tripathi, R. C., & Tripathi, B. J., (1983). Lens Morphology, aging and cataract. *J. Gerontology, 38*, 258–270.

132. Tyler, V. E., (1994). Herbs of choice. In: Tyler, V. E., (eds.), *The Therapeutic Use of Phytomedicinals* (p. 113). Pharmaceutical Product Press, New York.

133. Ulusu, N. N., & Tandogan, B., (2007). Purification and kinetic properties of glutathione reductase from bovine liver. *Molecular and Cellular Biochemistry, 303*, 45–51.

134. Update, (2015). *International Diabetes Federation (IDF)* (p. 13). <www.idf.org>, (accessed on 21 March 2016).

135. Vadivelan, R., Dipanjan, M., Umasankar, P., Dhanabal, S. P., Satishkumar, M. N., Antony, S., & Elango, K., (2011). Hypoglycemic, antioxidant and hypolipidemic activity of *Asparagus racemosus* on streptozotocin-induced diabetic in rats. *Advances in Applied Science Research, 2*, 179–185.

136. Valko, M., Izakovic, M., Mazur, M., Rhodes, C. J., & Telser, J., (2004). Role of oxygen radicals in DNA damage and cancer incidence. *Molecular and Cellular Biochemistry, 266*, 37–56.

137. Venkatachalam, T., Kumar, K. V., Selvi, K. P., Maske, O. M., Anbarasan, V., & Kumar, S. P., (2011). Antidiabetic activity of *Lantana camara* Linn fruits in normal and streptozotocin-induced diabetic rats. *Journal of Pharmacy Research, 4*, 1550–1552.

138. Venkatesh, S., Thilagavathi, J., & Sundar, S. D., (2008). Anti-diabetic activity of flowers of *Hibiscus rosasinensis*. *Fitoterapia., 79*, 79–81.
139. Vibin, M., Priya, S. G., Rooban, B. N., Sasikala, V., Sahasranamam, V., & Abraham, A., (2010). Broccoli regulates protein alterations and cataractogenesis in selenite Models. *Current Eye Research, 35*, 99–107.
140. Wiernsperger, N. F., (2003). Oxidative stress as a therapeutic target in diabetes: Revisiting the controversy. *Diabetes & Metabolism, 29*, 579–585.
141. Yamashiro, S., Noguchi, K., Matsuzaki, T., Miyagi, K., Nakasone, J., Sakanashi, M., et al., (2003). Cardioprotective effects of extracts from *Psidiumguajava* L and *Limoniumwrightii*, Okinawan medicinal plants, against ischemia-reperfusion injury in perfused rat hearts. *Pharmacology, 67*, 128–135.
142. Yoshino, K., Miyauchi, Y., Kanetaka, T., Takagi, Y., & Koga, K., (2009). Anti-diabetic activity of a leaf extract prepared from *Salacia reticulata* in mice. *Bioscience, Biotechnology and Biochemistry, 73*, 1096–1104.
143. Young, S. C., & Yun, H. C., (2005). Water extract of *Aralia elata* prevents cataractogenesis *in vitro* and *in vivo*. *Journal of Ethnopharmacology, 101*, 49–54.

CHAPTER 4

HEALTH BENEFITS OF *EUGENIA UNIFLORA* L.: A REVIEW

S. SYAMA, M. K. PREETHA, L. R. HELEN, A. VYSAKH, and M. S. LATHA*

Corresponding author. E-mail: mslathasbs@yahoo.com.

ABSTRACT

The main properties of *Eugenia uniflora* are antiseptic and astringent. Surinam berries reduce inflammation and enhance the functioning of the lungs. *Eugenia uniflora* can be used both as food and also as folk medicine. Secondary metabolites like tannins, flavonoids, and terpenoids are present in the plant. Brazilians use the juice of this plant to ferment into wine or vinegar and in some cases to prepare distilled liquor. Seeds of *Eugenia* cannot be eaten. *Eugenia uniflora*, which is endemic to Brazil, also can be found in special or dissimilar vegetation types and ecosystems.

4.1 INTRODUCTION

Plants are rich sources of phytochemicals, which can be used for diabetes mellitus, inflammation, pancreatitis, cancer, viral infection, etc. Medicinal plants are the primary life-supporting system for rural as well as for tribal communities. In many developing countries, a large portion of the population relies on traditional medicines to meet their health care needs. Herbal medicines have a promising future because most of the plants with medicinal properties have not yet been fully investigated. There are also a lot of developments in preventive therapies of cancer treatment.

In India, herbal wealth constitutes more than 8000 species. The growing field of herbal medicine provides significant income to the

Indian economy. Plants that are rich in secondary metabolites like alkaloids, flavonoids, glycosides, coumarins, and steroids are called medicinal plants.

4.1.1 SOME IMPORTANT MEDICINAL HERBS

Black pepper, aloe, sandalwood, ginseng, bayberry, red clover, cinnamon, and safflower are used to heal wounds. Turmeric is used for inhibiting the growth of microorganisms. Herbs such as marshmallow root and leaf are used as an antacid. *Zingiber officinale* and *Ocimum sanctum* are used in several herbal expectorants.

Several phytochemicals are used in cancer treatment, for example, flavones such as apigenin, which is cytotoxic to breast cancer cells (MCF7). Anti-cancer effects of *Curcuma longa* L. have been effectively studied in case of breast cancer, colon cancer, brain tumor, and lung metastases. Crocetin from saffron and cyanidins from grapes are also phytochemicals with anticancer properties.

4.1.2 EUGENIA UNIFLORA

Eugenia uniflora has a good source of essential nutrients that are needed for human health. Exotic fruits like *Eugenia* have nutritional importance and contain compounds that have therapeutic effects on certain human diseases [23].

Eugenia uniflora is commonly known as Pitanga Surinam or Brazilian cherry. It is a tropical shrub that is inhabitant to South America and belongs to the family Myrtaceae. It is a fruit-bearing tree that grows to a height of 6–12 m. In the Brazilian Atlantic forest, *Eugenia uniflora* is ecologically very important [15]. The local name of the fruit is pitanga, which is a fleshy indehiscent one with two seeds in it. The fruit is rich in vitamin c and nutrients. Tannins, flavonoids, mono, and sesquiterpenes constitute the chemical composition. The first Brazilian plantation of pangas of commercial scale is in Bonito, Pernambuco (Figure 4.1). Pitanga or Surinam cherry acts like a fruit fly host.

This chapter focuses on the role of *Eugenia uniflora* (Brazilian cherry) for human health.

FIGURE 4.1   (See color insert.) *Eugenia uniflora* fruits.

4.2   PHYSICAL CHARACTERISTICS OF *EUGENIA UNIFLORA*

Other names of *Eugenia uniflora* are: stopper cherry, cayennecherry, Surinam cherry, Florida cherry, pitanga, cereza de Surinam, cerise de cayenne. Dewandaru Guinda Hong Ziguo, Mayom–farang, pendanga, pitanga de Praia, Kirschmyrte, Surinam kirsche, etc.

Eugenia uniflora grows best in soil with abundant sunlight and drainage. It has peeling bark with thin and tan colored multiple stems. Propagation of this plant is by seed or by cuttings. Usually, *Eugenia uniflora* is disturbed by caterpillars. The seed of this plant is toxic and resinous. The pollen of *Eugenia* may cause an allergic reaction. *Eugenia* blooms during the months of August to November, and ripening of the fruits is during the months of October-January.

Pitanga cherry is sweet and acidic in nature with exotic flavor. During the maturation process the green colored epicarp of fruit changes to yellow, orange, red, and black. The ripened fruit appears in black color. Taste of Black cherry is better than the red cherry. The pitanga fruit matures within 3 weeks after flowering of the plant. These cherries are bitter in taste with high acidity and are balanced by high carbohydrate content. Characteristics of different parts of *Eugenia uniflora* are shown in Table 4.1.

TABLE 4.1   Characteristics of Different Parts of *Eugenia uniflora*

Fruit	Nature	Round
	Distance end to end	Less than 1.27 cm
	Coat	Fleshy
	Shade	Dark red
	Feature	Suited for human consumption; attracts birds
Flower	Colour	White
	Characteristic	Pleasant fragrance; spring flowering
Folio	Venation	Pinnate
	Leaf foliage	Less than 5.08 cm
	Tint	Purple or red
	Fall paint	No fall color change
	Fall quality	Not showy
	Shape	Ovate

4.3   POTENTIAL USES OF *EUGENIA UNIFLORA*

4.3.1   FOOD

Eugenia fruits can be eaten as raw. The leaves of *Eugenia* are used as an alternative for tea. Unripen fruits are used for making chutneys and relishes. Once it is ripe, the fruits are used for making jams, jellies, sherbets, pies, and juices.

4.3.2   MEDICINE

A mixture containing different parts such as leaves, roots, and fruits of *Eugenia uniflora* is drunk shortly before childbirth. The fruits are used to reduce the blood pressure. Some compounds present in stems and leaves have antimicrobial effects. Syrup made from this fruit is used to treat influenza.

4.3.3   OTHER USES

Eugenia uniflora is usually cultivated in orchards for its fruits. It is also an ornamental tree. The wood of *Eugenia uniflora* is used for making tool handles and agricultural implements. The fruits of *Eugenia uniflora* can be consumed as raw or can be used for preparing pickles, salads, custard pudding, and juices.

4.4 GEOGRAPHICAL DISTRIBUTION

Eugenia uniflora or Pitanga Surinam is mainly found in places like Africa, Amazon, Argentina, Central America, El Salvador, Zimbabwe, Indonesia, Malaysia, Samoa, Pacific, Panama, South America. USA, Suriname, Thailand, Uganda, Uruguay, India, Philippines, Mozambique, Sierra Leone, St Lucia, Puerto Rico, Australia, Ethiopia, Bahamas, Fiji, Guyana, Indochina, Haiti, Guinea- Bissau, etc.

The leaf of Pitanga contains substances with biological effects, which are beneficial for human health [17]. *Eugenia uniflora* acts as significant colonizing type in distressed areas. *Eugenia* acts as food for the local fauna. Some reports have described that *Eugenia* behaves as shade lenient [20]. Light intensity and soil flooding have a role in changing the leaf gas exchange, chlorophyll fluorescence and pigment indexes of *Eugenia uniflora* L. *Eugenia uniflora* acts as an ecological model because it grows in the areas of large and medium levels of rainfall. It is found in various types of vegetation and ecosystems [11].

Pitanga Surinam originates from Surinam, Guyana, French Guiana, parts of Uruguay and southern parts of Brazil. *Eugenia uniflora* has already recommended for heterogeneous reforestation in order to recover the degraded areas, to create permanent preservation areas, and also for providing food for birds. Normally fruit contains 69% pulp, 31% seeds and has a humidity of 85%. *Eugenia uniflora* has a long taproot that helps it to survive in a drought. Pharmacological properties of *Eugenia uniflora* leaves are due to the presence of high amount of phenolic compounds [8].

Surinam cherry becomes spoiled at room temperature; therefore, it should be refrigerated for further use. The seeds of *Eugenia uniflora* contain lecithin with antibacterial activity [22]. In Nigeria, *Eugenia uniflora* leaves are used as ethnomedicine for malaria, and in Brazil, leaves are spread on the floor because of its insect repellent property. The seeds of *Eugenia uniflora* cannot be eaten. The spicy fragrance of the leaves of *Eugenia uniflora* can cause respiratory discomfort to the allergic person.

4.5   IMPORTANCE OF *EUGENIA UNIFLORA* L.

Pitanga can be used to treat inflammation. Pitanga fruit contains vitamin C, which helps to prevent cardiac issues and also protect against stomach cancer, colon cancer, rectum cancer, mouth cancer, and lung cancers. High intake of Pitanga results in an increased supply of blood to ocular areas. Vitamin A present in Pitanga prevents the body from toxins and free radicals.

4.6   PHYTOCHEMISTRY OF *EUGENIA UNIFLORA L.*

Tannins, flavonoids, mono, and sesquiterpenes are responsible for the chemical composition of pitanga. The volatile profile of the leaves *Eugenia uniflora* with different fruit color biotypes have been recognized. This was done by using headspace solid-phase microextraction along with gas chromatography-quadrupole mass spectrometry (HS-SPME/GC-MS) [19]. The influence of various seasons on the chemical composition of essential oil from the leaves of *Eugenia uniflora* was reported [4]. From the 15 constituents identified, 80–95% is volatile, and 78–93% of essential oil is composed of oxygenated sesquiterpenes, during the wet season it increases up to 85–93%.

For the phytochemical screening, the leaves, stem, and root of *Eugenia uniflora* were collected. Then the leaves were properly dried, crushed, and powdered. Standard phytochemical methods are used for the phytochemical analysis. In the leaves, anthracenes, balsams, alkaloids, and volatile oils were absent. The leaves contain flavonoid, saponin, saponin glycosides, tannins, and phenols. The stem of *Eugenia uniflora* contains flavonoids, tannins, and phenols. The root contains only saponins [6].

The phytochemical screening of the different extracts (Petroleum ether, chloroform, ethyl acetate, acetone, cold, and hot water extracts) revealed the presence of various primary and secondary metabolites, and thus it has the potential to act as a source of useful drugs. The hot water extract has more phytoconstituents than other extracts [4]. In another study, polyphenolic compounds from pitanga leaves were extracted by different processes that are, first by a one-step process that uses water, ethanol or supercritical CO_2 and next to a two-step process that uses supercritical CO_2 which is followed by either ethanol or water. From all these extracts, total flavonoids, total polyphenolic, and free radical scavenging activity were determined. For one-step extraction, the results revealed that the yield of extraction is increased with solvent polarity. But in the case of 2 step process, the results showed that when the matrix was extracted previously with extracting the phenolic compounds from the *Eugenia uniflora* using beta -carotene bleaching, method or DPPH radical scavenging method have high antioxidant activities [3]. Eight compounds were isolated for the first time from methanolic extracts of *Eugenia uniflora* leaves. The structures of these compounds were determined by a combination of chemical and spectroscopic analysis. Their antibacterial, anti-leishmania, antifungal, scavenging assay by DPPH and MTT assay for cytotoxic activity were also determined [26].

Oils from the leaves of *Eugenia uniflora* L. possess antibacterial and antifungal activities. Free radical scavenging activities *of Eugenia uniflora* leaves essential oil were determined by DPPH (2,2-diphenyl dipicrylhy-drazyl radical scavenging capacity) ABTS (2, 2'-azinobis (3-ethylbenzo-thiazolinessulphonic acid Scavenging assay) and FRAP (Ferric Reducing Antioxidant Power assay) assays. Essential oil also reduced lipid peroxidation in mice kidney. Acute oral administration of essential oil did not produce toxicological or lethal effects in mice. The essential oil has antifungal activity against *Candida lipolytica* and *Candida guilliermondii* and also exhibits antibacterial activity against *Staphylococcus aureus* and *Listeria monocytogenes*. Thus the essential oil from *Eugenia uniflora* has pharmacological properties and also has application in phytomedicine [31]. Qualitative analysis of secondary metabolites of hot aqueoushot extracts of *Eugenia uniflora* leaves was assessed [7].

Primary metabolites (total soluble carbohydrates, proteins, total amino acids) and secondary metabolites (flavonoids, total phenolics, and tannins) were estimated using standard protocols. Results indicated that the hot aqueoushot extracts of the leaves of *Eugenia uniflora* showed the presence of primary and secondary metabolites. This justifies the utilization of *Eugenia uniflora* L. species for the treatment of various diseases in traditional medicine. The characteristic flavor of *Eugenia uniflora* is due to the presence of ketones and some of the identified sesquiterpenes [16].

In order to obtain natural extracts, the sequential extraction of *Eugenia uniflora* L. leaves was conducted, and results were compared with conventional extraction methods and fixed bed methods, and then the performance of extraction was evaluated for the yield of total flavonoid content and the total concentration of phenolics content. Supercritical CO_2, ethanol, and water were used (60°C and 400 bar) in the fixed bed sequential extraction. The result indicated that the combination process was advantageous when compared to respective aqueous or ethanolic extractions.

The sequential extraction has importance as it produces fractions. The yields of the extraction were higher, and the extracts have higher and more concentrated active polar compounds. The main effect of prior extraction with supercritical carbon dioxide produced extracts with a high content of phenolic compounds [10]. The leaf and branch of *Eugenia uniflora* L. contain flavonoid, saponin, and tannin. Flavonoids contained in leaf extract of *Eugenia uniflora* L. are myricetin, quercetin, gallocatechin, and quercitrin. Phytochemical studies with *Eugenia* species revealed the presence of flavonoids, tannins, terpenoids, and essential oils.

4.7   TRADITIONAL IMPORTANCE OF *EUGENIA UNIFLORA* L.

Eugenia uniflora is used by farmers because of its versatility of the fruit. The cherry is used in the Brazilian cosmetic industry for the production of shampoos, face creams, perfumes, and hair conditioners and also for making bath soaps. Pitanga is used for making wine, jellies, ice creams, liquors, and soft drinks [18]. The essential oil from *Eugenia uniflora* is active against two-gram positive Staphylococcus *epidermis* and *Streptococcus equi*. The development of *Klebsiella, Pseudomonas aeruginosa, Staphylococcus aureus* were inhibited by the lecithin from *Eugenia uniflora*. Juice of *Eugenia uniflora* has been used as an antipyretic, antidiabetic, anti-rheumatic, and as a diuretic. *Eugenia uniflora* leaf extract is a potential source of natural antioxidant and effective against free radical-mediated diseases [5]. *Eugenia uniflora* L. red fruits are used for hypertensive and digestive problems. In some places, the fruits are considered diuretic, hypotensive, anti-inflammatory, anti-diarrheal, and antimicrobial—*Eugenia uniflora* fruit act as a significant source of phenolic compounds [17].

Several diseases like bronchitis, chest cold, cough, hypertension, diarrhea, hepatic issues, painful urination, and gastrointestinal disorders can be treated using extracts of *Eugenia uniflora*. Some reports suggest that to ease the process of childbirth *Eugenia uniflora* is ingested. There is also evidence regarding its diuretic and insect repelling properties [12].

Three macrocyclic ellagitanninsimers (Oenothein B, Eugeniflorins D1, and D2) and the flavonol myricitrin have been isolated from *Eugenia uniflora* by phytochemical studies. Studies have indicated that oenothein B was the one responsible for the inhibition of 60 KDa heat shock protein HSP 60 and the (1,3)-beta-glucan synthase transcript of *Paracoccidioides brasiliensis* [20]. *Eugenia uniflora* leaves and branches contain flavonoids, gallocatechin, myricetin, quercitrin, quercetin, and myricitrin that are effective against cancer [2]. Four tannins were isolated from the active fractions of *Eugenia uniflora*, and all them were tested for the inhibition of EBV DNA Polymerases. The result of these tests showed 50% inhibitory concentration. Guarani Indians in Brazil use *Eugenia uniflora* as a tonic stimulant for the treatment of symptoms related to mood disorders and sadness [2].

4.8 OTHER TRADITIONAL HEALTH BENEFITS OF *EUGENIA UNIFLORA*

The syrup obtained from *Eugenia uniflora* helps to cure influenza, and the fruit reduces blood pressure. Surinam leaf along with lemongrass is used as a remedy to lower the body temperature. In Brazil, *Eugenia uniflora* leaf combination is used as an astringent and also for stomach pain [27]. The mixture of *Eugenia uniflora* is used against cold, cough, and chest cold and is consumed before the childbirth. The leaves of *Eugenia uniflora* is used to reduce the uric acid level and also used against rheumatism, stomach diseases, yellow fever and for hypertension. The bark of *Eugenia uniflora* is used to treat edema and eye infections.

4.9 NUTRITIONAL IMPORTANCE OF EUGENIA UNIFLORA

The leaves of *Eugenia uniflora* are rich in tannins and flavanols. Young leaves were differentiated from older leaves by their high content of Zn and nitrogen. Mature leaves contained high levels of calcium and iron. Young leaves and adult leaves had moderate amounts of phenolics. Adult leaves were again divided into two groups. One is that had the highest levels of all phenols and Mn, Cu in small amounts, the second group showed opposite amounts. There is a balance that exists between micronutrients such as Mn, Cu, and Zn that may influence competition among the different classes of phenolic compounds [27]. Major nutrients present in *Eugenia uniflora* is vitamin C, vitamin A, carbohydrate, vitamin B_2, and magnesium. For example, 173g of raw *Eugenia uniflora* gives us 157.1g of moisture, 12.96g of carbohydrate, 0.69g of total fat, 57 calories and 0.86g of ash.

Eugenia uniflora or Pitanga contains a high amount of carotenoids especially beta carotene, delta carotene, beta-cryptoxanthin, gamma-carotene, lycopene, and rubixanthin. Of these, lycopene is the major carotenoid (32%) and is the functional ingredient in nutraceutical products. Fruits of Pitanga contain 20–25% monosaturated and 29–32% polyunsaturated fatty acids. The highest content of barium (Ba) was found in *Eugenia uniflora* [25].

Surinam cherry also contains proteins, fats carbohydrates, calcium, phosphorous, and vitamins like riboflavin, thiamine, iron, niacin, vitamin A and vitamin C. About 93% of the cherry is water according to the results from previous studies. Vitamin B complex is also present in *Eugenia uniflora*. For making strawberry cakes, *Eugenia uniflora* is often used as an alternate.

4.10   ANTIOXIDANT ACTIVITY

Endogenous antioxidants include catalase, glutathione peroxidase, and superoxide dismutase. Exogenous antioxidants include vitamins A, E, and C from our diet. Both the exogenous and endogenous antioxidants have an important role in controlling the cell damage and also in oxidative stress. Free radicals are neutralized by the phenolic compounds from *Eugenia uniflora* that in turn controls the oxidative stress [28]. *Eugenia uniflora* extracts are highly active in the (DPPH) antioxidant assay [24]. Fruits of *Eugenia uniflora* L. also possess antioxidant activity. In a study, different fractions of *Eugenia uniflora* fruit extracts were tested for antioxidant activity by DPPH activity [17]. Alcoholic *Eugenia uniflora* fruit pulp extract exhibited antioxidant activity and is useful in the treatment of diseases that are associated with carbon tetrachloride-induced lipid peroxidation. *Eugenia uniflora* is a plant with a rich source of tremendous phytochemicals and antioxidant vitamins having different pharmacological activities [21].

The fruit pulp was an excellent source of carotenoids and (flavonoids) phenolic compounds having high antioxidant activities. *Eugenia uniflora* is rich in the antioxidants like beta-cryptoxanthin, lycopene, rubixanthin, and gamma-carotene.

4.11   ANTIHYPERTENSIVE ACTIVITY

In North-Eastern Argentina, the use of *Eugenia uniflora* as a hypotensive agent was tested in normotensive rats. Here intraperitoneal administration of the *Eugenia uniflora* extract was found to reduce blood pressure in rats. In order to test the origin of the hypotensive activity, vasorelaxant ACE and Alpha-adrenergic antagonistic was performed, and the results showed the hypotensive effect of *Eugenia uniflora*. This hypotensive effect was mediated by the vasodilating activity. Week diuretic effect was also found, which is related to increased renal blood flow [2].

4.12   ANTINOCICEPTIVE ACTIVITY

A study was conducted to discuss the antinociceptive, hypothermic profile and chemical composition of *Eugenia uniflora* leaves essential oil. Only after the isolation of atractylone and 3-furanoeudesmene the characterization of oil was performed. Pentane fraction of *Eugenia uniflora* essential oil, sesquiterpenes, atractylone along with 3-furanoeudesmene prevented the abdominal constrictions caused by acetic acid. Increased latency time in hot plate test shows the hypothermic effect. *Eugenia uniflora* essential oil, as well as Furano sesquiterpenes, were found to possess hypothermic effect. These results lead to the conclusion that a centrally mediated mechanism is responsible for induced hyperthermia [1].

4.13   ANTITUMOR ACTIVITY

The antiproliferative effect of *Eugenia uniflora* ethanolic leaves extract was experienced on the T47D (breast cancer cell line). The phenomenon of apoptosis was observed using acridine orange-ethidium bromide double staining method. Cytotoxic effect of *Eugenia uniflora* extract on T47D cells showed an IC_{50} value of 65 µg/ml. Apoptosis was also induced by *Eugenia uniflora* ethanol (EEU) extract of 50 µg/ml and 100 µg/ml. Thin layer chromatography revealed the presence of phenolic, flavonoid, and saponin compounds, which are responsible for the antiproliferative effect. This study on ethanolic extract of *Eugenia uniflora* leaves *showed* an antiproliferative effect on breast cancer cells (T47D). Further research is required to identify the molecular method of action and its selectivity on normal cells [13].

4.14   ANTI-INFLAMMATORY ACTIVITY

Eugenia uniflora or pitanga juice exhibited anti-inflammatory effect by reducing the PG-LPS-stimulated release of IL–8, which act as a common biomarker of gingival inflammation [30].

4.15   ANTI-GINGIVITIS PROPERTIES

Eugenia uniflora dentifrice showed anti-gingivitis properties in children aged between 10–12 years. Thus, it may be a potentially able and safe product to be used alternatively in preventive dental practices [14].

4.16   FUTURE TRENDS

- *Eugenia uniflora* may be an alternative therapeutic for oral candidiasis in the future [29].
- It is important to enhance the production of these fruits since the bioactive fraction of compound isolated from this can be used as a mild antioxidant for the preservation and storage of foods.

4.17   SUMMARY

Eugenia uniflora has nutritional, medicinal, and other uses. About 550 *Eugenia uniflora* species are found generally in subtropical and tropical South America. Atractylone and curzerene are main active constituents found in Pitanga essential oil. All the Bioactive compounds present in *Eugenia uniflora* are dependable for its anti-inflammatory, sedative, and diuretic properties. Phenolic compounds present in *Eugenia uniflora* exhibit various properties like anticancer, anti-aging, and anti-inflammatory. Flavonoids present in this plant act as anti-inflammatory, anti-viral, and also as an antitumoral agent. *Eugenia uniflora* is also used as a source of timber. When the leaves are crushed, they produce pungent oil. This pungent oil acts as an insect repellent, and therefore it has applications in agroforestry.

KEYWORDS

- antinociceptive
- antioxidants
- astringent
- candidiasis
- *Eugenia uniflora*
- flavonoids
- phenols
- pitanga
- secondary metabolites
- Surinam cherry

REFERENCES

1. Amorim, A. C. L., Lima, C. K. F., Hovell, A. M. C., Miranda, A. L. P., & Rezende, M., (2009). Antinociceptive and hypothermic evaluation of the essential leaf oil and isolated terpenoids from *Eugenia uniflora* L. (Brazilian Pitanga). *Phytomedicine, 16,* 923–928.
2. Colla, A. R., Machado, D. G., Bettio, L. E., Colla, G., Magina, M. D., Brighente, I. M., et al., (1999). Pharmacological basis for the empirical use of *Eugenia uniflora* L. (Myrtaceae) as antihypertensive. *J. Ethnopharmacol., 66*(1), 33–39.
3. Correa, H. A. M., Magalhaes, P. M., Queiroga, C. L., Pexioto, C. A., Oliveira, A. L., & Cabral, F. A., (2011). Extracts from pitanga (*Eugenia uniflora* L.) leaves: Influence of extraction process on antioxidant properties and yield of phenolic compounds. *Journal of Supercritical Fluids, 55,* 998–1006.
4. Costa, D. P., Santos, S. C., Seraphin, J. C., & Ferri, P. H., (2009). Seasonal variability of essential oils of *Eugeniauniflora* leaves. *J. Braz. Chem. Soc., 20*(7), 1287–1293.
5. Daniel, G., & Krishnakumari, S., (2016). *In vitro,* free radical scavenging activity of *Eugenia uniflora* (L.) leaves. *Journal of Medicinal Plants Studies, 4*(4), 25–29.
6. Daniel, G., & Krishnakumari, S., (2015). Screening of *Eugenia uniflora* (L). leaves in various solvents for qualitative phytochemical constituents. *Int. J. Pharma. BioSci., 6*(1), 1008–1015.

7. Daniel, G., & Krishnakumaris, S., (2015). Qualitative analysis of primary and secondary metabolites in an aqueous hot extract of *Eugenia uniflora*. (L.). *Asian Journal of Pharmaceutical and Clinical Research, 8*(1), 334–338.

8. Esene, O., Uchechukwu, R., Husseini, S. J., & Asuqu, S., (2011). Proximate and phytochemical analysis of leaf, stem, and root of *Eugenia uniflora* (Surinam or Pitanga cherry). *J. Nat. Prod. Plant Resource, 1*(4), 1–4.

9. Famuyiwa, F. G., & Adebajo, A. C., (2012). Larvicidal properties of *Eugenia uniflora* leaves. *Agric. Biol. J. N. Am., 3*(10), 400–405.

10. Garmus, T. T., Paviani, L. C., & Cabral, F. A. S., (2014). Extracts from Pitanga leaves (*Eugenia uniflora* L.) with sequential extraction in fixed bed using supercritical carbon dioxide ethanol and water as solvents. *The Journal of Supercritical Fluids, 86*, 4–14.

11. Guzman, F., Almero, M. P., Korbes, A. P., Morais, G. L., & Margis, R., (2012). Identification of MicroRNAs from *Eugenia uniflora* by high-throughput sequencing. *Bioinformatics, 7*(11), e49811.

12. Iqbal, M., (2013). *NMR-Based Metabolomics to Identify Bioactive Compounds in Herbs and Fruits* (p. 182). Natural products laboratory (NPL), Institute of biology (IBL), Faculty of Science, Leiden University, Leiden.

13. Ismiyati, N., & Febriansyah, R., (2012). Antiproliferative effect of ethanolic extract *Eugenia uniflora* Lam. *Indonesian Journal of Cancer Chemoprevention, 3*(2), 371–376.

14. Jovito, V. C., Freires, I. A., Ferreira, D. A. H., Paulo, M. Q., & De Castro, R. D., (2016). *Eugenia uniflora* dentifrice for treating gingivitis in children. *Braz. Dent. J., 27*(4), 387–392.

15. Kuhn, A. W., Tedesco, M., Laughinghouse, H. D., & Flores, F. C., (2015). Mutagenic and antimutagenic effects of *Eugenia uniflora* L. by the *Allium cepa* L. test. *International Journal of Cytology, Cytosystematics, and Cytogenetics, 68*(1), 25–30.

16. Malaman, F. S., Moraes, L. A., West, C., Ferreira, N. J., & Oliveira, A. L., (2011). Supercritical fluid extracts from the Brazilian cherry (*Eugenia uniflora* L.). *Food Chemistry, 124*, 85–92.

17. Massarioli, A. P., Oldoni, T. L. C., Moreno, I. A. M., Rocha, A. A., & De Alencar, S. M., (2013). Antioxidant activity of different pitanga (*Eugenia uniflora* L.) fruit fractions. *Journal of Food, Agriculture & Environment, 11*(1), 288–293.

18. Melo, R. M., Corre, V. F. S., Amorim, A. C. L., Miranda, A. L. P., & Rezende, C. M., (2007). Identification of impact aroma compounds in *Eugenia uniflora* L. (Brazilian pitanga) leaf essential oil. *J. Braz. Chem. Soc., 1*(18), 179–183.

19. Mesquita, P. R. R., Nunes, E. C., & Bastos, L. P., (2017). Discrimination of *Eugenia uniflora* L. biotypes based on volatile compounds in leaves using HS-SPME/GC–MS and chemometric analysis. *Microchemical Journal, 130*, 79–87.

20. Mielke, S. M., & Schaffer, B., (2009). Leaf gas exchange, chlorophyll fluorescence and pigment indexes of *Eugenia uniflora* L. in response to changes in light intensity and soil flooding. *Tree Physiology, 30*, 45–55.

21. Nwanneka, N. O., Bioma, N. U., Elijah, J. P., & Joseph, O. C., (2011). Antioxidant effect of *Eugenia uniflora* pulp extract on the pro-and antioxidant status of carbon tetrachloride (CCL_4)-induced hepatotoxicity in albino rats. *J. C. Biomed. Phar. Res., 1*(4), 157–163.

22. Oliveira, M. D. L., & Andrade, N. S., (2008). Purification of a lectin from *Eugenia uniflora* L. seeds and its potential antibacterial activity. *Letters in Applied Microbiology, 46,* 0266–8254.

23. Oliveira, A. L., Lopes, R. B., Cabral, F. A., & Eberlin, M. N., (2006). Volatile compounds from pitanga fruit (*Eugenia uniflora* L.). *Food Chemistry, 99,* 1–5.

24. Reynertson, K. A., Basile, M. J., & Kennelly, E. J., (2005). Antioxidant potential of seven myrtaceous fruits. *Ethnobotany Research and Applications, 3,* 25–35.

25. Rodolfo, D. M., Lucilaine, S., & Francisconi, S. C., (2011). *Inorganic Constituents in Herbal Medicine by Neutron Activation Analysis* (p. 213). International Nuclear Atlantic Conference –INAC 2011.INAC 2011, Belo Horizonte, MG, Brazil.

26. Samy, M. N., Sugimoto, S., Matsunami, K., Otsuka, H., & Kamel, M. S., (2014). Bioactive compounds from the leaves of *Eugenia uniflora. Journal of Natural Products, 7,* 37–47.

27. Santos, R. M., Fortes, G. A. C., Ferri, P. H., & Santos, S. C., (2011). Influence of foliar nutrients on phenol levels in leaves of *Eugenia uniflora. Brazilian Journal of Pharmacognosy, 21*(4), 581–586.

28. Schumacher, N. S. G., & Colomeu, T. C., (2015). Identification and antioxidant activity of the extracts of *Eugenia uniflora* leaves. *Antioxidants, 4*(4), 662–680.

29. Silva-Rocha, W. P., Souza, L. B. F. C., Ferreira, M. R. A., Soares, L. A. L., Svidzisnki, T. I. E., Milan, E. P., & Chaves, G., (2011). Effect of the crude extract of *Eugenia uniflora* in morphogenesis in *Candida albicans* from the oral cavity of kidney transplant recipients. *Gustavo C. Brazil, 20,* 24–28.

30. Soares, D. J., Walker, J., & Pignitter, M., (2014). Pitanga (*Eugenia uniflora* L.) fruit juice and two major constituents thereof exhibit anti-inflammatory properties in human gingival and oral gum epithelial cells. *Food Funct., 5,* 2981–2988.

31. Victoria, F. N., Lenardao, E. J., Savegnagoc, L., Perin, G., & Jacob, R. G., (2012). Essential oil of the leaves of *Eugenia uniflora* L.: Antioxidant and antimicrobial properties. *Food and Chemical Toxicology, 50,* 2668–2674.

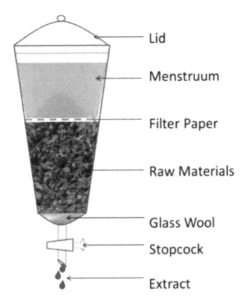

FIGURE 1.3   Schematic diagram of percolator.

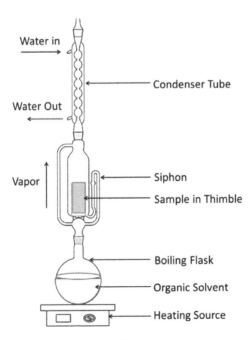

FIGURE 1.4   Schematic diagram of a Soxhlet apparatus.

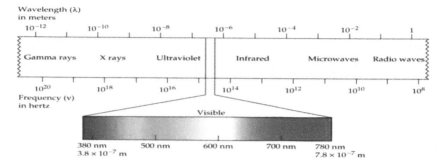

FIGURE 1.5 Electromagnetic spectrum.

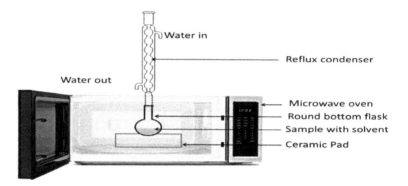

FIGURE 1.6 Schematic diagram of microwave-assisted extraction.

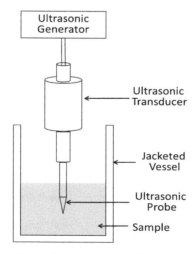

FIGURE 1.8 Batch type ultrasonic probe for extraction.

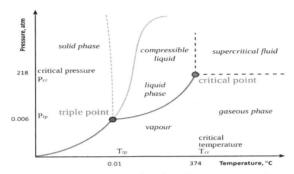

FIGURE 1.11   Water phase diagram as a function of temperature and pressure.

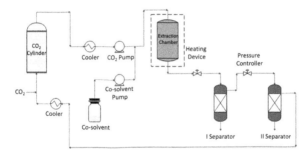

FIGURE 1.12   Experimental set up of supercritical fluid extraction.

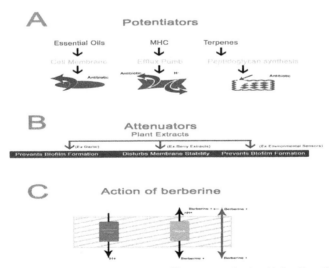

FIGURE 2.2   Action of plant extracts on efflux pump of microbial cell wall synthesis pathway: (a) Potentiators; (b) Attenuators; and (c) Berberine; Adapted and modified from Ref. [188].

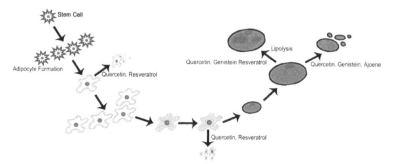

FIGURE 2.3 Action of some plant-derived compounds in triggering/inhibiting adipogenesis pathways; modified and adapted from [105].

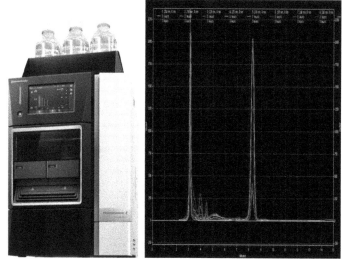

FIGURE 2.4 HPLC instrument (Courtesy Shimadzu Co. Japan): Left: Overlaid chromatogram of plant extract *A. scholaris* (right).

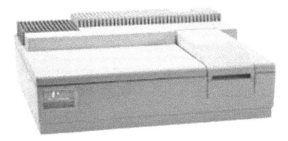

FIGURE 2.6 UV-VISIBLE spectrophotometer [Courtesy Perkin-Elmmer].

FIGURE 2.7   Freeze dryer [http://www.chem.iastate.edu/faculty/Wenyu_Huang/labtour; http://www.labogene.com/ScanVac].

FIGURE 2.8 CO_2 incubator [Courtesy Eppendorf India Ltd.].

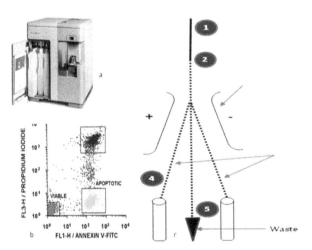

FIGURE 2.9   Flow cytometer [Courtesy Bio-RAD]; **Legend:** (a) Flow cytometer; (b) Flow chart of flow cytometer steps: (1) Particle pass through laser and produce images, (2) Particles are partitioned to droplets, (3) Pass through an electric field, (4) Charged particles are pulled through each side. (5) Uncharged particles are passed through as waste); and (c) Report of flow cytometer analysis of viable, apoptotic, and necrotic cell separation.

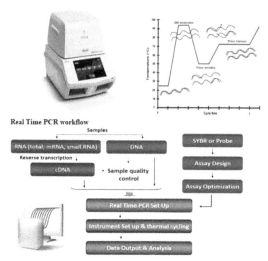

FIGURE 2.10a   Real-time PCR workflow [Courtesy Bio-RAD].

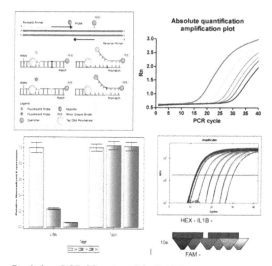

FIGURE 2.10b   Real-time PCR [Courtesy Bio-RAD].

FIGURE 4.1   *Eugenia uniflora* fruits.

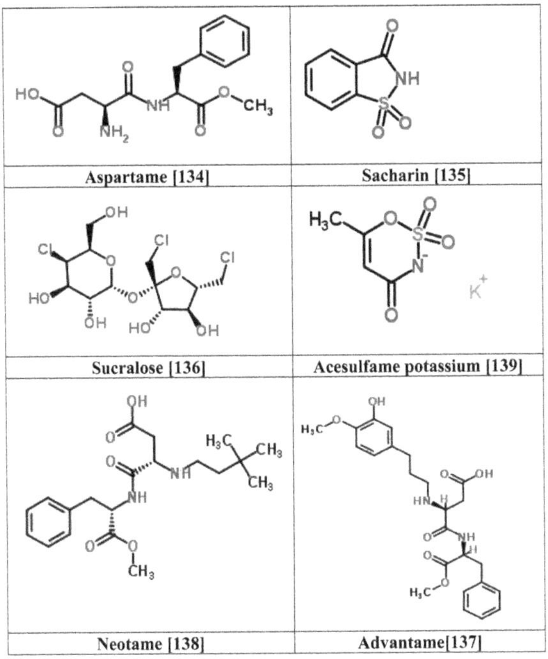

Aspartame [134]	Sacharin [135]
Sucralose [136]	Acesulfame potassium [139]
Neotame [138]	Advantame[137]

FIGURE 6.1 This sweetener was approved to be safe by FDA in 1981 (Aspartame [134]; Saccharin [135]; Sucralose [136]; Acesulfame potassium [139]; Neotame [138]; Advantame [137]).

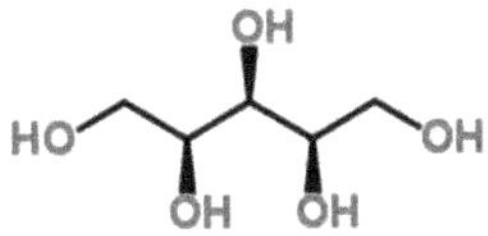

FIGURE 6.3 Structure of xylitol.

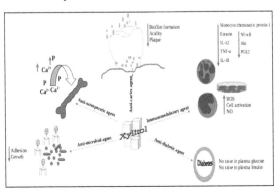

FIGURE 6.4 Multifaceted action of xylitol.

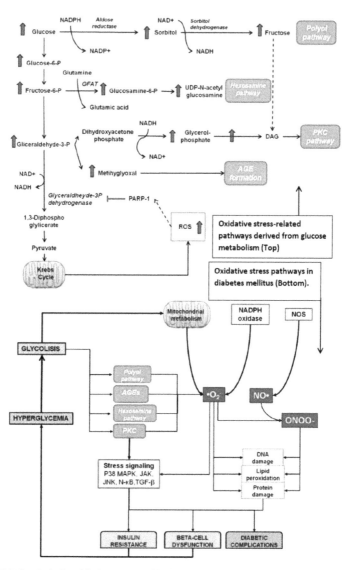

FIGURE 7.2 Relationship between oxidative stress and diabetes mellitus. **Note:** Consequences of up-regulation of polyol pathway are: Increased AR activity leads to increased sorbitol levels causing osmotic stress. Depletion of NADPH reduces glutathione reductase and catalase activity and increased the action of sorbitol dehydrogenase results in the increased supply of NADH to NADH oxidase; both results in increased ROS generation. Osmotic and oxidative stress leads to secondary diabetic complications. [https://www.intechopen.com/books/oxidative-stress-and-chronic-degenerative-diseases-a-role-for-antioxidants/oxidative-stress-in-diabetes-mellitus-and-the-role-of-vitamins-with-antioxidant-actions].

CHAPTER 5

FUNCTIONAL FOODS AND NUTRITIONAL EPIGENETICS: REVIEW ON SCOPE, ROLE, AND HEALTH BENEFITS

K. C. DHANYA* and ADITYA MENON

*Corresponding author. E-mail: dhanuchandra@gmail.com.

ABSTRACT

Epigenetics includes the inheritable manipulations in gene expression towards a phenotype without any alteration in the underlying DNA sequence. This is a change in phenotype without any change in genotype. While genetics of an individual is permanent, epigenetics is variable and is subjected to several factors such as environment, age, lifestyle, and diseased conditions. Among these factors, dietary habits play an important role in deciding the epigenetic fate and are termed as nutritional epigenetics. Functional foods, beyond providing the basic nutrition, may offer a potentially positive effect on health by acting as epigenetic modulators. This modulatory action might be utilized both as a preventive as well as a therapeutic approach for the prevention or cure of various disease conditions such as metabolic disorders, cancer, and chronic inflammatory reactions, where epigenetic changes play a vital role. This interconnection between epigenetics and foods or rather functional food substantiate the proposal by Hippocrates: *Let food be thy medicine and medicine be thy food.*

5.1 INTRODUCTION

In any multicellular organism, all cells contain the same genetic information in the form of DNA, only a few genes (functional unit of genomic

DNA) are activated in each group of cells while all the rest of the genes in cells are in a 'Switched-Off' stage. This makes sure that even though all the cells in an organism have the same genotype, different cellular phenotypes are present and this enables them to carry out different physiological functions. The molecular mechanism that allows this differentiation was named 'epigenetic landscape' by Conrad Waddington [80, 232, 310].

Genetics is a branch of biology that studies genes, its variation and heredity. The molecular information stored in a cell in the form of DNA is converted to messenger RNA (mRNA) through the process of transcription, and the process of translation translates the nucleic acid level information from mRNA for the synthesis of corresponding protein. The overall process is known as the 'Central dogma.' It represents three processes namely: DNA Replication, Transcription, and Translation. Gene expression is the sum total of transcription, post-transcriptional modifications, translation, and post-translational modifications (PTMs) of protein. Gene expression translates the genetic information stored in the language of nucleic acid to the language of the polypeptide (protein), causing the genotype to be expressed into its corresponding phenotype. Thus theoretically, all cells in a multicellular organism should be alike since each, and every cell in an organism carry essentially the same genotype. In reality, this is not the case due to Epigenetics.

The cells having the same genotype may end up into cells of different phenotype at the end of cellular differentiation, and this is due to the modifications or changes in gene expression rather than changes in the genetic code. This modulation of gene expression that results in different cell types and development comes under Epigenetics. This means that in addition to the hereditary genetic information and its expression to a phenotype through various stages of molecular events (such as transcription, translation, and post-transcriptional/PTMs), there is an additional mechanism that controls the gene expression and heredity. This inherited information, which is not coded in the DNA, is termed as epigenetic information [97]. Epigenetics can be explained as the study of alterations or modifications in gene expression, which are not caused by any change in cellular DNA sequences and are heritable, both mitotically and meiotically [313].

The molecular mechanisms, which are believed to drive epigenetic changes, are cytosine base modifications in DNA, modifications in histone molecules and nucleosome position alterations. They regulate cellular processes like gene expression, micro RNA expression, interactions between DNA and Proteins, X Chromosome inactivation, cellular differentiation, embryogenesis, etc. [232]. Epigenetic modifications result in imprinting, which is the silencing of one of the two alleles of a chromosome via modifications such as: methylation or acetylation. The epigenome is different in each cell even in a single organism, since it changes with time and upon exposure to various environmental conditions, unlike the genome [136, 286]. The epigenome is influenced by environmental conditions, variable depending upon the stage of development when the exposure happens, and the epigenetic changes that are induced may remain as such for long periods and influence the phenotypic outcome [97].

Conrad Waddington coined "epigenetics" to link genetics and developmental biology, and the literal meaning of this word is "in addition to changes in genetic sequence." Epigenetics involves any process that alters gene activity without altering the DNA sequence, and it leads to modifications in DNA in a way so as to be transmitted to daughter cells [326]. In other words, epigenetics is the sum total of all the mechanisms needed to unfold the genotype into the phenotype [29].

Alternatively, epigenetics involves any heritable change that may happen during the expression of genes where the changes are not due to any modification in the DNA sequence [31, 36, 243]. Epigenetics acts as a bridge between phenotype and genotype; how the final outcome of a chromosome may get changed without any variation in the underlying DNA sequence. This includes any potentially stable, heritable change in expressions of genes or phenotype that happens without any change in Watson-Crick base pairing of DNA. Several regulatory steps are involved in the precise functioning of gene expression. Modifications in the epigenome, such as methylation of DNA and modification of histones manipulate the expression of genes during transcription of DNA to RNA and translation of RNA to protein [97, 104].

This chapter outlines different mechanisms (Methylation of DNA bases, Chromatin remodeling by histone variants and by histone modifications, Noncoding RNA mediated changes in gene expression, etc.) of epigenetic alterations for the regulation and maintenance of a mammalian genome. The chapter also focuses on various disease conditions that arise due to the epigenetic modifications and the various factors that influence such as: age, environment, disease state, lifestyle, and diet. This chapter elaborates the influence of various functional foods and phytochemicals on various epigenetic mechanisms and the potential of such agents in maintaining the normal epigenetics or reversing disease prone epimutations. The chapter includes discussion on an epigenetic diet for medicinal and therapeutic purposes.

5.2   DIFFERENT MECHANISMS OF EPIGENETIC MODIFICATIONS

There are several mechanisms, cross-talks, and interactions between various pathways, which are responsible for the induction and maintenance of epigenetic status. These include modification of the nucleotide bases and chromatin proteins that package DNA [340]. The epigenetic modifications (either their presence or absence) at particular chromosomal regions results in various epigenetic effects.

Epigenetic processes are essential for many functions, but if they occur in an improper manner, can cause adverse health or behavioral effects. Three types of epigenetic modifications are: DNA methylation, modification of histones and nucleosome positioning, even though the interplay and interaction among epigenetic factors and positive/negative feedback mechanisms decide the final outcome [232].

5.2.1   DNA METHYLATION

Methylation of DNA bases occurs at CpG islands, which are regions where CpG dinucleotides cluster together. These areas will have about 200 bases, where 50% of the bases will be Guanine and Cytosine. CpG islands are rarely observed in the mammalian genome (~1%). In human beings, CpG islands are observed in approximately 60% of the gene promoters [232], whereas methylation may result in gene silencing.

In eukaryotic genomes, the most commonly observed epigenetic modification is methylation of cytosine to form 5-methylcytosine (5mC), which has a significant function of controlling gene expression via epigenetic mechanisms. Cells methylate or demethylate DNA thereby influencing the expression of specific genes [148]. DNA methyltransferase transfer methyl group to DNA bases, where S-adenosyl methionine (SAM) acts as the source of methyl group [67, 148, 210].

The four types of DNA methyltransferases (DNMTs) are: DNMT1, DNMT2, DNMT3a, and DNMT3b [324]. In DNA, cytokines are methylated [210] to 5-methylcytosine by a reversible covalent modification and about 3% of cytosine in human DNA can have this modification. (Figure 5.1).

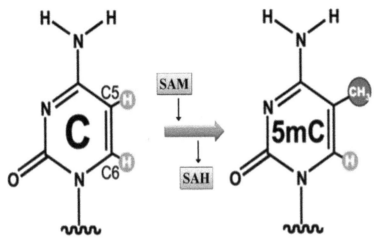

FIGURE 5.1   Conversion of cytosine to 5-methylcytosine. Reprinted from Maresca, A., Zaffagnini, M., Caporali, L., Carelli, V., & Zanna, C., (2015). DNA methyltransferase 1 mutations and mitochondrial pathology: Is mtDNA methylated? *Frontiers in Genetics*, 690. (Creative Commons Attribution License (CC BY).

The structure of the DNA helix major groove gets changed by the methylation of cytosine at the 5th position. This promotes binding of DNA binding proteins resulting in modified gene expression. The modifications in cytosine and protein binding at specific locations in DNA act as epigenetic markers that are heritable as they tend to be copied in the daughter DNA molecules after DNA replication. Methylation at CpGs in promoters prevents initiation of transcription and thus results in gene silencing, and this is predominantly observed in X Chromosome inactivation, Gene imprinting and in Parasitic DNAs. This is due to the methylation caused changes in the interaction between DNA and transcriptional proteins resulting in either an increased or a decreased transcription rate.

DNA methylation cause inhibition of the expression of genes, since methylated DNA is capable of recruiting methyl CpG binding domain proteins (MBD) that further causes recruitment of histone modifying and chromatin remodeling molecules and also by preventing the interaction between DNA and DNA binding proteins. Gene expression is favored by unmethylated CpG islands [76, 163, 180]. While DNA methylation at promoter region causes inhibition of transcription, gene body methylation enhances the expression of the gene [114].

Chromosomal instability and gene translocations are prevented, and thus chromosomal integrity is maintained by DNA methylation of CpGs present in repetitive elements which function to prevent the reactivation of endoparasitic DNA sequences [76]. Gene expression in eukaryotes is controlled and regulated via cis-acting (DNA sequence) and trans-acting factors (molecules that interact with DNA) at the transcriptional level. While an increase in the promoter site methylation results in reduced gene expression, the rate of methylation in the transcribed area of DNA variably influences the gene expression [125, 271]. DNA methylation causes transcriptional repression by two mechanisms:

- Directly interfering with specific transcription factor binding at promoter; and
- Specific transcriptional repressors directly bind to methylated DNA.

DNA methylation can change rate of expression of various genes via modifying histones thereby altering chromatin structure [62].

5.2.2   HISTONE MODIFICATION

Histones and their modification play a key role in causing epigenetic changes. Different types of histones such as H2A, H2B, H3 and H4 act as core histones which assemble in dimeric and tetrameric forms and are wrapped around in a 1.65 turn by 147 bp DNA to form nucleosome, the basic structural unit of chromatin. In between two nucleosomes, an average 50 bp DNA (linker DNA) is bound with Histone H1 (Linker Histone) [157].

5.2.3   POST-TRANSLATIONAL HISTONE MODIFICATIONS

The core histones undergo several PTMs [10, 157, 166], such as: acetylation, methylation, phosphorylation, etc. Other types of modifications such as sumoylation, ubiquitination, ADP ribosylation, deimination, noncovalent proline isomerizations, etc. also occur in histone H3. Histone modifications play significant functions in regulation of DNA repair, alternative splicing, DNA replication and chromosome condensation [127, 183]. Actively transcribed euchromatin is highly acetylated and has trimethylated H3K4, H3K36, and H3K79.

Heterochromatin is having low level of acetylation, and elevated levels of methylation in histones tend to be transcriptionally inactive [170]. High levels of methylation in histones in promoter (H3K4me3, H2BK5ac, H3K27ac, and H4K20me1) and gene body (H4K20me1 and H3K79me1) are associated with active transcription [141].

- Histone acetylation, which is the most frequently occurring modification, carried out by Histone acetyltransferase, almost invariably results in activation of transcription.
- Histone deacetylation by histone deacetylases (HDAC) correlates with transcriptional repression.
- Little is known about histone phosphorylation and its influence on gene expression.
- Histone methylation is by the enzyme methyltransferases which modify one single lysine or arginine on a single histone and cause either activation or repression of transcription [21].
- Histone demethylation by histone demethylase enzyme antagonizes methylation.
- Histone deamination involves the conversion of an arginine in H3 and H4 to a citrulline, and this process has the potential to antagonize the active effect of arginine methylation since citrulline prevents arginines from being methylated [61].
- Several other enzymatic reactions are also involved in histone modification such as: action by Histone kinases or phosphatases, Ubiquitylation, Proline Isomerization, ADP Ribosylation, Sumoylation, etc. [157]. Some other enzymes affect changes in nucleosome arrangement or composition, associated with ATP hydrolysis.

5.2.4   HISTONE VARIANTS

Histone variants and core histones show differences in few amino acids only and are incorporated into chromatin independently of DNA replication [170, 287]. Histone variants regulate gene expression and nucleosome positioning [349]. They play a critical role in nucleosome stability. Some can both activate and repress transcription. Among various histone variants, histone H2A family is the most abundant [34, 45].

5.2.5   CHROMATIN REMODELING

Chromatin is the state in which DNA is packaged within the cell. Nucleosomes are basic unit of chromatin [202]. Histones help to package DNA as chromatin to be contained in the nucleus and also have role in regulating gene expression. Nucleosome packaging of DNA affects transcription and thereby regulate gene expression. Tightly folded chromatin tends to be less or not expressed while open chromatin is easily expressed.

Nucleosome positioning determines the rate of transcription factor binding with target DNA and functions to shape methylation landscape [51]. Chromatin remodeling controls gene expression by modifying chromatin arrangement to allow interaction between DNA and transcription factors. Histone variants can regulate gene expression via positioning nucleosome for chromatin remodeling [349]. Chromatin remodeling mainly occurs by:

- **Covalent modifications** of histones by specific enzymes such as histone acetyltransferases (HATs), deacetylases, methyltransferases, and kinases. The covalent modifications status of histones and composition and arrangement of nucleosomes determine the epigenetic status that either enhances or prevents gene expression [97].
- **ATP dependent chromatin remodeling** complexes, which restructure and realign nucleosomes [292] regulate gene expression via repositioning or restructuring nucleosomes. Hydrolyzing ATP by the common ATPase domain repositions nucleosomes along the DNA, thus opening up DNA sequences for gene expression [312].

5.2.6   NONCODING RNA (NCRNA)

Short interfering RNA (siRNA), microRNA (miRNA), PIWI-interacting RNA (piRNA), repeat associated small interfering RNA (rasiRNA), etc. are examples for small Noncoding RNA (ncRNA), which are synthesized from a large precursor RNA molecule.

The siRNAs being similar in size to miRNAs can base-pair with their target mRNAs thereby directing them for degradation and thus cause repression of translation. siRNAs also play a significant role in transcriptional gene silencing (especially silences transposable elements) through siRNA guided methylation of homologous DNA in gene. Mature miRNAs if base paired with target mRNA in an imperfect manner can cause inhibition of translation and if the base pairing is perfect it will cause degradation of mRNA via the RISC complex. The piRNAs are associated with PIWI family proteins in gametocytes. The siRNAs correspond to piRNAs. These RNA control activity of transposable elements in the germline cells and are essential for germline viability. Long ncRNAs (lncRNAs) can increase or decrease transcription either by altering chromatin configuration by altering the recruitment of RNA Polymerase II [97, 185].

5.3 EFFECTS OF EPIGENETIC ON VARIOUS DISEASES

Traditionally, it was thought that genetics and environment have combined effect on individual variation in susceptibility towards diseases due to single nucleotide polymorphisms. Currently evidences show that DNA methylation and chromatin modification are subjective to environmental factors that play an imperative role in disease susceptibility [71].

Epigenetics or epigenetic variation is the reason for the inter-individual variation in gene expression thereby creating discrepancies in disease predisposition. These epigenetic changes are inherited mitotically as well as transgenerationally [238]. This has brought about a comparison wherein, the genome is considered as the hardware of the computer and the epigenome as the software to control operations by computer [71]. Epigenome records memories or cues regarding the lifestyle, behaviors, dietary exposures and other social aspects thus providing an interface between genome and environment. Epigenetic mechanisms control gene expression all the way through a person's life span. Immediately following fertilization, rapid DNA demethylation and histone modification happen in the paternal genome while the maternal genome undergoes gradual demethylation [205]. Embryonic methylation is initiated so that each cell has a unique epigenetic pattern which should be preserved for appropriate gene expression. Disturbances in this epigenetic pattern lead to congenital disorders or predispose an individual to acquired diseases such as cancer, neurodegenerative disorders, autoimmune diseases, etc. [232].

Throughout person's life, epigenetic changes build up as a part of normal aging. Over time this will cause certain genes to switch on or off. Sometimes some genes may undergo changes that may result in more or less age-associated diseases like cancer and diabetes [3, 75, 255]. There are a number of diseases with an epigenetic origin. Several diseases and syndromes have abnormal DNA methylation or other modifications. Examples of such diseases are: Silver-Russell Syndrome, Beckwith-Wiedemann Syndrome [293], Angelmann, and Prader-Willi Syndromes [187], Fragile X Syndrome [310], brain disorders such as Autism, Schizophrenia, Rett's Syndrome, etc. [107, 263, 310]. Epigenetic causes have been implicated in various physiological alterations and diseases like obesity, heart disease, various cancers, Rubinstein-Taybi Syndrome, ATR-X Syndrome, Coffin-Lowry Syndrome, etc. [73].

5.3.1   EPIGENETIC MODIFICATIONS IN CANCER

Besides the presence of mutations, cancerous cells also possess severely distorted epigenetic characters. Epigenetic modifications such as histone modifications, changes in DNA methylation, etc. functions to promote carcinogenesis and cause progression of cancer [47, 267].

As proposed by Chen et al. [47], cancer formation is a vicious cycle in which epigenetic modifiers undergo mutation to alter the epigenetics causing altered gene expression [198]. The cancer epigenome is having changes as follows:

DNA Methylation: Cancer cells show changes in DNA methylation pattern [26]. Both Hypomethylation and hypermethylation are observed. Hypomethylation usually found in repeated DNA sequences whereas hypermethylation are solely pertaining to CpG islands [74].

Hypermethylation in Cancer: Hypermethylation has been noted as a salient feature of cancer cells and genes that functions in cell cycle regulation (Rb, p14ARF), differentiation, apoptosis (DAPK, TMS1), DNA repair (BRCA1, MGMT), angiogenesis, metastasis, etc. are the major ones that undergo hypermethylation. Tissue-specific hypermethylation gain of promoter CpG islands is consistently observed in various cancers [77, 249]. Such tissue-specific cancers, which show hypermethylation that are not chance events, are termed CIMP (CpG island methylator phenotype) [296]. Several genes critical in breast cancer development have shown to be hypermethylated thereby losing its expression. Examples are steroid receptor gene, cell adhesion genes, and matrix metalloproteinase inhibitors [336]. Hypermethylation is allied with leukemia and hematologic disorders. Though the frequency of hypermethylation has been consistently observed and characterized in various tumors, yet this hypermethylation cannot always mean a bad prognosis. For instance, highly methylated cases of acute myeloid leukemia (AML) show better prognosis [159].

Hypomethylation in Cancer: The second type of methylation pattern observed in most of malignancies is hypomethylation [85, 150]. Many cancers have lower m^5C content, and a huge majority of them are both highly metastatic and malignant [74]. It has been correlated with metastatic hepatocellular cancer, prostate tumors, cervical cancer and hematological malignancy like Chronic B cell lymphocytic leukemia.

Histone Modification: In cancer, the most important histone modification is the reduction in monoacetylation at H4K16 [91] mediated by HDACs which is either overexpressed or mutated [251, 347]. Several cancer types (such as leukemia or cancer of colon, uterus or lung) have translocations, mutations or deletions in HATs and related genes [204]. Histone modification is not yet clinically accepted as a biomarker for cancer, but many researchers have found out their importance in prognosis of various cancers [48].

Nucleosome positioning: Chromatin remodeling has been observed to be associated with cancer such as: primary non-small cell lung cancers, choroid plexus carcinomas, medulloblastomas, etc., although the molecular mechanisms underlying the mechanism remain unclear. Nucleosome remodeling also causes transcriptional repression by promoter hypermethylation [197, 277]. Chromatin translocation has been implicated as the causative factor for the cancer of white blood cells, AML and its subtype, acute promyelocytic leukemia (APML) [331].

5.3.2   EPIGENETIC MODIFICATIONS IN NEURAL DEVELOPMENT DISORDERS

Mutations in epigenetic genes results in certain neurodevelopmental disorders. The important step in neural development is the stage at which neural cells lose their multipotency and exit mitosis. Epigenetic factors play an important role in this mitotic exit of the neural cells [333]. The factors are:

Methylation: An X linked neurological disease, Rett syndrome, results from mutation in MeCP2, which recruit HDACs to methylated DNA, both up-regulation and down-regulation of MeCP2 in the brain cause neurodevelopmental disorders [44, 116, 302, 303].

Histone modifications: Dysfunction of a HAT is the main reason for the autosomal dominant disorder, Rubinstein-Taybi Syndrome [6].

Loss of function mutations: A rare X linked neuro disorder, Coffin-Lowry syndrome results from mutation in RSK2, which is a serine/threonine protein kinase that function to change chromatin structure and help the H3 residue acetylation [53].

Nucleosome positioning: Mutation to ATRX gene (a member of snf2 family gene that help in the formation of heterochromatin, chromosome alignment, and cohesion; maintain X chromosome inactivation, etc.) causes an X-linked disease and ATRX syndrome [262, 303].

5.3.3 EPIGENETICMODIFICATIONSINNEURO-DEGENERATIVE DISEASES

Epigenetic alterations have role in the following neurodegenerative diseases:

DNA Methylation: In most of the neurological diseases, there are hyper- and hypomethylated sites. In Fragile X syndrome patients, FMR1 promoter hypermethylation has been described [96]. Other literature only permethylated promoters include: SMN2 in spinal muscular atrophy, FXN in Friedreich's ataxia, neprilysin in Alzheimer's disease, etc. [303].

Hypomethylation: In Parkinson's patients, the *Substantia nigra*, due to promoter hypomethylation, overexpresses tumor necrosis factor alpha (TNFα) and this result in the induction of apoptosis of neuronal cells [228]. Multiple Sclerosis patients have hypomethylation in the promoter region of PADI2 [303]. Prader-Willi and Angelman Syndromes result from abnormal methylation of DNA in imprinting control region at 15q11-q13 that results in loss of paternally or maternally expressed genes resulting in Prader-Willi syndrome or Angelman syndrome, respectively [250].

Histone Modification: Histone hypoacetylation, is the most frequently observed change among other modifications leading to neurological disorders. In amyotrophic lateral sclerosis (ALS), there occurs histone hypoacetylation. ALS patients over express FUS protein which induces histone hypoacetylation [303].

Other examples for neurological diseases having hypoacetylation are: Parkinson's disease. Friedreich's ataxia and Huntington's disease. In Alzheimer's disease and in epilepsy, there are alterations in acetylation and phosphorylation of histones. Histone demethylation is observed to be associated with X linked mental retardation while H3K9 hyper trimethylation is involved in Friedreich's ataxia and Huntington's disease [4, 154].

Nucleosome positioning: In congenital myotonic dystrophy, the amplification of CTG repeats acts as a strong nucleosome-positioning signal for creating a closed chromatin domain [162] and has implications in various neuronal malignancies.

5.3.4   *EPIGENETIC MODIFICATIONS AND AUTOIMMUNE DISEASES*

Breakdown of immune tolerance is the basal reason for autoimmune diseases. Following types of epigenetic modifications are observed in this category:

DNA methylation: The ICF syndrome (immunodeficiency, centromeric instability, facial anomalies) result from marked DNA hypomethylation due to a mutation in the DNMT3B [135]. Rheumatoid arthritis and Systemic lupus erythematosus (SLE) are presented with global hypomethylation. SLE patients have hypomethylations in various genes such as PRF1, LCN2, IFGNR2, CD154, MMP14, CD70, AIM2 CSF3R, and 18S and 28S ribosomal RNA gene promoter [132]. In rheumatoid arthritis, hypermethylated (DR3) and hypomethylated (L1 and IL6) sites are present [133, 142].

Histone modification: In rheumatoid arthritis, there is a reduced activity of HDACs [126]. The characteristic pattern of histone modifications are also observed in patients with Type I diabetes [199]. Nucleosomes act as autoantigens in SLE, and histone modifications make apoptotic nucleosomes more immunogenic leading to autoantibody production [307].

Nucleosome positioning: Asthma, Primary biliary cirrhosis, Type 1 diabetes, Crohn's disease, etc. have single-nucleotide polymorphism (in 17q12-q21) that leads to alterations in nucleosome distribution [307].

Apart from the diseases listed above, epigenetic alterations may be found in other disorders, such as: cardiovascular diseases (CVDs) [109, 207, 299], metabolic disorders [284], myopathies [344] and in kids born through assisted reproductive treatment [331]. Diseases develop when wrong epigenetic modification happen at the wrong time at the wrong place, and this reflects the magnitude of epigenetics in normal development [77]. For instance, DNA methylation has an obvious role in cancer as hypermethylation of genes (p14ARF, p16INK4a, MGMT, etc.) occur during carcinogenesis [79].

To conclude, via epigenetic reprogramming an organism develops an alternate and unique phenotype due to the interaction of its genotype with the environment. The three distinct and intertwined mechanisms of epigenetic modification (such as DNA methylation, histone modifications, noncoding RNA, etc.) regulate the "epigenome" [49, 78, 206]. These processes have an effect on transcript stability, DNA folding, the positioning of nucleosomes, chromatin condensation and nuclear organization and decide if a gene is to be silenced or expressed. They control genomic stability, development, X chromosome inactivation, etc. [290]. Disruption of the epigenome, also known as induction of epimutations, [119] has an essential role in disease development [71, 103, 134]. Therefore, the multifaceted interaction involving the genetic system and epigenetic marks embossed in the genetic system by exogenous or endogenous factors plays a huge role in disease susceptibility [131].

5.4   FACTORS THAT INFLUENCE EPIGENETIC CHANGES

Epigenetics is the interaction between nature (genes) and nurture (environment) and is concerned with how environmental factors influence gene expression.

The health of an individual is the interplay between the genetics and environmental factors. The influence of environmental and lifestyle on the epigenome of an individual has been established. Nutrition influences the development of and/or continuance of epigenetic mechanisms, and it persistently alters the mammalian phenotype [316]. Epigenetic modifications are vulnerable to environmental influences, and this is observed to be maximum during early development [226]. Therefore, nutrition influences on epigenetic gene regulation link early nutrition and later metabolism and susceptibility to chronic diseases [318].

Epigenetic mechanisms are influenced by aging, lifestyle, nutritional factors (such as methyl donors like folate), inorganic contaminants (like arsenic), airborne polycyclic aromatic hydrocarbons, drugs (such as cocaine), endocrine disruptors (such as BPA), phytoestrogens, chemical fungicides/pesticides, etc. Some studies also have shown that behavioral patterns such as maternal nursing behavior or depression influence DNA methylation [273].

5.4.1   LIFESTYLE

The epigenetic landscape is known to be influenced by lifestyle factors such as environmental pollutants, diet, obesity, smoking, alcohol consumption, physical activity, psychological stress and working shifts [8, 264, 345]. Regular exercise has also shown to modulate the miRNA profile of circulating neutrophils, some of which are involved in inflammatory pathways [236]. Cigarette smoke condensate reduces certain histone modifications [191] similar to changes that can be found in lung cancer tissue. Also, a study on children exposed to prenatal maternal smoking had shown hypomethylation in DNA repetitive elements in the buccal cells.

Glucocorticoid receptor gene hypermethylation was found in suicide victims who had childhood abuse [195]. Whereas, positive early social experience helps to adapt to stress later in life via epigenetic modifications [321, 323]. In animals, maternal stress and nurturing stability altered the epigenome of the progeny [319, 320, 321]. Briefly, mothers with low nurture tendency had offspring with high levels of methylation of CpG in the promoter of glucocorticoid receptor gene and offspring with methylation were not able to manage well under stress [320, 323]. Regarding the influence of shift work on epigenetics, recent studies showed that chronological regulators may induce chromatin remodeling [212] through a histone-acetyltransferase activity. There is variation in DNA methylation pattern of inflammatory genes in night shift workers [33].

5.4.2   AGE

The epigenome is vulnerable to deregulation during various stages of life such as gestation, neonatal period, puberty, old age, etc. Its sensitivity towards the environment is especially high during embryogenesis because during this period the DNA is synthesized at a high rate and during this time the base DNA methylation pattern and chromatin structure is established. It is also known that certain stable epigenetic programming is reversible in adulthood when there are exposures to HDAC inhibitors, various nutrients, contaminants, etc. [65, 319]. It was observed that in yellow agouti mice, supplementation of methyl donors such as choline, folic acid, betaine or genistein) [316] shifts coat color and reduce obesity, diabetes, and cancer in the offspring [71].

Even though epigenetic marks are observed to be stable in somatic cells, yet the epigenome undergoes extensive reprogramming (demethylation and remethylation) during two developmental time periods, during gametogenesis and in early preimplantation embryos [205]. This reprogramming function to erase preceding paternal/maternal imprints and acquired epigenetic marks. There are epigenetic changes due to various nutritional and environmental factors, *in utero* and these have a role in development and disease susceptibility [24]. For example, synthesis of SAM, which is necessary for DNA methylation, is dependent on methyl donors and cofactors from the diet. So environmental factors that may influence early nutrition and thus availability of SAM can cause alterations in methylation of epigenetically important regions in the genome and thereby may influence adult phenotype [227].

Aging, development of neurologic diseases, autoimmunity, cancer, etc. are found to be associated with alterations, both increase, and decrease, in DNA methylation [244]. Decrease in the levels of methylated cytosines which promote chromosomal rearrangements and instability is observed in some tissues during aging, while in some other tissues increased methylations may be the predisposing factor for cancer [239]. Aging tissue is also known to have a lower global DNA methylation level, and the accumulated methylation pattern varies during development, even in the case of monozygotic twins with identical genetic sequences [90].

5.4.3 ENVIRONMENTAL FACTORS

Monozygotic twins have differences in their epigenome, which is due to the differences in the level of environmental exposures. Abnormal epigenetic programming is the reason why there is increased occurrence of developmental disorders and decreased viability in cloned embryos and stem cells after somatic cell nuclear transfer [204, 205].

Epigenetics is at play when cloned animals fail to be exactly identical to the donor with whom they share an identity in genotype, and they develop diseases differently from their donor [246]. Even though monozygotic twins may be considered as human clones and are genomically identical, there are differences in their epigenome and thus there occur differences in susceptibility towards development of diseases such as cancer, autoimmune diseases, etc. [90, 140, 132]. Differential environmental exposures and thereby different levels of epigenetic modifications result in differences in phenotypes and disease susceptibility in monozygotic twins and in genetically identical animal models. Epigenetic mechanisms are considered to be responsible for changes in phenotype as a function of diet, age, xenobiotic exposure, behavior, etc. [71].

Various environmental factors such as xenobiotics, low dose radiation, and behavior directly influence epigenome by altering chromatin remodeling and methylation and results in changes in the subsequent gene expression pattern. These environments driven epigenetic alterations are observed in fetal, post-natal, and adult stages.

5.4.3.1 EXPOSURE TO XENOBIOTIC CHEMICALS

Epigenotype changes occur following exposure to environmental xenobiotics. For example, exposure to arsenite, *in vivo,* caused decreased methylation in adult mice genome while there was gene-specific hypomethylation (in oncogene promoter, Ha-ras) when the animals were exposed to sodium arsenite along with a methyl-deficient diet [319].

Heavy metals (such as Cadmium, Nickel, and Lead [231, 260, 269]) also interact with epigenome. Nickel exposure cause increased histone methylation decreased histone acetylation and subsequently result in decreased gene expression [46, 336]. Chromium interacts with histone acetyltransferase and HDAC and results in alterations in gene expression [325]. Endocrine active chemicals cause epigenetic changes after either *in utero* or adult exposures. Li et al. observed hypomethylation in mice exposed to diethylstilbestrol (DES) *in utero* or perinatally [171, 172]. Individuals who had exposure to DES *in utero* during the first trimester developed reproductive disorders, and clear-cell adenocarcinoma (CCA) of the vagina and these diseases showed epigenetic transgenerational inheritance in granddaughters [214, 215].

Bisphenol A (BPA) present in many plastics those used for making baby bottles, food containers and dental composites also causes changes in the epigenome. Estradiol and BPA when exposed neonatally caused changes in DNA methylation pattern in a gene-specific pattern in rat prostate [117]. Exposure to a plant-based phytoestrogen (genistein) cause alterations in DNA methylation pattern, and it possesses favorable health effect, such as chemo-prevention of prostate and breast cancer and reduced deposition of adipose. It also causes several negative health effects such as decreased reproductive fitness [65, 165, 209, 211]. Genistein induce hypermethylation and thereby reduced the incidence of diabetes, obesity, and cancer in the offspring [71].

5.4.3.2   LOW DOSE RADIATION

The bystander effect or the radiation-induced bystander effect is an event when cells which happen to be near to ionizing radiation exposed cells develop radiation-induced damages, even though they themselves are not exposed to radiation. The unirradiated cells also exhibit irradiated effects. The development of chromosomal and genomic instability in these cells neighboring the radiation inflicted cells may be due to epigenetic changes [266]. Genomic instability may be arising from a genetic or epigenetic mutation in DNA repair system [86, 305]. A genome-wide dysregulation of epigenome may also result in a mutator phenotype [30].

5.4.3.3   NUTRITIONAL FACTORS

Maternal nutritional status and dietary pattern of mother determine fetal epigenetic reprogramming those finally impact adult diseases. It was observed that when maternal diet was supplemented with methyl donors such as Vitamin B12, folic acid, choline, betaine, etc., there was an elevated level of methylation at 7 Avypseudoexon1A (PS1A) CpG site resulting in shifting of the coat color distribution in offspring [58, 317].

It was also observed that early postnatal nutritional status has an influence on the epigenetic level regulation of the Igf2 gene which is reported in several cancers [82, 86]. Similarly, postweaning diet caused changes in Igf2 expression [314]. Therefore, it is to be understood that the stage of development during which the nutritional exposure happened is important in determining the effects of nutrition on the epigenome.

5.5   INTRODUCTION TO FUNCTIONAL FOODS

Functional food is defined as whole foods and either fortified or enriched or enhanced foods or dietary molecules which can off era health and physiological advantage further than being just food that provides nutrients. The word 'functional food' was first used in Japan to define processed food with additive ingredients, which supports and aids specific bodily functions in addition to nutritive value.

Foods for Specified Health Use (FOSHU) come under Japanese Ministry of Health and Welfare and all the products approved by it are targeted for healthy people and people in a preliminary stage of a disease or a borderline condition [257]. After Japan, functional foods have already become part of the dietary landscape of many other countries. For example,

- In Canada, functional foods are "those which may appear similar to or may be a conventional food, is used as a part of a normal diet and is confirmed to have physiological advantages and/or reduce chronic disease risk beyond basic nutritional function."
- According to European Commission Concerted Action on Functional Food Science in Europe (FUFOSE), a food is functional food only if it has a favorable effect on one or more bodily activities and is still a food and not a dietary supplement.
- The American Dietetic Association (ADA) defines food as functional food which may be whole food and fortified, enriched or enhanced and has a beneficial effect on health when consumed as a part of diet on a regular basis at effective level.

Though the definition of functional food varies, the main sources of them are pretty much limited to plants, animals, and microorganisms. The incorporation of bioactive compounds, which are produced by any of these sources, like fiber, flavonoids, carotenoids, minerals, fatty acids, phytoestrogens, prebiotics, and probiotics, vitamins, and soy protein in our daily diet can reduce risks and even cure certain diseases. Table 5.1 provides a summary of different categories of functional foods, sources, and effects.

5.5.1   FUNCTIONAL FOODS FROM PLANT SOURCES

The review by Block et al. [32] consisted of over 200 epidemiological studies detailing the relation of plant-based diet and lower cancer frequencies. The National Academy of Sciences has also acknowledged the

TABLE 5.1 List of Functional Food Components, Sources, and Potential Health Benefits

Class/Examples	Sources	Health effects	Reference
Carotenoids			
B Carotene	Carrot, pumpkin, sweet potatoes, cantaloupe, spinach, tomatoes	Neutralizes free radicals. Bolsters cellular antioxidant defense. Anticancer activity. Prevents Chronic Heart Diseases (CHD).	[40, 56, 219]
Zeaxanthin Lutein	Kale, spinach, collard, eggs, corn, citrus fruits, asparagus, carrot, broccoli	Maintain eye health.	[158, 245]
Lycopene	Tomato, pink or red grapefruit, watermelon	Maintain prostate health.	[100, 327]
Dietary-Fibers			
Insoluble fiber	Wheat and corn bran, fruit skin	Maintain digestive health. Reduce risk of cancer. Control of diabetes.	[121, 139, 190, 196]
Soluble Fiber	Husk of Psyllium seed, pea, apple, beans, citrus fruits	Reduce the risk of CHD and cancer	[224, 274, 333]
	Cereal grains, brown rice oatmeal	Reduce risk of CHD and cancer Maintain blood glucose.	[19, 194, 240, 253]
Fatty Acids			
Mono-unsaturated fatty acid (MUFA)	Tree nuts, canola oil, olive oil, avocados	Reduces risk of CHD.	[99, 272]
Poly-unsaturated fatty acids (PUFAs) -ALA	Walnuts, flax seed	Reduce risk of CHD. Maintain health of heart, eye, and mental function	[22, 66, 101, 152, 254]

TABLE 5.1 (Continued)

Class/Examples	Sources	Health effects	Reference
PUFAs –DHA/EPA	Salmon, tuna, fish oil	Reduce risk of CHD. Maintain f health of eye and mental function	[14, 118, 283]
Conjugated linoleic acid (CLA)	Lamb, Beef, some cheese	Boosts immune system	[28, 295]
Flavonoid			
Anthocyanins–Malvidin, Cyanidin Delphinidin, Pelargonidin	Cherries, Berries, red grapes	Cellular antioxidant defense. Supports maintenance of healthy brain function.	[12, 341, 339]
Flavanols–epicatechins. Catechins, Epigallocatechin	Tea, chocolate, grapes, cocoa, apple	Maintain health of heart	[68, 89, 339]
Procyanidins and Proanthocyanidins	Cranberries, cocoa, apples, peanuts, grapes, cinnamon, tea, strawberries, chocolate	Maintain health of heart and urinary tract	[27, 120, 122, 123, 138]
Flavanones–Hesperetin, Naringenin	Citrus fruits	Neutralizes free radicals	[89, 106, 115]
Flavonols–Quercetin, Isorhamnetin, Kaempferol, Myricetin	Onion, tea, apples, broccoli	Scavenges free radicals	[129, 201, 229]
Isothiocyanate			
Sulforaphane	Cruciferous vegetables–Cauliflower, broccoli, cabbage, kale, etc.	Enhance detoxification Boost cellular antioxidant defense.	[145, 203, 247, 304]
MINERALS			
Calcium	Sardine, spinach, yogurt, low-fat dairy product	Decrease osteoporosis	[81, 234]

TABLE 5.1 (Continued)

Class/Examples	Sources	Health effects	Reference
Magnesium	Spinach, pumpkin seeds, halibut, almonds, brazil nuts, beans	Maintain normal muscle and nerve function, immunity, and bone health.	[125, 288, 294, 298]
Potassium	Potato, beans, low-fat dairy products, banana, leafy greens	Reduce blood pressure and stroke.	[147, 328]
Selenium	Fish, whole grains, red meat, liver, garlic, eggs	Scavenges free radicals Maintain immune and prostate health.	[25, 52, 153]
Phenolic Acid			
Ferulic acid Caffeic acid	Apple, pear, coffee citrus fruits, whole grains	Cellular antioxidant defense. Maintenance of eye and heart health.	[5, 161, 213, 218, 256]
Plant Stanol/Sterol			
Free Stanols/Sterols/ Sterol esters	Soy, corn, wheat, fortified foods and beverages	Reduce risk of CHD	[108, 130, 167, 230]
Polyol			
Xylitol, Mannitol, Sorbitol, Lactitol	Chewing gums and other food applications	Reduces risk of dental caries	[182, 225]
Prebiotics			
Inulin, Polydextrose Fructo oligosaccharides	Whole grains, honey, onions, garlic, leeks, banana	Maintenance of digestive health. Supports absorption of calcium.	[37, 275]
Probiotics			
Yeast, Bifidobacteria Lactobacilli, etc.	Yogurts and other cultured dairy and non-dairy foods	Maintenance of digestive and immune health. Benefits are strain specific.	[98, 105, 278]

TABLE 5.1 *(Continued)*

Class/Examples	Sources	Health effects	Reference
		Phytoestrogens	
Isoflavones–Daidzein, Genistein	Soybean, soy-based foods	Maintenance of health of bone, immune system, brain, and supports menopausal health	[146, 156, 300, 346]
Lignans	Flax seeds, rye, triticale, nuts, lentils, cauliflower, broccoli, carrot	Maintenance of health of heart and immune system.	[2, 50]
		Soy Protein	
Soy protein	Soybeans, soy-based foods like milk, yogurt, cheese, and tofu	Reduce risk of CHD. May improve bone health.	[9, 233]
		SULFIDES/THIOLS	
Diallyl sulfide, Allyl-methyl trisulfide	Garlic, onions, leeks, scallions	Enhance detoxification Maintenance of health of heart, immune system and digestive system	[20, 92, 113, 208]
Dithiolthiones	Cruciferous vegetables	Enhance detoxification Maintenance of health of immune system	[144, 164]

positive effect of vegetables and fruits in reducing the possibility of both heart diseases and cancer. This effect was later attributed to all components in plants, which do not add to the nutritive value that is now known as 'phytochemicals' [280].

Phytochemicals are non-nutritive chemicals, which impart color and other organoleptic properties to the plant. Phytochemicals are produced by plants for self-defense, but recent research has also shown them to be effective in humans for improving health and protecting against various diseases [177]. More than 5000 phytochemicals have been isolated and characterized, but a large percentage still remains unknown.

Human cells are exposed to various oxidizing agents some of which are essential for the normal functioning of the cell. Conservation of the delicate balance between the oxidants and antioxidants is the cornerstone in preventing many diseases [193]. Overproduction of oxidants can lead to oxidative stress, for example in various chronic bacterial, parasitic or viral infections [176], which can result in inflicting damage to different biomolecules like proteins, DNA, and lipids thereby increasing the risk of diseases [69, 242, 276, 308]. The phytochemicals with antioxidant properties like phenolics and carotenoids can protect the cells from these oxidative assaults and minimize and prevent the risk of chronic diseases [88].

Phytochemicals from food items (like green tea, cocoa, wheat, rye, oats, rice, and other grains) have been implicated to lower blood pressure, reduce inflammation, increase HDL cholesterol, decrease LDL oxidation, dilate blood vessels, etc., as well as to decrease the blood clotting, which translates to reducing the risks of CVDs. This effect has been attributed to the phytochemicals like carotenoids anthocyanins, lignans, phenolic acids, phytosterols, etc., present in these plants [13, 84, 282].

Just like the effect of phytochemicals in reducing CVDs, many of the chemicals of plant origin are capable of reducing the risk of cancer occurrence [155, 188]. Carcinogenesis, though being a multistep process, is tightly linked to the oxidative damages caused to cellular components [54, 178]. Dietary antioxidants have been extensively studied and found to reduce the oxidative stress and furthermore prevent the frequency of induction of cancer [35, 41, 220].

5.5.2 FUNCTIONAL FOODS FROM ANIMAL SOURCES

Animals also provide physiologically active components, which can be classified under functional food. The example of animal-based functional food is Omega-3 fatty acids. They are polyunsaturated fatty acids (PUFA) of which docosahexaenoic acid (DHA) and eicosapentaenoic acid (EPA) are two important forms apart from alpha-linolenic acid (ALA). Fish and fish products are the major sources of DHA and EPA, with functions ranging from serving as the essential components for development of brain and retina in infants. There are other major effects of omega-3 fatty acids which was observed and hypothesized such as:

- better retina function;
- good effects on digestion;
- improves the efficiency of the immune system;
- increasing bile output for effective digestion;
- prevents depression;
- prevents diabetes;
- prevents premature birth;
- reduces allergic diseases;
- reduces blood pressure;
- reduces inflammation symptoms;
- reduces the blood cholesterol and triglyceride levels; and
- reduces the risk of gastro intestinal tumors.

Clinical investigations have shown the role of omega-3 fatty acids in various processes [59]. Various studies have shown the role of omega-3 fatty acids in chronic conditions such as Cancer, Psoriasis, Rheumatoid arthritis, Cognitive dysfunction, Crohn's disease, etc. of which the most promising effect was found in Cardiac diseases [248]. Conjugated linoleic acid (CLA) is another of the functional compound that is mainly obtained from ruminant fat and dairy products; thus beef and cow milk are its main sources. CLA has attracted attention when the anticarcinogenic activity was discovered [55, 128]. CLA has also been found to modulate the immune system under preclinical experiments.

5.5.3 PROBIOTICS

Initially, the word probiotics was used as an antonym for antibiotics and later it was redefined as viable microbes that are favorable to health [259]. Nobel laureate and a microbiologist, Metchnikoff indicates that the lactic acid bacteria (LAB) contributed to the lifespan of Bulgarian peasants [93]. Extensive research on various probiotic microorganisms have provided evidences for their effects in preventing cancer, allergy, hypertension, etc., and maintaining intestinal tract function, immune system, urogenital health, cholesterol lowering, etc. [261]. The ease of maintenance and culturing of these microorganisms have resulted in development and marketing of various probiotic formulations and food additives.

5.6 DIETARY REGULATION OF GENE FUNCTION

Epigenetic characters may undergo changes throughout lifespan, mediated by environmental exposure or nutritional status and influence appearance, behavior, response to stress, susceptibility to various diseases, and even duration of existence [289]. This effect is most prominent during early life. The honeybee (*Apis mellifera*) is an example of the influence of nutrition during early life in the induction of alternative phenotypes [95]. Even though female bee, the queen bee, are genetically identical to all bees, the queen is distinct from worker bees in morphology, ability to reproduce, activities, and life span. This difference in phenotype developed when the genetically identical larvae got exposed to royal gelly, a mixture of vitamins, proteins, amino acids, steroids, fatty acids, lipids, hormones, etc. [160, 184].

Changes in epigenetic patterns can influence the gene expression in a cell without causing genomic mutations, but instead cause various epigenetic alterations that induce a different pattern of gene products resulting in changes in disease risk, response to stress situations and metabolism [173, 175, 195]. It is a well-established fact that diet plays a serious role in modeling the epigenetic pattern of an individual and next generation. The Dutch Hunger Winter Cohort study had shown that mothers who suffered severe undernourishment during the first trimester had offspring who are more prone to cardiovascular disease [221, 252]. Also, it was observed in Overkalix, Sweden, during the 19th century that there occurred a difference in risk for early cardiovascular death in grandchildren depending upon the paternal nutritional status during the prepubertal growth period [329]. Atherogenesis is found to have a link with various epigenetic changes [149, 342]. Epigenetic reprogramming is also implied to be a causative factor for obesity, hypercholesterolemia, high glucose and hyperhomocysteinemia which are significant cardiovascular risk factors [330, 338].

5.6.1 EPIGENETIC HEALTH AND FUNCTIONAL FOODS

The idea of "epigenetic diet" has come from the fact that dietary bioactive food components cause alterations in epigenetic patterns and thereby can change gene expression [111, 222].

The idea of epigenetic diet proposes that early disease intervention or prevention may be possible via modifying the epigenome through the diet. Diet rich in molecules such as Insulin, folate, flavanol, glucose, etc. influence the methyl metabolism and availability of SAM (SAM- the universal methyl donor) resulting in DNA methylation changes. The availability of SAM is dependent on vitamins Folate, B6, B12, choline, and betaine and amino acids cysteine, methionine, glycine, and serine. If SAM is not available up to levels needed, Methylation of DNA and histones are altered [217]. The research to decipher the mechanisms of action of each of these molecules is currently ongoing [43, 301].

Consumption of alcohol results in wastage of methionine and choline, and it modify availability of Vitamin B, thereby reduce the amount of SAM available for DNA methylation and thus can change epigenetic patterns. [192]. The methylation pattern of a cell is affected by Epigallocatechin–3-gallate (EGCG), the polyphenol presents in green tea. It has been found to reduce global DNA methylation and thus expression level in cancer cell lines [83]. In recent years several other nutrients are being identified to be able to modulate the epigenome. An example is Selenium, a mineral found in vegetables and grains when grown in soil rich with selenium. It is observed that reduced levels of selenium cause decreased DNA methylation [63, 335]. It has been observed that the red carotenoid lycopene can cause demethylation in breast cancer cell lines [151, 174].

Sulforaphane is present in Broccoli sprouts. It causes inhibition of HDAC. Other food components such as butyrate, reserve ratrol, genistein, and curcumin also have gained interest in the epigenetic nutritional field

[237, 347]. It is a known fact that flavanol-rich diets are able to reduce cardiovascular risk [87, 186, 268]. Both hyper- and hypo-methylation of specific DNA locus have been demonstrated and act as promoter of genes involved in atherosclerosis. Another important aspect of one's epigenetic health is that it depends on maternal diet. Maternal transmission of epigenetic character is influenced by mother's diet, thereby affecting the phenotype or disease risk of offspring. Recently, there were research findings which showed that exposure to various dietary molecules during different stages of life (before conception, during pregnancy, Lactation, childhood, Puberty, Pre Menopause, post Menopause) determine the epigenetic pattern [39]. Epigenetic modifications during embryonic growth is most prominent since these get transmitted from the few embryonic stem cells over consecutive mitotic divisions to maximum number of cells than those which are occurring during later stages of development in adult stem cells or somatic cells [3].

Maternal diet rich in Vitamin B is allied with reduced risk of cancer in offspring, such as colorectal cancer [60, 258]. Maternal supplementation of folic acid 2–3 months before and after conception is important for methylation of DNA in the offspring [279]. Poor maternal nutrition is associated with increased risk towards diabetes type 2 [17]. Maternal supplementation of choline was found to increase histone modifications such as histone–3 lysine–9 demethylation and histone–3 lysine–27 trimethylation in rat offspring [64], due to increased histone methyltransferase expression. Maternal protein-restriction in rats has an influence in epigenetics of metabolism in offspring. With a maternal diet having low protein, the fat and carbohydrate metabolism in the pup was observed to be altered [173]. Similarly, low, and high protein diets for mother pigs affected global DNA methylation in liver and skeletal muscle of the offspring [11].

5.6.2 FUNCTIONAL FOODS AS EPIGENETIC THERAPEUTICS

Nutri-epigenomics is the epigenome level remodeling by various nutritional factors that happens lifelong. Phytochemicals present in diet are having important role in achieving the balance between dietary balance and health, in other words, nature, and man.

Cardioprotective, chemopreventive, neuroprotective and anti-inflammatory activities are possessed by phytochemicals found in vegetables, fruits, and beverages. So, rather than simply considering food as a base material for generating energy and biomolecules via metabolism, we now consider food also as a conditioning environment to shape the epigenome/genome, and thereby the process of metabolism. Thus diet determines stress response, metabolism, immune system homeostasis and the over al physiological functioning of the body [124, 200, 241, 285].

As discussed earlier, nutritional imbalances and metabolic disturbances in mother during various stages of development have an effect on epigenome and thus the fitness of progeny. This pattern is hereditary and will be inherited through generations [38, 57, 322]. This idea is hypothesized as developmental origins of health and disease (DOHAD) [296]. According to this hypothesis, epigenetic influences such as maternal lifestyle, nutrition, environment, etc. effect development during various stages and have an imperative role in the cause of several noncommunicable diseases like diabetes, cardiovascular disease, cancer, cognitive decline disorders, etc. [15, 16, 23, 136].

Soy polyphenol genistein (GEN) intake have shown to change the DNA methylation pattern in the offspring and decrease risk of diseases such as obesity, diabetes, cancer, etc., throughout generations [70, 72, 315]. Restriction or supplementation with betaine, choline, folic acid, methionine, vitamin B12, etc. in maternal diet is shown to influence the DNA methylation patterns in the offspring, in experimental models [270].

5.7   FUTURE DIRECTIONS AND THERAPEUTIC INTERVENTIONS

Epigenetic pattern tends to change naturally with aging. DNA methylation decrease throughout the genome [42] and changes in histone modifications also decreases. These epigenetic pattern alterations may be contributing to the physiology of aging or development of age-related diseases. With a better information of the age-associated epigenetic signature changes, we may be able to implement a proper dietary or pharmaceutical intervention.

Nutri epigenetics may be considered as a novel and natural method for modulating the epigenetic patterns causative of particular diseases. Further research is very much needed to identify between a "healthy" epigenetic patterns from a 'diseased' epigenetic pattern so that an easier diagnosis and better prognosis is possible [169, 181].

Since the timing of epigenetic changes have a profound influence in an individual and also in the next progenies, the timing of an intervention (a nutrient or pharmaceutical) will also be important. An example of an epigenetic intervention is the use of DNMT or HDAC inhibitors in cancer therapy for the restoration of tumor suppressor gene expression. In future targeted epigenetic modifications may be employed for the treatment or prevention of more diseases [289]. From studies, we now know the complementary and contributing role the epigenetic changes play along with gene mutations in various disease development and progression. Various inflammatory disorders like metabolic diseases (obesity, diabetes type 2), rheumatoid arthritis, cardiovascular disease, cancer, etc. are examples [102, 202, 223, 265, 309]. Nutrients and dietary routines play a significant role in the prevention and treatment of several of these diseases [18, 343]. It is known that lifestyle and diet strongly regulate the epigenetic manipulation of genes (leptin in controlling appetite, insulin receptor in glucose homeostasis, TNFα that control obesity and inflammation) [168, 235, 291, 281].

Several available dietary patterns (low glycemic index diet, Mediterranean diet, low carbohydrate intake vegetarian diet, etc.) could be customized to meet cultural and personal inclinations and respective calorie requirements for proper epigenetic management so that various goals such as weight control, diabetes control, cardiometabolic management, prevention of various metabolic diseases, etc. [94] can be achieved. Since adverse epigenetic modifications due to lifestyle are observed to be due to influences *in utero* during pregnancy and during early post-natal life, proper epigenetic management should be adopted for those periods [1]. Epigenetic changes may precede disease pathology. Therefore, epigenetic marks may be utilized as diagnostic indicators for risk for a disease, and they may also act as prognostic indicators [143].

The present knowledge on the role of nutrition in epigenetics is based on the comprehensive relationship between the following three components: nutrition, epigenetics, and aging. Nutritional interventions for shaping the epigenome will have a profound effect when applied during critical windows of development (e.g., embryonic period, fetal stage and prepubertal period) [216]. It has become clear that health is influenced by epigenetic programming and thus dietary intervention at specific developmental points could carry out both programming and reprogramming of epigenome which can in turn act as an effective tool to improve health [110].

To conclude, the phenotypic status arises from the complicated and continuous interactions between genes and environment during the ancestral, past, present conditions. These gene-environment interactions are accountable for the lifelong amendments of the epigenome. Any disruption in epigenome or the development of epimutations can alter immune function and results in certain disease conditions as well as unhealthy aging [306]. To prevent or to reverse adverse epigenetic alterations, therapeutic, and/or epigenetic nutritional approaches are to be devised and practiced. A better understanding of nutritional epigenomics will help to achieve the ultimate goal of controlled and predictable beneficial manipulation of human health at all stages of life through dietary interventions.

5.8 SUMMARY

The final phenotypic outcome of any organism is the combined result of nature and nurture. While nature refers to the genes and hereditary factors, nurture is the environmental factors that influence the expression of genes. Genetics of an individual comes under nature while epigenetics corresponds to the changes in the expression of genetics due to nurture. This chapter outlined different epigenetic mechanisms such as methylation of DNA, modification of histones, chromatin remodeling, Histone variants, Noncoding RNAs, etc. in the regulation and maintenance of mammalian genome. The chapter also outlined the various disease conditions that arise due to the epigenetic modifications and the various factors such as: age, environment, disease state, lifestyle, and diet.

KEYWORDS

- alpha-linolenic acid
- chromatin remodeling
- epigenetic therapeutics
- genistein
- nutritional epigenetics
- PIWI-interacting RNAs
- repeat-associated RNAs

REFERENCES

1. Adamo, K. B., Ferraro, Z. M., & Brett, K. E., (2012). Can we modify the intrauterine environment to halt the intergenerational cycle of obesity? *International Journal of Environmental Research and Public Health, 9*(12), 1263–1307.

2. Adlercreutz, H., Satu-Maarit, H., & Penalvo-Garcia, J., (2004). Phytoestrogens, cancer and coronary heart disease. *BioFactors (Oxford, England), 22*(1–4), 229–236.

3. Aguilera, O., Fernandez, A. F., Munoz, A., & Fraga, M. F., (2010). Epigenetics and environment: A complex relationship. *Journal of Applied Physiology, 109*(1), 243–251.

4. Al-Mahdawi, S., Pinto, R. M., Ismail, O., Varshney, D., Lymperi, S., Sandi, C., Trabzuni, D., & Pook, M., (2007). The Friedreich ataxia GAA repeat expansion mutation induces comparable epigenetic changes in human and transgenic mouse brain and heart tissues. *Human Molecular Genetics, 17*(5), 735–746.

5. Alam, M. A., & Conrad, S., (2013). Ferulic acid improves cardiovascular and kidney structure and function in hypertensive rats. *Journal of Cardiovascular Pharmacology, 61*(3), 240–249.

6. Alarcón, J. M., Gaël, M., Khalid, T., Svetlana, V., Shunsuke, I., Kandel, E. R., & Barco, A., (2004). Chromatin acetylation, memory, and LTP are impaired in CBP+/− mice. *Neuron, 42*(6), 947–959.

7. Alberts, B., Johnson, A., Lewis, J., Raff, M., Roberts, K., & Walter, P., (2002). Chromosomal DNA and its packaging in the chromatin fiber. In: *Molecular Biology of the Cell* (p. 1616). Garland Science, New York.

8. Alegría-Torres, Jorge, B., Andrea, B., & Valentina, (2011). Epigenetics and lifestyle. *Epigenomics, 3*(3), 267–277.

9. Alekel, D., Lee, A., St Germain, P., Charles, T., Hanson, K. B., Stewart, J. W., & Toda, T., (2000). Isoflavone-rich soy protein isolate attenuates bone loss in the lumbar spine of perimenopausal women 1–4. *American Society for Clinical Nutrition, 72*(3), 844–852.

10. David, A. C., Berger, S. L., Cote, J., Dent, S., Jenuwien, T., Kouzarides, T., et al., (2007). New nomenclature for chromatin-modifying enzymes. *Cell, 131*(4), 633–636.

11. Altmann, S., Murani, E., Schwerin, M., Metges, C. C., Wimmers, K., & Ponsuksili, S., (2012). Maternal dietary protein restriction and excess affects offspring gene expression and methylation of non-SMC subunits of condensing i in liver and skeletal muscle. *Epigenetics, 7*(3), 239–252.

12. Andres-Lacueva, C., Shukitt-Hale, Barbara, G., Rachel, L., Jauregui, O., Lamuela-Raventos, Rosa, M., & Joseph, J. A., (2005). Anthocyanins in aged blueberry-fed rats are found centrally and may enhance memory. *Nutritional Neuroscience, 8*(2), 111–120.

13. Andújar, I., Recio, M. C., Giner, R. M., & Rios, J. L., (2012). Cocoa polyphenols and their potential benefits for human health. *Oxidative Medicine and Cellular Longevity, 2012*, 1–23.

14. Anonymous, (2004). EPA and DHA enriched omega-3 supplement for the treatment of dry eye, meibomianitis and xerostomia. US Patent 20040076695-A1.

15. Anway, M. D., Cupp, A. S., Uzumcu, M., & Skinner, M. K., (2005). Epigenetic transgenerational actions of endocrine disruptors and male fertility. *Science, 308* (5727), 1466–1469.
16. Anway, M. D., & Skinner, M. K., (2006). Epigenetic transgenerational actions of endocrine disruptors. *Endocrinology, 147*(6), 43–49.
17. Attig, L., Gabory, A., & Junien, C., (2010). Nutritional developmental epigenomics: Immediate and long-lasting effects. *Proceedings of the Nutrition Society, 69*(2), 221–231.
18. Attig, L., Vigé, A., & Gabory, A., (2013). Dietary alleviation of maternal obesity and diabetes: Increased resistance to diet-induced obesity transcriptional and epigenetic signatures. *PLoS ONE, 8*(6), e66816.
19. Aune, D., Chan, D. S. M., Lau, R., Vieira, R., Greenwood, D. C., Kampman, E., & Norat, T., (2011). Dietary fiber, whole grains, and risk of colorectal cancer: Systematic review and dose-response meta-analysis of prospective studies. *BMJ, 343*, d6617.
20. Banerjee, S. K., & Maulik, S. K., (2002). Effect of garlic on cardiovascular disorders: A review. *Nutrition Journal, 1*, 1–14.
21. Bannister, A. J., & Kouzarides, T., (2005). Reversing histone methylation. *Nature, 436*(7054), 1103–1106.
22. Barceló-Coblijn, G., & Murphy, E. J., (2009). Alpha-linolenic acid and its conversion to longer chain n−3 fatty acids: Benefits for human health and a role in maintaining tissue n−3 fatty acid levels. *Progress in Lipid Research, 48*(6), 355–374.
23. Barker, D. J., & Martyn, C. N., (1992). The maternal and fetal origins of cardiovascular disease. *Journal of Epidemiology and Community Health, 46*(1), 8–11.
24. Bateson, P., Barker, D., Brock, T. C., & Deb, D., (2004). Developmental plasticity and human health. *Nature, 430*(6998), 419–421.
25. Battin, E. E., Perron, N. R., & Brumaghim, J. L., (2006). The central role of metal coordination in selenium antioxidant activity. *Inorganic Chemistry, 45*(2), 499–501.
26. Baylin, S. B., & Herman, J. G., (2000). DNA hypermethylation in tumorigenesis: Epigenetics joins genetics. *Trends in Genetics, 16*(4), 168–174.
27. Bearden, M. M., Pearson, D. A., Rein, D., Chevaux, K. A., Carpenter, D. R., Keen, C. L., & Schmitz, H., (2000). Potential cardiovascular health benefits of procyanidins present in chocolate and cocoa. Chapter 19, In: Parliament, T. H. (ed.), *Caffeinated Beverages Health Benefits, Physiological Effects, and Chemistry* (pp. 177–186). American Chemical Society, Philadelphia–PA.
28. Belury, M. A., (2002). Dietary conjugated linoleic acid in health: Physiological effects and mechanisms of action. *Annual Review of Nutrition, 22*(1), 505–531.
29. Berger, S. L., Kouzarides, T., Shiekhattar, R., & Shilatifard, A., (2009). An operational definition of epigenetics. *Genes & Development, 23*(7), 781–783.
30. Bestor, T. H., Xu, G. L., & Bourchis, D., (1999). Chromosome instability and immunodeficiency syndrome caused by mutations in a DNA methyltransferase gene. *Nature, 402*(6758), 187–191.
31. Bird, A., (2007). Perceptions of epigenetics. *Nature, 447*(7143), 396–398.
32. Block, G., Patterson, B., & Subar, A., (1992). Fruits, vegetables, and cancer prevention: A review of the epidemiological evidence. *Nutrition and Cancer, 18*(1), 1–29.

33. Bollati, V., & Baccarelli, A., (2010). Epigenetic effects of shiftwork on blood DNA methylation. *Chronobiology International*, *27*(5), 1093–1104.

34. Bonisch, C., & Hake, S. B., (2012). Histone H2A variants in nucleosomes and chromatin: More or less stable. *Nucleic Acids Research*, *40*(21), 10719–10741.

35. Borek, C., (2004). Dietary antioxidants and human cancer. *Integrative Cancer Therapies*, *3*(4), 333–341.

36. Bossdorf, O., Richards, C. L., & Pigliucci, M., (2007). Epigenetics for ecologists. *Ecology Letters*, *11*, 106–115.

37. Brownawell, A. M., Caers, W., Gibson, G. R., Kendall, C. W. C., Lewis, K. D., Ringel, Y., & Slavin, J. L., (2012). Prebiotics and the health benefits of fiber: Current regulatory status, future research, and goals. *Journal of Nutrition*, *142*(5), 962–974.

38. Burdge, G. C., & Lillycrop, K. A., (2010). Nutrition, epigenetics and developmental plasticity: Implications for understanding human disease. *Annual Review of Nutrition*, *30*(1), 315–339.

39. Burdge, G. C., Lillycrop, K. A., & Jackson, A. A., (2009). Nutrition in early life, and risk of cancer and metabolic disease: Alternative endings in an epigenetic tale. *British Journal of Nutrition*, *101*(5), 619–630.

40. Burton, G. W., & Ingold, K. U., (1984). Beta-carotene: An unusual type of lipid antioxidant. *Science (New York)*, *224*(4649), 569–573.

41. Byers, T., & Perry, G., (1992). Dietary carotenes, vitamin C, and vitamin E as protective antioxidants in human cancers. *Annual Review of Nutrition*, *12*(1), 139–159.

42. Calvanese, V., Lara, E., Kahn, A., & Fraga, M. F., (2009). The role of epigenetics in aging and age-related diseases. *Ageing Research Reviews*, *8*(4), 268–276.

43. Casero, R. A., & Marton, L. J., (2007). Targeting polyamine metabolism and function in cancer and other hyperproliferative diseases. *Nature Reviews Drug Discovery*, *6*(5), 373–390.

44. Chahrour, M., Jung, S. Y., Shaw, C., Zhou, X., Wong, S. T. C., Qin, J., & Zoghbi, H. Y., (2008). MeCP2, a key contributor to neurological disease, activates and represses transcription. *Science*, *320*(5880), 1224–1229.

45. Cheema, M., & Ausió, J., (2015). The structural determinants behind the epigenetic role of histone variants. *Genes*, *6*(3), 685–713.

46. Chen, H., Ke, Q., Kluz, T., Yan, Y., & Costa, M., (2006). Nickel ions increase histone H3 lysine 9 dimethylation and induce transgene silencing. *Molecular and Cellular Biology*, *26*(10), 3728–3737.

47. Chen, Q. W., Zhu, X. Y., Li, Y. Y., & Meng, Z. Q., (2014). Epigenetic regulation and cancer: Review. *Oncology Reports*, *31*(2), 523–532.

48. Chervona, Y., & Costa, M., (2012). Histone modifications and cancer: Biomarkers of prognosis. *American Journal of Cancer Research*, *2*(5), 589–597.

49. Cheung, P., & Lau, P., (2005). Epigenetic regulation by histone methylation and histone variants. *Molecular Endocrinology*, *19*(3), 563–573.

50. Cho, J. Y., Kim, A. R., Yoo, E. S., Baik, K. U., & Park, M. H., (1999). Immunomodulatory effect of arctigenin, a lignan compound, on tumor necrosis factor-alpha and nitric oxide production, and lymphocyte proliferation. *The Journal of Pharmacy and Pharmacology*, *51*(11), 1267–1273.

51. Chodavarapu, R. K., Feng, S., Bernatavichute, Y. V., Chen, P., Stroud, H., Yu, Y., & Hetzel, J. A., (2010). Relationship between nucleosome positioning and DNA methylation. *Nature, 466*(7304), 388–392.

52. Clark, L. C., Dalkin, B., Krongrad, A., & Combs, G. F., (1998). Decreased incidence of prostate cancer with selenium supplementation: Results of a double-blind cancer prevention trial. *British Journal of Urology, 81*(5), 730–734.

53. Clayton, A. L., Rose, S., Barratt, M. J., & Mahadevan, L. C., (2000). Phosphoacetylation of histone H3 on c-Fos- and c-Jun-associated nucleosomes upon gene activation. *The EMBO Journal, 19*(14), 3714–3126.

54. Collins, A. R., (1999). Oxidative DNA damage, antioxidants, and cancer. *BioEssays, 21*(3), 238–246.

55. Collomb, M., Schmid, A., Sieber, R., Wechsler, D., & Ryhänen, E. L., (2006). Conjugated linoleic acids in milk fat: Variation and physiological Effects. *International Dairy Journal, 16*(11), 1347–1361.

56. Cook, N. R., Albert, C. M., Gaziano, J. M., Zaharris, E., & MacFadyen, J., (2007). Randomized factorial trial of vitamins C and E and beta-carotene in the secondary prevention of cardiovascular events in women. *Archives of Internal Medicine, 167*(15), 1610–1618.

57. Cooney, C. A., (2006). Germ cells carry the epigenetic benefits of grandmother's diet. *Proceedings of the National Academy of Sciences of the United States of America, 103*(46), 17071–17072.

58. Cooney, C. A., Dave, A. A., & Wolff, G. L., (2002). Maternal methyl supplements in mice affect epigenetic variation and DNA methylation of offspring. *The Journal of Nutrition, 132*(8), 2393–2400.

59. Crawford, M., (2000). Placental delivery of arachidonic and docosahexaenoic acids: Implications for the lipid nutrition of preterm infants. *The American Journal of Clinical Nutrition, 71*(1), 275–284.

60. Cui, H., Cruz-Correa, M., & Giardiello, F. M., (2003). Loss of IGF2 imprinting: Potential marker of colorectal cancer risk. *Science, 299*(5613), 1753–1755.

61. Cuthbert, G. L., Daujat, S., & Snowden, A. W., (2004). Histone determination antagonizes arginine methylation. *Cell, 118*(5), 545–553.

62. Das, P. M., (2004). DNA methylation and cancer. *Journal of Clinical Oncology, 22*, 4632–4642.

63. Davis, C. D., Uthus, E. O., & Finley, J. W., (2000). Dietary selenium and arsenic affect DNA methylation *in vitro* in caco-2 cells and *in vivo* in rat liver and colon. *The Journal of Nutrition, 130*(12), 2903–2909.

64. Davison, J. M., Mellott, T. J., Kovacheva, V. P., & Blusztajn, J. K., (2009). Gestational choline supply regulates methylation of histone H3, expression of histone methyltransferases G9a (Kmt1c) and Suv39h1 (Kmt1a) and DNA methylation of their genes in rat fetal liver and brain. *Journal of Biological Chemistry, 284*(4), 1982–1989.

65. Day, J. K., Bauer, A. M., & DesBordes, C., (2002). Genistein alters methylation patterns in mice. *The Journal of Nutrition, 132*(8), 2419–2423.

66. Deckelbaum, R. J., & Torrejon, C., (2012). The omega-3 fatty acid nutritional landscape: Health benefits and sources. *Journal of Nutrition, 142*(3), 587–591.

67. Delcuve, G. P., Rastegar, M., & Davie, J. R., (2009). Epigenetic control. *Journal of Cellular Physiology, 219*(2), 243–250.

68. Desideri, G., Kwik-Uribe, C., Grassi, D., Necozione, S., Ghiadoni, L., Mastroiacovo, D., et al., (2012). Benefits in cognitive function, blood pressure, and insulin resistance through cocoa flavanol consumption in elderly subjects with mild cognitive impairment: The cocoa, cognition, and aging (CoCoA) study. *Hypertension, 60*(3), 794–801.

69. Dhalla, N. S., Temsah, R. M., & Netticadan, T., (2000). Role of oxidative stress in cardiovascular diseases. *Journal of Hypertension, 18*(6), 655–673.

70. Dolinoy, D. C., & Jirtle, R. L., (2008). Environmental epigenomics in human health and disease. *Environmental and Molecular Mutagenesis, 49*(1), 4–8.

71. Dolinoy, D. C., Weidman, J. R., & Jirtle, R., (2007). Epigenetic gene regulation: Linking early developmental environment to adult disease. *Reproductive Toxicology, 23*, 297–307.

72. Dolinoy, D. C., Weidman, J. R., Waterland, R. A., & Jirtle, R. L., (2006). Maternal genistein alters coat color and protects any mouse offspring from obesity by modifying the fetal epigenome. *Environmental Health Perspectives, 114*(4), 567–572.

73. Egger, G., Liang, G., Aparicio, A., & Jones, P. A., (2004). Epigenetics in human disease and prospects for epigenetic therapy. *Nature, 429*(6990), 457–463.

74. Ehrlich, M., (2002). DNA methylation in cancer: Too much, but also too little. *Oncogene, 21*(35), 5400–5413.

75. Ellis, L., Atadja, P. W., & Johnstone, R. W., (2009). Epigenetics in cancer: Targeting chromatin modifications. *Molecular Cancer Therapeutics, 8*(6), 1409–1420.

76. Esteller, M., (2007). Epigenetic gene silencing in cancer: The DNA hypermethylome. *Human Molecular Genetics, 16*(R1), 50–59.

77. Esteller, M., (2002). CpG Island hypermethylation and tumor suppressor genes: Booming present, a brighter future. *Oncogene, 21*(35), 5427–5440.

78. Esteller, M., (2005). Aberrant DNA methylation as a cancer-inducing mechanism. *Annual Review of Pharmacology and Toxicology, 45*(1), 629–656.

79. Esteller, M., (2007). Cancer epigenomics: DNA methylomes and histone-modification maps. *Nature Reviews Genetics, 8*(4), 286–298.

80. Esteller, M., (2008). Epigenetics in evolution and disease. *The Lancet, 372*, S90–S96.

81. Ettinger, B., Genant, H. K., & Cann, C. E., (1987). Postmenopausal bone loss is prevented by treatment with low-dosage estrogen with calcium. *Annals of Internal Medicine, 106*(1), 40–45.

82. Falls, J. G., Pulford, D. J., Wylie, A. A., & Jirtle, R. L., (1999). Genomic imprinting: Implications for human disease. *The American Journal of Pathology, 154*(3), 635–647.

83. Fang, M. Z., Wang, Y., Ai, N., Hou, Z., Sun, Y., Lu, H., Welsh, W., & Yang, C. S., (2003). Tea polyphenol (-)-Epigallocatechin–3-Gallate inhibits DNA methyltransferase and reactivates methylation-silenced genes in cancer cell lines. *Cancer Research, 63*(22), 7563–7570.

84. Fardet, A., (2010). New hypotheses for the health-protective mechanisms of whole-grain cereals: What is beyond fiber? *Nutrition Research Reviews, 23*(1), 65–134.

85. Feinberg, A. P., & Vogelstein, B., (1983). Hypomethylation distinguishes genes of some human cancers from their normal counterparts. *Nature, 301*(5895), 89–92.

86. Feinberg, A. P., & Tycko, B., (2004). The History of cancer epigenetics. *Nature Reviews Cancer, 4*(2), 143–153.

87. Fisher, N. D. L., & Hollenberg, N. K., (2006). Aging and vascular responses to flavanol-rich cocoa. *Journal of Hypertension, 24*(8), 1575–1580.

88. Fraga, C. G., Oteiza, P. I., Litterio, M. C., & Galleano, M., (2012). Phytochemicals as antioxidants: Chemistry and health effects. *Acta Horticulturae, 939,* 63–69.

89. Fraga, C. G., Galleano, M., Verstraeten, S. V., & Oteiza, P. I., (2010). Basic biochemical mechanisms behind the health benefits of polyphenols. *Molecular Aspects of Medicine, 31*(6), 435–445.

90. Fraga, M. F., Ballestar, E., Paz, M. F., & Ropero, S., (2005). Epigenetic differences arise during the lifetime of monozygotic twins. *Proceedings of the National Academy of Sciences of the United States of America, 102*(30), 10604–10609.

91. Fraga, M. F., Ballestar, E., Villar-Garea, A., & Boix-Chornet, M., (2005). Loss of acetylation at Lys16 and trimethylation at Lys20 of histone H4 is a common hallmark of human cancer. *Nature Genetics, 37*(4), 391–400.

92. Fukao, T., Hosono, T., Misawa, S., Seki, T., & Ariga, T., (2004). The effects of allyl sulfides on the induction of phase II detoxification enzymes and liver injury by carbon tetrachloride. *Food and Chemical Toxicology, 42*(5), 743–749.

93. Fuller, R., (1992). History and development of probiotics. In: Dordrecht (ed.), *Probiotics* (pp. 1–8). Springer, Netherlands.

94. Gaal, L. F., & Maggioni, A. P., (2014). Overweight, obesity, and outcomes: Fat mass and beyond. *The Lancet, 383*(9921), 935–936.

95. Gabor, M., George, L., & Maleszka, R., (2011). Epigenomic communication systems in humans and honey bees: From molecules to behavior. *Hormones and Behavior, 59*(3), 399–406.

96. Gheldof, N., Tabuchi, T. M., & Dekker, J., (2006). The active FMR1 promoter is associated with a large domain of altered chromatin conformation with embedded local histone modifications. *Proceedings of the National Academy of Sciences, 103*(33), 12463–12468.

97. Gibney, E. R., & Nolan, C. M., (2010). Epigenetics and gene expression. *Heredity, 105*(1), 4–13.

98. Gill, H., & Prasad, J., (2008). Probiotics, immunomodulation, and health benefits. In: *Bioactive Components of Milk* (pp. 423–454). Springer, New York, New York.

99. Gillman, M. W., & Cupples, L. A., (1997). Inverse association of dietary fat with development of ischemic stroke in men. *JAMA, 278*(24), 2145–2150.

100. Giovannucci, E., Rimm, E. B., Liu, Y., Stampfer, M. J., & Willett, W. C., (2002). Prospective study of tomato products, lycopene, and prostate cancer risk. *Journal of the National Cancer Institute, 94*(5), 391–398.

101. Glubi, T., & Maggioni, A. P., (2008). Effect of N–3 polyunsaturated fatty acids in patients with chronic heart failure (the GISSI-HF trial): A randomized, double-blind, placebo-controlled trial. *The Lancet, 372*(9645), 1223–1230.

102. Gluckman, P. D., Hanson, M. A., Cooper, C., & Thornburg, K. L., (2008). Effect of in utero and early-life conditions on adult health and disease. *The New England Journal of Medicine, 359*(1), 61–73.

103. Godfrey, K. M., Lillycrop, K. A., Burdge, G. C., Gluckman, P. D., & Hanson, M. A., (2007). Epigenetic mechanisms and the mismatch concept of the developmental origins of health and disease. *Pediatric Research, 61*(5 Part–2), 5R–10R.

Plant Secondary Metabolites for Human Health

104. Goldberg, A. D., Allis, C. D., & Bernstein, E., (2007). Epigenetics: A landscape takes shape. *Cell*, *128*(4), 635–638.
105. Gorbach, S. L., (2000). Probiotics and gastrointestinal health. *The American Journal of Gastroenterology*, *95*(1), S2–S4.
106. Graf, B. A., Milbury, P. E., & Blumberg, J. B., (2005). Flavonols, flavones, flavanones, and human health: Epidemiological evidence. *Journal of Medicinal Food*, *8*(3), 281–290.
107. Gräff, J., & Mansuy, I. M., (2009). Epigenetic dysregulation in cognitive disorders. *European Journal of Neuroscience*, *30*(1), 1–8.
108. Gylling, H., & Miettinen, T. A., (2005). The Effect of plant stanol- and sterol-enriched foods on lipid metabolism, serum lipids and coronary heart disease. *Annals of Clinical Biochemistry*, *42*(4), 254–263.
109. Hang, C. T., Yang, J., Han, P., Cheng, H. L., Shang, C., Ashley, E., Zhou, B., & Chang, C. P., (2010). Chromatin regulation by Brg1 underlies heart muscle development and disease. *Nature*, *466*(7302), 62–67.
110. Hanley, B., Dijane, J., & Fewtrell, M., (2010). Metabolic imprinting, programming and epigenetics: A review of present priorities and future opportunities. *British Journal of Nutrition*, *104*(S1), 1–25.
111. Hardy, T. M., & Tollefsbol, T. O., (2011). Epigenetic diet: Impact on the epigenome and cancer. *Epigenomics*, *3*(4), 503–518.
112. Hayek, M. G., Han, S. N., Wu, D., Watkins, B. A., Meydani, M., Dorsey, J. L., Smith, D. E., & Meydani, S. N., (1999). Dietary conjugated linoleic acid influences the immune response of young and old C57BL/6NCrlBR mice. *The Journal of Nutrition*, *129*(1), 32–38.
113. Hayes, M. A., Rushmore, T. H., & Goldberg, M. T., (1987). Inhibition of hepatocarcinogenic responses to 1, 2-Dimethylhydrazine by diallyl sulfide: A component of garlic oil. *Pub. Med.-NCBI*, 1155–1157.
114. Hellman, A., & Chess, A., (2007). Gene body-specific methylation on the active X chromosome. *Science*, *315*(5815), 1141–1143.
115. Heo, H. J., Kim, D. O., & Shin, S., (2004). Effect of antioxidant flavanone, naringenin, from *Citrus Junos* on neuroprotection. *Journal of Agricultural and Food Chemistry*, *52*(6), 1520–1525.
116. Hite, K. C., Adams, V. H., & Hansen, J. C., (2009). Recent advances in MeCP2 structure and function. *Biochemistry and Cell Biology*, *87*(1), 219–227.
117. Ho, S. M., Tang, W. Y., De Frausto, Jessica, B., & Prins, G. S., (2006). Developmental exposure to estradiol and bisphenol: A increases susceptibility to prostate carcinogenesis and epigenetically regulates phosphodiesterase type 4 variant 4. *Cancer Research*, *66*(11), 5624–5632.
118. Hodge, W., Barnes, D., Schachter, H. M., Pan, Y., & Lowcock, E. C., (2005). *Effects of Omega-3 Fatty Acids on Eye Health: Summary*. AHRQ Publication No. 05-E008–1, 117 pages, e-doc.
119. Holliday, R., (1991). Mutations and epimutations in mammalian cells. *Mutation Research*, *250*(1 & 2), 351–363.
120. Hollman, P. C. H., (2001). Evidence for health benefits of plant phenols: Local or systemic effects? *Journal of the Science of Food and Agriculture*, *81*(9), 842–852.

121. Howarth, N. C., Saltzman, E., & Roberts, S. B., (2001). Dietary fiber and weight regulation. *Nutrition Reviews, 59*(5), 129–139.

122. Howell, A. B., (2002). Cranberry proanthocyanidins and the maintenance of urinary tract health. *Critical Reviews in Food Science and Nutrition, 42*(3), 273–278.

123. Howell, A. B., (2007). Bioactive compounds in cranberries and their role in prevention of urinary tract infections. *Molecular Nutrition & Food Research, 51*(6), 732–737.

124. Huang, J., Plass, C., & Gerhauser, C., (2011). Cancer chemoprevention by targeting the epigenome. *Current Drug Targets, 12*(13), 1925–1956.

125. Hubbard, J. I., Jones, S. F., & Landau, E. M., (1968). On the mechanism by which calcium and magnesium affect the release of transmitter by nerve impulses. *The Journal of Physiology, 196*(1), 75–86.

126. Huber, L. C., Stanczyk, J., Jüngel, A., & Gay, S., (2007). Epigenetics in inflammatory rheumatic diseases. *Arthritis & Rheumatism, 56*(11), 3523–3531.

127. Huertas, D., Sendra, R., & Muñoz, P., (2009). Chromatin dynamics coupled to DNA repair. *Epigenetics, 4*(1), 31–42.

128. Ip, C., Scimeca, J. A., & Thompson, H. J., (1994). Conjugated linoleic acid: A powerful anticarcinogen from animal fat sources. *Cancer, 74*(3), 1050–1054.

129. Ishige, K., Schubert, D., & Sagara, Y., (2001). Flavonoids protect neuronal cells from oxidative stress by three distinct mechanisms. *Free Radical Biology & Medicine, 30*(4), 433–446.

130. Plat, J., & Mensink, R. P., (2001). Effects of plant sterols and stanols on lipid metabolism and cardiovascular risk. *Nutrition, Metabolism, and Cardiovascular Diseases: NMCD, 11*(1), 31–40.

131. Jaenisch, R., & Bird, A., (2003). Epigenetic regulation of gene expression: How the genome integrates intrinsic and environmental signals. *Nature Genetics, 33*(3s), 245–254.

132. Javierre, B. M., Fernandez, A. F., Richter, J., & Al-Shahrour, F., (2010). Changes in the pattern of DNA methylation associate with twin discordance in systemic lupus erythematosus. *Genome Research, 20*(2), 170–179.

133. 133.Javierre, B. M., Esteller, M., & Ballestar, E., (2008). Epigenetic connections between autoimmune disorders and haematological malignancies. *Trends in Immunology, 29*(12), 616–623.

134. Jiang, Y. H., Bressler, J., & Beaudet, A. L., (2004). Epigenetics and human disease. *Annual Review of Genomics and Human Genetics, 5*(1), 479–510.

135. Jin, B., Tao, Q., Peng, J., Soo, H. M., & Wu, W., (2007). DNA methyltransferase 3B (DNMT3B) mutations in ICF syndrome lead to altered epigenetic modifications and aberrant expression of genes regulating development, neurogenesis and immune function. *Human Molecular Genetics, 17*(5), 690–709.

136. Jirtle, R. L., & Skinner, M. K., (2007). Environmental epigenomics and disease susceptibility. *Nature Reviews Genetics, 8*(4), 253–262.

137. Jones, P. A., (1999). The DNA methylation paradox. *Trends in Genetics: TIG, 15*(1), 34–37.

138. Joshi, S. S., Kuszynski, C. A., & Bagchi, D., (2001). The cellular and molecular basis of health benefits of grape seed proanthocyanidin extract. *Current Pharmaceutical Biotechnology, 2*(2), 187–200.

139. Kaczmarczyk, M. M., Miller, M. J., & Freund, G. G., (2012). The health benefits of dietary fiber: Beyond the usual suspects of type 2 *diabetes mellitus*, cardiovascular disease and colon cancer. *Metabolism, 61*(8), 1058–1066.

140. Kaminsky, Z. A., Tang, T., Wang, S. C., & Ptak, C., (2009). DNA methylation profiles in monozygotic and dizygotic twins. *Nature Genetics, 41*(2), 240–245.

141. Karlic, R., Chung, H. R., Lasserre, J., Vlahovicek, K., & Vingron, M., (2010). Histone modification levels are predictive for gene expression. *Proceedings of the National Academy of Sciences, 107*(7), 2926–2931.

142. Karouzakis, E., Gay, R. E., Gay, S., & Neidhart, M., (2009). Epigenetic control in rheumatoid arthritis synovial fibroblasts. *Nature Reviews Rheumatology, 5*(5), 266–272.

143. Kelly, T. K., De Carvalho, D. D., & Jones, P. A., (2012). Epigenetic modifications as therapeutic targets. *Nature Biotechnology, 28*(10), 1069–1078.

144. Kensler, T. W., Curphey, T. J., Maxiutenko, Y., & Roebuck, B. D., (2000). Chemoprotection by organosulfur inducers of phase 2 enzymes: Dithiolethiones and dithiins. *Drug Metabolism and Drug Interactions, 17*(1–4), 3–22.

145. Kensler, T. W., Egner, P. A., & Agyeman, A. S., (2012). Keap1-Nrf2 signaling: A target for cancer prevention by sulforaphane. In: *Natural Products in Cancer Prevention and Therapy* (pp. 163–177). Springer, Berlin Heidelberg.

146. Khaodhiar, L., Ricciotti, H. A., Li, L., Pan, W., Schickel, M., Zhou, J., & Blackburn, G. L., (2008). Daidzein-rich isoflavone aglycones are potentially effective in reducing hot flashes in menopausal women. *Menopause (New York, N.Y.), 15*(1), 125–132.

147. Khaw, K. T., Barrett, C., & Elizabeth., (1987). Dietary potassium and stroke-associated mortality. *New England Journal of Medicine, 316*(5), 235–240.

148. Kim, J. K., Samaranayake, M., & Pradhan, S., (2009). Epigenetic mechanisms in mammals. *Cellular and Molecular Life Sciences, 66*(4), 596–612.

149. Kim, M., Long, T. I., Arakawa, K., Wang, R., Yu, M. C., & Laird, P. W., (2010). DNA methylation as a biomarker for cardiovascular disease risk. *PLoS ONE, 5*(3), https://doi.org/10.1371/journal.pone.0009692.

150. Kim, Y. I., Giuliano, A., Hatch, K. D., Schneider, A., Nour, M. A., Dallal, G. E., Selhub, J., & Mason, J. B., (1994). Global DNA hypomethylation increases progressively in cervical dysplasia and carcinoma. *Cancer, 74*(3), 893–899.

151. King-Batoon, A., Leszczynska, J. M., & Klein, C. B., (2008). Modulation of gene methylation by genistein or lycopene in breast cancer cells. *Environmental and Molecular Mutagenesis, 49*(1), 36–45.

152. Kinsella, J. E., Lokesh, B., & Stone, R. A., (1990). Dietary N–3 polyunsaturated fatty acids and amelioration of cardiovascular disease: Possible mechanisms. *The American Journal of Clinical Nutrition, 52*(1), 1–28.

153. Klein, E. A., Thompson, I. M., Tangen, C. M., & Crowley, J. J., (2011). Vitamin E and the risk of prostate cancer: The selenium and vitamin E cancer prevention trial (SELECT). *JAMA, 306*(14), 1549–1556.

154. Kleine-Kohlbrecher, D., & Christensen, J., (2010). Functional link between the histone demethylase PHF8 and the transcription factor ZNF711 in x-linked mental retardation. *Molecular Cell, 38*(2), 165–178.

155. Knekt, P., Järvinen, R., Seppänen, R., Hellövaara, M., Teppo, L., Pukkala, E., & Aromaa, A., (1997). Dietary flavonoids and the risk of lung cancer and other malignant neoplasms. *American Journal of Epidemiology*, *146*(3), 223–230.

156. Kouki, T., Kishitake, M., Okamoto, M., Oosuka, I., Takebe, M., & Yamanouchi, K., (2003). Effects of neonatal treatment with phytoestrogens, genistein and daidzein, on sex difference in female rat brain function: Estrous cycle and lordosis. *Hormones and Behavior*, *44*(2), 140–145.

157. Kouzarides, T., (2007). Chromatin modifications and their function. *Cell*, *128*(4), 693–705.

158. Krinsky, N. I., Landrum, J. T., & Bone, R. A., (2003). Biologic mechanisms of the protective role of Lutein and Zeaxanthin in the eye. *Annual Review of Nutrition*, *23*(1), 171–201.

159. Kroeger, H., Jelinek, J., & Estecio, M. R. H., (2008). Aberrant CpG-Island methylation in acute myeloid leukemia is accentuated at relapse. *Blood*, *112*(4), 1366–1373.

160. Kucharski, R., Maleszka, J., Foret, S., & Maleszka, R., (2008). Nutritional control of reproductive status in honeybees via DNA methylation. *Science*, *319*(5871), 1827–1830.

161. Kumaran, K. S., & Prince, P. S. M., (2010). Caffeic acid protects rat heart mitochondria against isoproterenol-induced oxidative damage. *Cell Stress and Chaperones*, *15*(6), 791–806.

162. Kumari, D., & Usdin, K., (2009). Chromatin remodeling in the noncoding repeat expansion diseases. *The Journal of Biological Chemistry*, *284*(12), 7413–7417.

163. Kuroda, A., Rauch, T. A., & Todorov, I., (2009). Insulin gene expression is regulated by DNA methylation. *PLoS ONE*, *4*(9), e6953.

164. Kwak, M. K., (2003). Modulation of gene expression by cancer chemopreventive dithiolethiones through the Keap1-Nrf2 pathway: Identification of novel gene clusters for cell survival. *Journal of Biological Chemistry*, *278*(10), 8135–8145.

165. Lamartiniere, C. A., Cotroneo, M. S., Fritz, W. A., Wang, J., Mentor-Marcel, R., & Elgavish, A., (2002). Genistein chemoprevention: Timing and mechanisms of action in murine mammary and prostate. *The Journal of Nutrition*, *132*(3), 552S–558S.

166. Latham, J. A., & Dent, S. Y. R., (2007). Cross-regulation of histone modifications. *Nature Structural & Molecular Biology*, *14*(11), 1017–1024.

167. Law, M. R., (2000). Topic in review: Plant sterol and stanol margarines and health. *Western Journal of Medicine*, *173*(1), 43–47.

168. Lehnen, H., Zechner, U., & Haaf, T., (2013). Epigenetics of gestational *diabetes mellitus* and offspring health: The time for action is in early stages of life. *Molecular Human Reproduction*, *19*(7), 415–422.

169. Levenson, V. V., & Melnikov, A. A., (2012). DNA methylation as clinically useful biomarkers–light at the end of the tunnel. *Pharmaceuticals*, *5*(12), 94–113.

170. Li, B., Carey, M., & Workman, J. L., (2007). The role of chromatin during transcription. *Cell*, *128*(4), 707–719.

171. Li, S., Washburn, K. A., Moore, R., Uno, T., Teng, C., Newbold, R. R., McLachlan, J. A., & Negishi, M., (1997). Developmental exposure to diethylstilbestrol elicits demethylation of estrogen-responsive lactoferrin gene in mouse uterus. *Cancer Research*, *57*(19), 4356–4359.

172. Li, S., Hansman, R., Newbold, R., Davis, B., McLachlan, J. A., & Barrett, J. C., (2003). Neonatal diethylstilbestrol exposure induces persistent elevation of

c-Fos expression and hypomethylation in its Exon–4 in mouse uterus. *Molecular Carcinogenesis, 38*(2), 78–84.

173. Lillycrop, K. A., Phillips, E. S., Jackson, A. A., Hanson, M. A., & Burdge, G. C., (2005). Dietary protein restriction of pregnant rats induces and folic acid supplementation prevents epigenetic modification of hepatic gene expression in the offspring. *The Journal of Nutrition, 135*(6), 1382–1386.

174. Liu, A. G., & Erdman, J. W., (2011). Lycopene and Apo–10′-Lycopenal do not alter DNA methylation of GSTP1 in LNCaP Cells. *Biochemical and Biophysical Research Communications, 412*(3), 479–482.

175. Liu, D., Diorio, J., Tannenbaum, B., Caldji, C., Francis, D., Freedman, A., et al., (1997). Maternal care, hippocampal glucocorticoid receptors, and hypothalamic-pituitary-adrenal responses to stress. *Science* (New York), *277*(5332), 1659–1662.

176. Liu, R. H., & Hotchkiss, J. H., (1995). Potential genotoxicity of chronically elevated nitric oxide: Review. *Mutation Research, 339*(2), 73–89.

177. Liu, R. H., (2003). Health benefits of fruit and vegetables are from additive and synergistic combinations of phytochemicals. *The American Journal of Clinical Nutrition, 78*(3), 517S–520S.

178. Loft, S., & Poulsen, H. E., (1996). Cancer risk and oxidative DNA damage in man. *Journal of Molecular Medicine, 74*(6), 297–312.

179. López-Otín, C., Blasco, M. A., Partridge, L., Serrano, M., & Kroemer, G., (2013). The hallmarks of aging. *Cell, 153*(6), 1194–1217.

180. Lopez-Serra, L., & Esteller, M., (2008). Proteins that bind methylated DNA and human cancer: Reading the wrong words. *British Journal of Cancer, 98*(12), 1881–1885.

181. Lorincz, A. T., (2011). The promise and the problems of epigenetics biomarkers in cancer. *Expert Opinion on Medical Diagnostics, 5*(5), 375–379.

182. Loveren, C., (2004). Sugar alcohols: What is the evidence for caries-preventive and caries-therapeutic effects? *Caries Research, 38*(3), 286–293.

183. Luco, R. F., Pan, Q., Tominaga, K., Blencowe, B. J., Pereira-Smith, O. M., & Misteli, T., (2010). Regulation of alternative splicing by histone modifications. *Science, 327*(5968), 996–1000.

184. Maleszka, R., (2008). Epigenetic integration of environmental and genomic signals in honey bees: The critical interplay of nutritional, brain and reproductive networks. *Epigenetics, 3*(4), 188–192.

185. Malone, C. D., & Hannon, G. J., (2009). Small RNAs as guardians of the genome. *Cell, 136*(4), 656–668.

186. Manach, C., Mazur, A., & Scalbert, A., (2005). Polyphenols and prevention of cardiovascular diseases. *Current Opinion in Lipidology, 16*(1), 77–84.

187. Mann, M. R. W., & Bartolomei, M. S., (1999). Towards a molecular understanding of Prader–Willi and Angelman syndromes. *Human Molecular Genetics Review, 8*(10), 1867–1873.

188. Le Marchand, L., Murphy, S. P., Hankin, J. H., Wilkens, L. R., & Kolonel, L. N., (2000). Intake of flavonoids and lung cancer. *Journal of the National Cancer Institute, 92*(2), 154–160.

189. Maresca, A., Zaffagnini, M., Caporali, L., Carelli, V., & Zanna, C., (2015). DNA methyltransferase 1 mutations and mitochondrial pathology: Is mtDNA methylated? *Frontiers in Genetics*, 690.

190. Marlett, J. A., McBurney, M. I., & Slavin, J. L., (2002). American dietetic association: Position of the American dietetic association: Health implications of dietary fiber. *Journal of the American Dietetic Association, 102*(7), 993–1000.

191. Marwick, J. A., Kirkham, P. A., & Stevenson, C. S., (2004). Cigarette smoke alters chromatin remodeling and induces proinflammatory genes in rat lungs. *American Journal of Respiratory Cell and Molecular Biology, 31*(6), 633–642.

192. Mason, J. B., & Choi, S. W., (2005). Effects of alcohol on folate metabolism: Implications for carcinogenesis. *Alcohol, 35*(3), 235–241.

193. McCord, J. M., (1993). Human disease, free radicals, and the oxidant/antioxidant balance. *Clinical Biochemistry, 26*(5), 351–357.

194. McCullough, M. L., Robertson, A. S., Chao, A., Jacobs, E. J., Stampfer, M. J., & Jacobs, D. R., (2003). Prospective study of whole grains, fruits, vegetables and colon cancer risk. *Cancer Causes & Control: CCC, 14*(10), 959–970.

195. McGowan, P. O., Sasaki, A., D'Alessio, A. C., Dymov, S., Labonté, B., Szyf, M., Turecki, G., & Meaney, M. J., (2009). Epigenetic regulation of the glucocorticoid receptor in human brain associates with childhood abuse. *Nature Neuroscience, 12*(3), 342–348.

196. McIntosh, M., & Miller, C., (2001). Diet containing food rich in soluble and insoluble fiber improves glycemic control and reduces hyperlipidemia among patients with type 2 *diabetes mellitus. Nutrition Reviews, 59*(2), 52–55.

197. Medina, P. P., & Sanchez, C. M., (2008). Involvement of the chromatin-remodeling factor BRG1/SMARCA4 in human cancer. *Epigenetics, 3*(2), 64–68.

198. Meng, Z., Zhu, X. Y., Li, Y. Y., & Meng, Z. Q., (2013). Epigenetic regulation and cancer: Review. *Oncology Reports, 31*(2), 523–532.

199. Miao, F., Smith, D. D., Zhang, L., Min, A., Feng, W., & Natarajan, R., (2008). Lymphocytes from patients with type 1 diabetes display a distinct profile of chromatin histone H3 lysine 9 dimethylation: An epigenetic study in diabetes. *Diabetes, 57*(12), 3189–3198.

200. Miceli, M., Bontempo, P., Nebbioso, A., & Altucci, L., (2014). Natural compounds in epigenetics: A current view. *Food and Chemical Toxicology, 73*, 71–83.

201. Middleton, E., (1998). Effect of plant flavonoids on immune and inflammatory cell function. *Advances in Experimental Medicine and Biology, 439*, 175–182.

202. Miller, R. L., & Ho, S. M., (2008). Environmental epigenetics and asthma. *American Journal of Respiratory and Critical Care Medicine, 177*(6), 567–573.

203. Misiewicz, I., Skupińska, K., Kowalska, E., Lubiński, J., & Kasprzycka, G. T., (2004). Sulforaphane-mediated induction of a phase 2 detoxifying enzyme NAD(P) H: Quinone reductase and apoptosis in human lymphoblastoid cells. *Acta Biochimica Polonica, 51*(3), 711–721.

204. Moore, S. D. P., (2004). Uterine leiomyomata with t(10, 17) disrupt the histone acetyltransferase MORF. *Cancer Research, 64*(16), 5570–5577.

205. Morgan, H. D., Santos, F., & Green, K., (2005). Epigenetic reprogramming in mammals. *Human Molecular Genetics, 14*(1), 47–58.

206. Morris, K. V., (2005). siRNA-mediated transcriptional gene silencing: The potential mechanism and a possible role in the histone code. *Cellular and Molecular Life Sciences, 62*(24), 3057–3066.

207. Movassagh, M., Choy, M. K., Goddard, M., Bennett, M. R., Down, T. A., & Foo, R. S. Y., (2010). Differential DNA methylation correlates with differential expression of angiogenic factors in human heart failure. *PLoS ONE, 5*(1), e8564.

208. Munday, R., & Munday, C. M., (2004). Induction of phase II detoxification enzymes in rats by plant-derived isothiocyanates: Comparison of allyl isothiocyanate with sulforaphane and related compounds. *Journal of Agricultural and Food Chemistry, 52*(7), 1867–1871.

209. Naaz, A., Yellayi, S., & Zakroczymski, M. A., (2003). The soy isoflavone genistein decreases adipose deposition in mice. *Endocrinology, 144*(8), 3315–3320.

210. Nafee, T. M., Farrell, W. E., Carroll, W. D., Fryer, A. A., & Ismail, K. M. K., (2007). Review article: Epigenetic control of fetal gene expression. *BJOG: An International Journal of Obstetrics & Gynaecology, 115*(2), 158–168.

211. Nagao, T., Yoshimura, S., Saito, Y., Nakagomi, M., Usumi, K., & Ono, H., (2001). Reproductive effects in male and female rats of neonatal exposure to genistein. *Reproductive Toxicology* (Elmsford, NY), *15*(4), 399–411.

212. Nakahata, Y., Grimaldi, B., Sahar, S., Hirayama, J., & Sassone, C. P., (2007). Signaling to the circadian clock: Plasticity by chromatin remodeling. *Current Opinion in Cell Biology, 19*(2), 230–237.

213. Nardini, M., Natella, F., Gentili, V., Felice, M. D., & Scaccini, C., (1997). Effect of caffeic acid dietary supplementation on the antioxidant defense system in rat: An *in vivo* study. *Archives of Biochemistry and Biophysics, 342*(1), 157–160.

214. Newbold, R. R., (2004). Lessons learned from perinatal exposure to diethylstilbestrol. *Toxicology and Applied Pharmacology, 199*(2), 142–150.

215. Newbold, R. R., Padilla, B. E., & Jefferson, W. N., (2006). Adverse effects of the model environmental estrogen diethylstilbestrol are transmitted to subsequent generations. *Endocrinology, 147*(6), S11–S17.

216. Niculescu, M. D., & Lupu, D. S., (2011). Nutritional influence on epigenetics and effects on longevity. *Current Opinion in Clinical Nutrition and Metabolic Care, 14*(1), 35–40.

217. Niculescu, M. D., & Zeisel, S. H., (2002). Diet, methyl donors and DNA methylation: Interactions between dietary folate, methionine and choline. *The Journal of Nutrition, 132*(8), 2333–2335.

218. Norata, G. D., Marchesi, P., Passamonti, S., Pirillo, A., Violi, F., & Catapano, A. Luigi., (2007). Anti-inflammatory and anti-atherogenic effects of cathechin, caffeic acid and trans-resveratrol in apolipoprotein E deficient mice. *Atherosclerosis, 191*(2), 265–271.

219. Omenn, G. S., Goodman, G. E., & Thornquist, M. D., (1996). Effects of a combination of beta carotene and vitamin A on lung cancer and cardiovascular disease. *New England Journal of Medicine, 334*(18), 1150–1155.

220. Pa, K., Jarvinen, R., Seppanen, R., & Rissanen, A., (1991). Dietary antioxidants and the risk of lung cancer. *American Journal of Epidemiology, 134*(5), 471–479.

221. Painter, R. C., Osmond, C., Gluckman, P., Hanson, M., Phillips, D. I. W., & Roseboom, T. J., (2008). Transgenerational effects of prenatal exposure to the Dutch famine on neonatal adiposity and health in later life. *BJOG: An International Journal of Obstetrics & Gynaecology, 115*(10), 1243–1249.

222. Park, L. K., Friso, S., & Choi, S. W., (2012). Nutritional influences on epigenetics and age-related disease. *Proceedings of the Nutrition Society*, *71*(1), 75–83.
223. Paul, D. S., & Beck, S., (2014). Advances in epigenome-wide association studies for common diseases. *Trends in Molecular Medicine*, *20*(10), 541–543.
224. Pereira, M. A., O'Reilly, E., Augustsson, K., Fraser, G. E., Goldbourt, U., Heitmann, B. L., & Hallmans, G., (2004). Dietary fiber and risk of coronary heart disease. *Archives of Internal Medicine*, *164*(4), 370–376.
225. Petersson, L. G., Birkhed, D., Gleerup, A., Johansson, M., & Jönsson, G., (1991). Caries-preventive effect of dentifrices containing various types and concentrations of fluorides and sugar alcohols. *Caries Research*, *25*(1), 74–79.
226. Petronis, A., (2001). Human morbid genetics revisited: Relevance of epigenetics. *Trends in Genetics: TIG*, *17*(3), 142–146.
227. Petronis, A., (2004). The Origin of schizophrenia: Genetic thesis, epigenetic antithesis, and resolving synthesis. *Biological Psychiatry*, *55*(10), 965–970.
228. Pieper, H. C., Evert, B. O., Kaut, O., Riederer, P. F., Waha, A., & Wüllner, U., (2008). Different methylation of the TNF-alpha promoter in cortex and Substantia Nigra: Implications for selective neuronal vulnerability. *Neurobiology of Disease*, *32*(3), 521–527.
229. Pietta, P. G., (2000). Flavonoids as antioxidants. *Journal of Natural Products*, *63*(7), 1035–1042.
230. Plat, J., Kerckhoffs, D. A., & Mensink, R. P., (2000). Therapeutic potential of plant sterols and stanols. *Current Opinion in Lipidology*, *11*(6), 571–576.
231. Poirier, L. A., & Vlasova, T. I., (2002). The prospective role of abnormal methyl metabolism in cadmium toxicity. *Environmental Health Perspectives*, 793–795.
232. Portela, A., & Esteller, M., (2010). Epigenetic modifications and human disease. *Nature Biotechnology*, *28*(10), 1057–1068.
233. Potter, S. M., Baum, J. A., Teng, H. Y., Stillman, R. J., Shay, N. F., & Erdman, J. W., (1999). Soy protein and isoflavones: Their effects on blood lipids and bone density in postmenopausal women. *American Journal of Clinical Nutrition*, *68*(6), 1375S–1379S.
234. Prince, R. L., Smith, M., Dick, I. M., Price, R. I., Webb, P. G., Henderson, N. et al., (1991). Prevention of postmenopausal osteoporosis. *New England Journal of Medicine*, *325*(17), 1189–1195.
235. Radford, E. J., Ito, M., Shi, H., Corish, J. A., Yamazawa, K., Isganaitis, E., et al., (2014). In-utero undernourishment perturbs the adult sperm methylome and intergenerational metabolism. *Science*, *345*(6198), 1255903.
236. Radom, A., Shlomit, Z., Frank, O., Stacy, G., Pietro, C., & Dan M., (2010). Evidence for microRNA involvement in exercise-associated neutrophil gene expression changes. *Highlighted Topic Epigenetics in Health and Disease J. Appl. Physiol.*, *109*, 252–261.
237. Rahman, I., & Chung, S., (2010). Dietary polyphenols, deacetylases and chromatin remodeling in inflammation. *Journal of Nutrigenetics and Nutrigenomics*, *3*(4–6), 220–230.
238. Rakyan, V. K., Blewitt, M. E., Druker, R., Preis, J. I., & Whitelaw, E., (2002). Metastable epialleles in mammals. *Trends in Genetics: TIG*, *18*(7), 348–351.
239. Rashid, A., Issa, J., & Pierre, J., (2004). CpG-Island methylation in gastroenterologic neoplasia: Maturing field. *Gastroenterology*, *127*(5), 1578–1588.

240. Rave, K., Roggen, K., Dellweg, S., Heise, T., & Dieck, H. T., (2007). Improvement of insulin resistance after diet with a whole-grain based dietary product: Results of a randomized, controlled cross-over study in obese subjects with elevated fasting blood glucose. *British Journal of Nutrition, 98*(5), 929–936.

241. Remely, M., Lovrecic, L., De la Garza, A. L., Migliore, L., Peterlin, B., Milagro, F. I., Martinez, A. J., & Haslberger, A. G., (2015). Therapeutic perspectives of epigenetically active nutrients. *British Journal of Pharmacology, 172*(11), 2756–2768.

242. Reuter, S., Gupta, S. C., Chaturvedi, M. M., & Aggarwal, B. B., (2010). Oxidative stress, inflammation, and cancer: How are they linked? *Free Radical Biology & Medicine, 49*(11), 1603–1616.

243. Richards, E. J., (2006). Inherited epigenetic variation: Revisiting soft inheritance. *Nature Reviews Genetics, 7*(5), 395–401.

244. Richardson, B., (2003). Impact of aging on DNA methylation. *Ageing Research Reviews, 2*(3), 245–261.

245. Richer, S., Stiles, W., & Statkute, L., (2004). Double-masked, placebo-controlled, randomized trial of lutein and antioxidant supplementation in the intervention of atrophic age-related macular degeneration: The veterans LAST (Lutein antioxidant supplementation trial) study. *Optometry (St. Louis, Mo.), 75*(4), 216–230.

246. Rideout, W. M., Eggan, K., & Jaenisch, R., (2001). Nuclear cloning and epigenetic reprogramming of the genome. *Science, 293*(5532), 1093–1098.

247. Riedl, M. A., Saxon, A., & Diaz, S. D., (2009). Oral sulforaphane increases phase ii antioxidant enzymes in the human upper airway. *Clinical Immunology, 130*(3), 244–251.

248. Rix, T. A., Christensen, J. H., & Schmidt, E. B., (2013). Omega-3 fatty acids and cardiac arrhythmias. *Current Opinion in Clinical Nutrition and Metabolic Care, 16*(2), 168–173.

249. Robertson, K. D., & Jones, P. A., (2000). DNA methylation: Past, present and future directions. *Carcinogenesis, 21*(3), 461–467.

250. Robertson, K. D., (2005). DNA methylation and human disease. *Nature Reviews Genetics, 6*(8), 597–610.

251. Ropero, S., Fraga, M. F., & Ballestar, E., (2006). Truncating mutation of HDAC2 in human cancers confers resistance to histone deacetylase inhibition. *Nature Genetics, 38*(5), 566–569.

252. Roseboom, T., De Rooij, S., & Painter, R., (2006). The Dutch famine and its long-term consequences for adult health. *Early Human Development, 82*(8), 485–491.

253. Rosén, L. A. H., Silva, L. O., Andersson, U. K., Holm, C., Östman, E. M., & Björck, I. M. E., (2009). Endosperm and whole grain rye breads are characterized by low post-prandial insulin response and a beneficial blood glucose profile. *Nutrition Journal, 8*(1), 42.

254. Ruxton, C. H. S., Reed, S. C., Simpson, M. J. A., & Millington, K. J., (2004). The health benefits of omega-3 polyunsaturated fatty acids: Review of the evidence. *Journal of Human Nutrition and Dietetics, 17*(5), 449–459.

255. Sadikovic, B., Al-Romaih, K., Squire, J., & Zielenska, M., (2008). Causes and consequences of genetic and epigenetic alterations in human cancer. *Current Genomics, 9*(6), 394–408.

256. Saija, A., Tomaino, A., Trombetta, D., De Pasquale, A., Uccella, N., Barbuzzi, T., Paolino, D., & Bonina, F., (2000). *In vitro* and *in vivo* evaluation of caffeic and ferulic acids as topical photoprotective agents. *International Journal of Pharmaceutics, 199*(1), 39–47.

257. Saito, M., (2007). Role of FOSHU (Food for specified health uses) for healthier life. *Yakugaku Zasshi: Journal of the Pharmaceutical Society of Japan, 127*(3), 407–416.

258. Sakatani, T., Kaneda, A., Iacobuzio, D. C. A., Carter, M. G., Witzel, S. D. B., Okano, H., et al., (2005). Loss of imprinting of Igf2 alters intestinal maturation and tumorigenesis in mice. *Science, 307*(5717), 1976–1978.

259. Salminen, S., Ouwehand, A., Benno, Y., & Lee, Y. K., (1999). Probiotics: How should they be defined? *Trends in Food Science & Technology, 10*(3), 107–110.

260. Salnikow, K., & Costa, M., (2000). Epigenetic mechanisms of nickel carcinogenesis. *Journal of Environmental Pathology, Toxicology and Oncology: Official Organ of the International Society for Environmental Toxicology and Cancer, 19*(3), 307–318.

261. Sanders, M. E., (1994). Lactic acid bacteria as promoters of human health. In: *Functional Foods* (pp. 294–322). Springer, Boston, MA.

262. De Sario, A., (2009). Clinical and molecular overview of inherited disorders resulting from epigenomic dysregulation. *European Journal of Medical Genetics, 52*(6), 363–372.

263. Schanen, N. C., (2006). Epigenetics of autism spectrum disorders. *Human Molecular Genetics, 15*(2), 138–150.

264. Schulz, W. A., Steinhoff, C., & Florl, A. R., (2006). Methylation of endogenous human retroelements in health and disease. *Current Topics in Microbiology and Immunology, 310*, 211–250.

265. Schwartz, & David, A., (2012). State of the art epigenetics and environmental lung disease. *Journal of Allergy and Clinical Immunology, 130*(6), 1243–1255.

266. Seymour, C. B., & Mothersill, C., (2004). Radiation-induced bystander effects–Implications for cancer. *Nature Reviews Cancer, 4*(2), 158–164.

267. Sharma, S., Kelly, T. K., & Jones, P. A., (2010). Epigenetics in cancer. *Carcinogenesis, 31*(1), 27–36.

268. Sies, H., Schewe, T., Heiss, C., & Kelm, M., (2005). Cocoa polyphenols and inflammatory mediators. *The American Journal of Clinical Nutrition, 81*(1), S304–S312.

269. Silbergeld, E. K., Waalkes, M., & Rice, J. M., (2000). Lead as a carcinogen: Experimental evidence and mechanisms of action. *American Journal of Industrial Medicine, 38*(3), 316–323.

270. Sinclair, K. D., Allegrucci, C., & Singh, R., (2007). DNA methylation, insulin resistance, and blood pressure in offspring determined by maternal periconceptional Vitamin B and methionine status. *Proceedings of the National Academy of Sciences, 104*(49), 19351–19356.

271. Singal, R., Wang, S. Z., Sargent, T., Zhu, S. Z., & Ginder, G. D., (2002). Methylation of promoter proximal-transcribed sequences of an embryonic globin gene inhibits transcription in primary erythroid cells and promotes formation of a cell type-specific methylcytosine binding complex. *The Journal of Biological Chemistry, 277*(3), 1897–1905.

272. Sirtori, C. R., Tremoli, E., Gatti, E., Montanari, G., Sirtori, M., Colli, S., et al., (1986). Controlled evaluation of fat intake in the Mediterranean diet: comparative activities

of olive oil and corn oil on plasma lipids and platelets in high-risk patients. *The American Journal of Clinical Nutrition, 44*(5), 635–642.

273. Skinner, M. K., Manikkam, M., & Guerrero, B. C., (2010). Epigenetic transgenerational actions of environmental factors in disease etiology. *Trends in Endocrinology and Metabolism: TEM, 21*(4), 214–222.

274. Slattery, M. L., Sorenson, A. W., Mahoney, A. W., French, T. K., Kritchevsky, D., & Street, J. C., (1988). Diet and colon cancer: Assessment of risk by fiber type and food source. *Journal of the National Cancer Institute, 80*(18), 1474–1480.

275. Slavin, J., (2013). Fiber and prebiotics: Mechanisms and health benefits. *Nutrients, 5*(4), 1417–1435.

276. Sosa, V., Moliné, T., Somoza, R., Paciucci, R., Kondoh, H., & Lleonart, M. E., (2013). Oxidative stress and cancer: An overview. *Ageing Research Reviews, 12*(1), 376–390.

277. Sporn, J. C., Kustatscher, G., Hothorn, T., Collado, M., Serrano, M., Muley, T., Schnabel, P., & Ladurner, A. G., (2009). Histone macroH2A isoforms predict the risk of lung cancer recurrence. *Oncogene, 28*(38), 3423–3428.

278. Stamatova, I., & Meurman, J. H., (2009). Probiotics: Health benefits in the mouth. *American Journal of Dentistry, 22*(6), 329–338.

279. Steegers, R. P., Obermann, B. S. A., Kremer, D., Lindemans, J., Siebel, C., Steegers, E. A., et al., (2009). Periconceptional maternal folic acid use of 400 μg per day is related to increased methylation of the IGF2 gene in the very young child. *PLoS ONE, 4*(11), e7845.

280. Steinmetz, K. A., & Potter, J. D., (1991). Vegetables, fruit, and cancer, II: Mechanisms. *Cancer Causes & Control: CCC, 2*(6), 427–442.

281. Susiarjo, M., & Bartolomei, M. S., (2014). You are what you eat, but what about your DNA? *Science, 345*(6198), 733–734.

282. Suzuki, Y., Miyoshi, N., & Isemura, M., (2012). Health-promoting effects of green tea. *Proceedings of the Japan Academy, Series B: Physical and Biological Sciences, 88*(3), 88–101.

283. Swanson, D., Block, R., & Mousa, S. A., (2012). Omega-3 fatty acids EPA and DHA: Health benefits throughout life. *Advances in Nutrition: An International Review Journal, 3*(1), 1–7.

284. Symonds, M. E., Sebert, S. P., Hyatt, M. A., & Budge, H., (2009). Nutritional programming of the metabolic syndrome. *Nature Reviews Endocrinology, 5*(11), 604–610.

285. Szarcvel, S., Katarzyna, N., Matladi, N., Haegeman, G. B., & Wim, V., (2010). Nature or nurture: Let food be your epigenetic medicine in chronic inflammatory disorders. *Biochemical Pharmacology, 80*(12), 1816–1832.

286. 286. Szyf, M., (2009). Early life, the epigenome and human health. *Acta Paediatrica, 98*(7), 1082–1084.

287. Talbert, P. B., & Henikoff, S., (2010). Histone variants–Ancient wrap artists of the epigenome. *Nature Reviews Molecular Cell Biology, 11*(4), 264–275.

288. Tam, M., Gómez, S., González-Gross, M., & Marcos, A., (2003). Possible roles of magnesium on the immune system. *European Journal of Clinical Nutrition, 57*(10), 1193–1197.

289. Tammen, S. A., Friso, S., & Choi, S. W., (2013). Epigenetics: The link between nature and nurture. *Molecular Aspects of Medicine, 34*(4), 753–764.

290. Tang, W., & Ho, S., (2007). Epigenetic reprogramming and imprinting in origins of disease. *Reviews in Endocrine & Metabolic Disorders*, 8(2), 173–182.

291. Teh, A. L., Pan, H., Chen, L., Ong, M., Dogra, S., Wong, J., MacIsaac, J. L., & Mah, S. M., (2014). The effect of genotype and in utero environment on interindividual variation in neonate DNA methylomes. *Genome Research*, 24(7), 1064–1074.

292. Teif, V. B., & Rippe, K., (2009). Predicting nucleosome positions on the DNA: Combining intrinsic sequence preferences and remodeler activities. *Nucleic Acids Research*, 37(17), 5641–5655.

293. Temple, I. K., (2007). Imprinting in human disease with special reference to transient neonatal diabetes and Beckwith-Wiedemann syndrome. *Endocrine Development*, 1, 2113–2123.

294. Terblanche, S., Noakes, T. D., Dennis, S. C., Marais, D., & Eckert, M., (1992). Failure of magnesium supplementation to influence marathon running performance or recovery in magnesium-replete subjects. *International Journal of Sport Nutrition*, 2(2), 154–164.

295. Thom, E., Wadstein, J., & Gudmundsen, O., (2001). Conjugated linoleic acid reduces body fat in healthy exercising humans. *The Journal of International Medical Research*, 29(5), 392–396.

296. Topol, E. J., (2014). Individualized medicine from Prewomb to tomb. *Cell*, 157(1), 241–253.

297. Toyota, M., Ahuja, N., Ohe-Toyota, M., Herman, J. G., Baylin, S. B., & Issa, J. P., (1999). CpG Island methylator phenotype in colorectal cancer. *Proceedings of the National Academy of Sciences of the United States of America*, 96(15), 8681–8686.

298. Tucker, K. L., Hannan, M. T., Chen, H., Cupples, L. A., Wilson, P. W., & Kiel, D. P., (1999). Potassium, magnesium, and fruit and vegetable intakes are associated with greater bone mineral density in elderly men and women. *The American Journal of Clinical Nutrition*, 69(4), 727–736.

299. Turunen, M. P., Aavik, E., & Ylä-Herttuala, S., (2009). Epigenetics and atherosclerosis. *Biochimica Biophysica Acta (BBA): General Subjects*, 1790(9), 886–891.

300. Uehara, M., (2013). Isoflavone metabolism and bone-sparing effects of daidzein-metabolites. *Journal of Clinical Biochemistry and Nutrition*, 52(3), 193–201.

301. Ulrich, C. M., & Reed, M. C., & Nijhout, H. F., (2008). Modeling folate, one-carbon metabolism, and DNA methylation. *Nutrition Reviews*, 66S, 27–30.

302. Urdinguio, R. G., Lopez-Serra, L., Lopez, N. P., Alaminos, M., Diaz-Uriarte, R., Fernandez, A. F., & Esteller, M., (2008). Mecp2-null mice provide new neuronal targets for ret syndrome. *PLoS ONE*, 3(11), e3669.

303. Urdinguio, R. G., Sanchez-Mut, J. V., & Esteller, M., (2009). Epigenetic mechanisms in neurological diseases: Genes, syndromes, and therapies. *The Lancet Neurology*, 8(11), 1056–1072.

304. Vauzour, D., Buonfiglio, M., Corona, G., Chirafisi, J., Vafeiadou, K., Angeloni, C., et al., (2010). Sulforaphane protects cortical neurons against 5- S -cysteinyl-dopamine-induced toxicity through the activation of ERK1/2, Nrf–2 and the upregulation of detoxification enzymes. *Molecular Nutrition & Food Research*, 54(4), 532–542.

305. Veigl, M. L., Kasturi, L., Olechnowicz, J., Ma, A. H., Lutterbaugh, J. D., Periyasamy, S., et al., (1998). Biallelic inactivation of hMLH1 by epigenetic gene silencing: Novel

mechanism causing human MSI cancers. *Proceedings of the National Academy of Sciences of the United States of America, 95*(15), 8698–8702.

306. Vel, S., Katarzyna, S., Declerck, K., Vidaković, M., & Berghe, W. V., (2015). From inflammation to healthy aging by dietary lifestyle choices: Is epigenetics the key to personalized nutrition? *Clinical Epigenetics, 7*(1), 33.

307. Verlaan, D. J., Berlivet, S., & Hunninghake, G. M., (2009). Allele-specific chromatin remodeling in the ZPBP2/GSDMB/ORMDL3 locus associated with the risk of asthma and autoimmune disease. *American Journal of Human Genetics, 85*(3), 377–393.

308. Vijayalakshmi, S. V., Padmaja, G., Kuppusamy, P., & Kutala, V. K., (2009). Oxidative stress in cardiovascular disease. *Indian Journal of Biochemistry & Biophysics, 4,* 6421–6440.

309. Villeneuve, L. M., & Natarajan, R., (2010). The role of epigenetics in the pathology of diabetic complications. *AJP: Renal Physiology, 299*(1), F14–F25.

310. Waddington, C. H., (2009). *An Introduction to Modern Genetics* (p. 310). Macmillan, N.Y.

311. Walter, E., Mazaika, P. K., & Reiss, A. L., (2009). Insights into brain development from neurogenetic syndromes: Evidence from fragile X Syndrome, Williams syndrome, Turner syndrome and velocardiofacial syndrome. *Neuroscience, 164*(1), 257–271.

312. Wang, G. G., Allis, C. D., & Chi, P., (2007). Chromatin remodeling and cancer, Part II: ATP-dependent chromatin remodeling. *Trends in Molecular Medicine, 13*(9), 373–380.

313. Waterland, R. A., Lin, J., Smith, C. A., & Jirtle, R. L., (2006). Post-weaning diet affects genomic imprinting at the insulin-like growth factor 2 (Igf2) locus. *Human Molecular Genetics, 15*(5), 705–716.

314. Waterland, R. A., (2006). Epigenetic mechanisms and gastrointestinal development. *The Journal of Pediatrics, 149*(5), S137–S142.

315. Waterland, R. A., (2009). Is epigenetics an important link between early life events and adult disease? *Hormone Research, 71.* 13–16.

316. Waterland, R. A., & Garza, C., (1999). Potential mechanisms of metabolic imprinting that lead to chronic disease. *Am. J. Clin. Nutr., 69,* 179–197.

317. Waterland, R. A., & Jirtle, R. L., (2003). Transposable elements: Targets for early nutritional effects on epigenetic gene regulation–II. *Molecular and Cellular Biology, 23*(15), 5300–5310.

318. Waterland, R. A., & Jirtle, R. L., (2004). Early nutrition, epigenetic changes at transposons and imprinted genes, and enhanced susceptibility to adult chronic diseases. *Nutrition, 20*(1), 63–68.

319. Weaver, I. C. G., Champagne, F. A., Brown, S. E., Dymov, S., Sharma, S., Meaney, M. J., & Szyf, M., (2005). Reversal of maternal programming of stress responses in adult offspring through methyl supplementation: Altering epigenetic marking later in life. *Journal of Neuroscience, 25*(47), 11045–11054.

320. Weaver, I. C. G., Meaney, M. J., & Szyf, M., (2006). Maternal care effects on the hippocampal transcriptome and anxiety-mediated behaviors in the offspring that are reversible in adulthood. *Proceedings of the National Academy of Sciences, 103*(9), 3480–3485.

321. Weaver, I. C. G., (2009). Shaping adult phenotypes through early life environments. *Birth Defects Research Part C: Embryo Today: Reviews, 87*(4), 314–326.

322. Weaver, I. C. G., (2007). Epigenetic programming by maternal behavior and pharmacological intervention–Nature versus nurture: Let's call the whole thing off. *Epigenetics*, *2*(1), 22–28.

323. Weaver, I. C. G., Cervoni, N., & Champagne, F. A., (2004). Epigenetic programming by maternal behavior. *Nature Neuroscience*, *7*(8), 847–854.

324. Weber, M., & Schübeler, D., (2007). Genomic patterns of DNA methylation: Targets and function of an epigenetic mark. *Current Opinion in Cell Biology*, *19*(3), 273–280.

325. Wei, Y. D., Tepperman, K., Huang, M., Sartor, M. A., & Puga, A., (2004). Chromium inhibits transcription from polycyclic aromatic hydrocarbon-inducible promoters by blocking the release of histone deacetylase and preventing the binding of p300 to chromatin. *The Journal of Biological Chemistry*, *279*(6), 4110–4119.

326. Weinhold, B., (2006). Epigenetics: The science of change. *Environmental Health Perspectives*, *114*(3), A160-A167.

327. Wertz, K., Siler, U., & Goralczyk, R., (2004). Lycopene: Modes of action to promote prostate health. *Archives of Biochemistry and Biophysics*, *430*(1), 127–134.

328. Whelton, P. K., He, J., Cutler, J. A., Brancati, F. L., Appel, L. J., Follmann, D., & Klag, M. J., (1997). Effects of oral potassium on blood pressure: Meta-analysis of randomized controlled clinical trials. *JAMA*, *277*(20), 1624–1632.

329. Whitelaw, E., (2006). Epigenetics: Sins of the fathers, and their fathers. *European Journal of Human Genetics*, *14*(2), 131–132.

330. Wierda, R. J., Geutskens, S. B., Jukema, J. W., Quax, P. H. A., & Van den Elsen, P. J., (2010). Epigenetics in atherosclerosis and inflammation. *Journal of Cellular and Molecular Medicine*, *14*(6a), 1225–1240.

331. Wilkins-Haug, L., (2009). Epigenetics and assisted reproduction. *Current Opinion in Obstetrics and Gynecology*, *21*(3), 201–206.

332. Wolffe, A. P., (2001). Chromatin remodeling: Why it is important in cancer. *Oncogene*, *20*(24), 2988–2990.

333. Wolk, A., Manson, J. E., Stampfer, M. J., Colditz, G. A., Hu, F. B., Speizer, F. E., et al., (1999). Long-term intake of dietary fiber and decreased risk of coronary heart disease among women. *JAMA*, *281*(21), 1998–2004.

334. Wynder, C., Hakimi, M. A., Epstein, J. A., Shilatifard, A., & Shiekhattar, R., (2005). Recruitment of MLL by HMG-domain protein iBRAF promotes neural differentiation. *Nature Cell Biology*, *7*(11), 1113–1117.

335. Xiang, N., Zhao, R., Song, G., & Zhong, W., (2008). Selenite reactivates silenced genes by modifying DNA methylation and histones in prostate cancer cells. *Carcinogenesis*, *29*(11), 2175–2181.

336. Yan, Y., Kluz, T., Zhang, P., Chen, H., & Costa, M., (2003). Analysis of specific lysine histone H3 and H4 acetylation and methylation status in clones of cells with a gene silenced by nickel exposure. *Toxicology and Applied Pharmacology*, *190*(3), 272–277.

337. Yang, X., Yan, L., & Davidson, N. E., (2001). DNA methylation in breast cancer. *Endocrine-Related Cancer*, *8*(2), 115–127.

338. Yang, X., Wang, X., Liu, D., Yu, L., Xue, B., & Shi, H., (2014). Epigenetic regulation of macrophage polarization by DNA methyltransferase 3b. *Molecular Endocrinology*, *28*(4), 565–574.

339. Yao, L. H., Jiang, Y. M., Shi, J., Tomás-Barberán, F. A., Datta, N., Singanusong, R., & Chen, S. S., (2004). Flavonoids in food and their health benefits. *Plant Foods for Human Nutrition (Dordrecht, Netherlands)*, *59*(3), 113–122.

340. Youngson, N. A., & Whitelaw, E., (2008). Transgenerational epigenetic effects. *Annual Review of Genomics and Human Genetics*, *9*(1), 233–257.

341. Zafra-Stone, S., Yasmin, T., Bagchi, M., Chatterjee, A., Vinson, J. A., & Bagchi, D., (2007). Berry anthocyanins as novel antioxidants in human health and disease prevention. *Molecular Nutrition & Food Research*, *51*(6), 675–683.

342. Zaina, S., Heyn, H., Carmona, F. J., Varol, N., Sayols, S., Condom, E., et al., (2014). DNA methylation map of human atherosclerosis. *Circulation: Cardiovascular Genetics*, *7*(5), 692–700.

343. Zamora-Ros, R., Forouhi, N. G., & Sharp, S. J., (2014). Dietary intakes of individual flavanols and flavonols are inversely associated with incident type 2 diabetes in European populations. *The Journal of Nutrition*, *144*(3), 335–343.

344. Zeng, W., De, G., & Jessica, C., (2009). Specific loss of histone H3 lysine 9 trimethylation and HP1γ/cohesin binding at D4Z4 repeats is associated with facioscapulohumeral dystrophy (FSHD). *PLoS Genetics*, 5(7), e1000559.

345. Zhang, F. F., Cardarelli, R., Carroll, J., Zhang, S., & Fulda, K. G., (2011). Physical activity and global genomic DNA methylation in a cancer-free population. *Epigenetics*, *6*(3), 293–299.

346. Zhang, R., Li, Y., & Wang, W., (1997). Enhancement of immune function in mice fed high doses of soy daidzein. *Nutrition and Cancer*, *29*(1), 24–28.

347. Zhang, Y., & Chen, H., (2011). Genistein, an epigenome modifier during cancer prevention. *Epigenetics*, *6*(7), 888–891.

348. Zhu, P., & Martin, E., (2004). Induction of HDAC2 expression upon loss of APC in colorectal tumorigenesis. *Cancer Cell*, *5*(5), 455–463.

349. Zilberman, D., Gehring, M., & Tran, R. K., (2007). Genome-wide analysis of Arabidopsis thaliana DNA methylation uncovers interdependence between methylation and transcription. *Nature Genetics*, *39*(1), 61–69.

PART III
Innovative Use of Plant-Based Drugs for Human Health

PLANT XYLITOL FOR HUMAN HEALTH: AN OVERVIEW ON BENEFITS AND POTENT IMMUNOMODULATOR

V. H. HARITHA, V. S. BINCHU, V. N. HAZEENA, and Y. ANIE*

*Corresponding author. E-mail: aniey@mgu.ac.in.

ABSTRACT

Modern lifestyle and intake of high-calorie food is a major cause of many metabolic and inflammatory disorders. Consuming nutraceuticals is an alternative to improve our health. Today, consumption of high-calorie sweeteners has increased at a distressing rate and to lessen the impact, use of sugars in confectionaries and other foods has been replaced by artificial sweeteners. However, artificial sweeteners, in the long run, have exhibited serious side effects and are therefore being replaced by natural sweeteners. Polyols like sorbitol, mannitol, and xylitol are potential natural sweeteners that are present in fruits and vegetables. Among these, xylitol seemed to be an excellent candidate in providing parenteral nutrition. It is also beneficial in the treatment of diabetes, pulmonary infection, otitis media, and osteoporosis. This molecule is very promising in preventing dental caries and possesses excellent immune modulating potential. This chapter reviews the immune-related activities of xylitol and its beneficial aspects to human health.

6.1 INTRODUCTION

Adequate energy from food and proper nutritional intake can facilitate human health and homeostasis. Apart from major energy sources (such as carbohydrates and fats), proteins, vitamins, and minerals also play an important role in maintaining our health. Our body gets these essential components from the food we eat. Hence, the right choice of diet and the quantity is very important in maintaining health and in preventing diseases. Ignoring the pyramid has resulted in a generation overburdened with serious health issues.

This chapter reviews the reported immune-related activities of xylitol and its beneficial aspects to humanity.

6.2 EFFECT OF PRESENT DAY FOOD HABITS ON HUMAN HEALTH

Globalization and modernization have pushed us to make a compromise in both the nutritional outlook and the concept of *healthy food for a healthy body*. Unhealthy junk foods have been welcomed in every home, and they have outpaced nutritional foods, fruits, and vegetables. Consequently, processed, and canned foods became an additional class in the food classification system. The transformation of a diet based on fresh and nutrient-rich foods to tasty fast food products with high calorie and no nutritional input has paved the way to the development of a society overwhelmed with malnutrition and/or lifestyle diseases.

Based on the type and extent of processing, food is classified into: unprocessed or minimally processed food, processed culinary ingredients, and ultra-processed food products [60].

6.2.1 UNPROCESSED OR MINIMALLY PROCESSED FOOD

The food that retains its natural state can be called as unprocessed foods, such as: Fresh fruits and vegetables, eggs from the farm, raw dairy, whole cereals, raw nuts, seeds, and honey. They are ingrained with low calorie. However, they are rich in dietary fibers, calcium, vitamins, and potassium with lower sodium and added sugars [113]. Unprocessed foods also exhibit high antioxidant capacity. When these products reach the market, they may undergo a minimum level of processing that ensures food safety, preservation, and nutritional backup [122]. The processing includes: cleaning, freezing, irradiation, vacuum packing, crushing, flaking, etc. [11, 60].

6.2.2   PROCESSED CULINARY INGREDIENTS

Processed culinary ingredients are extracted and purified ingredients of raw food materials. This processing may result in a great nutritional loss, but their caloric value remains the same. The method employed for the production of processed culinary ingredients include: physical methods (pressure, milling, refining) and chemical methods (hydrogenation and hydrolysis). Enzymes and additives are also used in the processing techniques. Proteins (milk and soy protein), carbohydrates (sugars, syrups, lactose, starch, flour), fat (animal fat and vegetable oils) and salts are commonly used as processed culinary ingredients. These products are not readily consumed but are used during cooking of unprocessed or minimally processed foods at home and restaurants or during the production of ultra-processed foods.

In this modern industrialized world, everybody is busy and to cope up with this limitation on time, and food industries have come up with the idea of ready-to-eat and ready-to-heat food products. They can also be called as ultra-processed food products. Packed, frozen, and powdered foods, desserts, and confectioneries, bread, soft drinks and many more fall under this category. The processed food products are usually marketed under different brand names with attractive packing and hyper-palatable taste. These foods are packed with calories but possessed with very minimal and/or no nutritional benefits.

6.2.3   EFFECTS OF PROCESSED FOODS ON HUMAN HEALTH

The obsession with junk foods, fast foods, and flavored foods has resulted in the high-calorie intake. Fried chips, cakes, pizzas, cookies, popcorn, soft drinks, ice creams, sausages, and burgers are examples of a few common fast foods. The pleasant taste, appealing appearance, convenience, and marketing strategies had made this generation additive towards these foods. Some negative effects of such foods are:

- Attraction of individuals to tasty foods has made people passionate for sweet dishes and confectionaries, which use a high amount of sugar, a silent killer. Sugar accelerates inflammatory disorders that lead to aging, tissue destruction and many pathophysiological conditions [39, 83].

- Carbonated drinks and spicy foods rich in fats and oils lead to acidity and gastritis.
- Flavors and colors added to food can evoke hypersensitivity reactions including asthma.
- High-calorie foods (rich in starch, sugar, and fat (saturated, trans-fat, and oxy-cholesterol)) affect the basal nutritional status and lead to the development of many metabolic diseases over time.
- In addition, the high fat intake can result in hypercholesterolemia, which can affect the vital function of heart, liver, and brain in the long run.
- Phosphate rich foods and soft drinks add on to the progression of heart, kidney, and bone-related diseases.
- Sodium rich fast foods can increase hypertension and also can affect normal renal functions.
- The sodium salt of glutamate (monosodium glutamate) in foods trigger the onset of many more health-related problems.

People consuming a high amount of sugar in the form of sucrose, high fructose corn syrup and fructose are always at the risk of metabolic syndromes including obesity, diabetes, cardiovascular disease and many more. In addition to the direct link of sugar with metabolic disorders, excess sugar consumption also affects normal renal function and may also result in fatty liver, hypertension, and dyslipidemia which are independent risk factors for the development of the above said metabolic diseases. Another chronic disease commonly seen accompanying with high sugar intake is cancer. Scientific studies suggested a strong correlation between the rise in blood sugar and cancer progression [141, 142]. Cancer cells feeding on excess sugar help in accelerating its angiogenic property, growth, and metastasis and contribute to disease progression.

Moreover, high sugar consumption also results in increased incidence of periodontal diseases. The oral bacteria like *Streptococcus mutans* feed on sucrose and other carbohydrates; ferment them to acids and lead to the development of dental caries [143]. Awareness of the ill effects of sugar raised concerns about its usage and triggered a search for some other components with the same sweetness but less calorie. This contemplation led to the era of natural and synthetic sugar substitutes.

6.3 ARTIFICIAL SWEETENERS

Sugar substitute, also called food additives, can be artificial sweeteners or natural sweeteners. The sweetness of artificial sweeteners (which is equal to or more than that of sugar) along with their low caloric value has made them popular among the public. Hence, the demand for sugar-free products containing artificial sweeteners has increased exponentially. The first generation and the second generation of artificial sweeteners are now available in the market. Aspartame, saccharin, sucralose, advantame, neotame, acesulfame potassium are some among the commonly used artificial sweeteners, which had been approved by Food *and Drug Administration (*FDA). Figure 6.1 shows the structural formula of selected sweeteners.

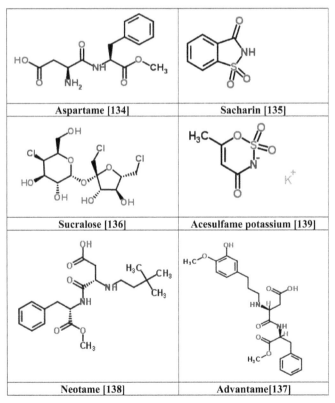

Aspartame [134]	Sacharin [135]
Sucralose [136]	Acesulfame potassium [139]
Neotame [138]	Advantame[137]

FIGURE 6.1   (See color insert.) This sweetener was approved to be safe by FDA in 1981 (Aspartame [134]; Saccharin [135]; Sucralose [136]; Acesulfame potassium [139]; Neotame [138]; Advantame [137]).

6.3.1 FDA APPROVED ARTIFICIAL SWEETENERS

6.3.1.1 ASPARTAME

L-α-aspartyl-L-phenylalanine methyl ester or aspartame is synthesized for food industry from L-aspartic acid and L-phenylalanine. Once consumed, it does not accumulate in the human body but metabolizes rapidly to its corresponding amino acids and methanol by digestive esterase and peptidases [16]. Phenylalanine and aspartic acid (the metabolic product of aspartame) is converted to tyrosine, oxaloacetate, and alanine while methanol breaks down to formaldehyde and formic acid.

Aspartame is familiar to the food industry under the brand names NutraSweet and Equal. This sugar substitute is 180 times sweeter than sucrose. The dominance of aspartame in the society can be attributed to its non-caloric nature, as they are mostly preferred by diabetic patients, who always crave to eat something sweet. Though aspartame is stable under drying and freezing, yet this sugar addictive loses its sweetness when heated. Hence it cannot be used in cooked and baked products.

In addition to FDA, Joint Food and Agriculture Organization of the United Nations (FAO)/WHO Expert Committee on Food Additives (JECFA), the Scientific Committee on Food of the European Commission and regulatory bodies in other countries have also approved the use of aspartame in the food industry. Aspartame is widely used in over 6000 different products [15] like beverages, chewing gums, confectionaries, soft drinks, etc. The daily approved intake of aspartame by FDA is 50 mg/kg body weight, and by European Food Safety Authority is 40 mg/kg. Some case studies and experimental reports have suggested carcinogenic effect [67, 95, 96, 97] and genotoxic effect [1, 37]; some other studies suggested that aspartame possesses no carcinogenic or genotoxic effects [24, 41, 47]. More research arguing the beneficial and negative effects of aspartame is still needed.

6.3.1.2 SACCHARIN

The o-sulfabenzamide; 2,3-dihydro–3-oxobenzisosulfonazole was the first discovered by Fahlberg and Ira Remsen of Johns Hopkins University. The use of saccharine was at peak during World War I. Later on, this sweetener slowly wiped out sucrose from the food industry; though it got popularized among diabetic persons.

This white crystalline solid is 300 times sweeter than sucrose but has a metallic after-taste especially at high concentrations. Since it is an inert and heat stable compound, it can blend with other food ingredients. Once consumed, saccharin remains un-metabolized, and most of it is excreted as such through urine. In the market, saccharin is known under the following brand names: Sweet and Low, Sweet Twin and Necta Sweet. The acid salt of saccharin with low solubility is used in the cosmetic and pharmaceutical industries while the water-soluble sodium and calcium salts are being used in the food and beverage industries. The acceptable daily intake of saccharine is about 5 mg/kg body weight.

The extensive studies on saccharin have ended up with different controversial conclusions. Saccharin and o-toluene sulphonamide (an impurity in saccharin) were shown to induce benign and malignant neoplasm in experimental rats [13, 81]. In 1977, FDA put a ban on saccharin based on these findings. But due to massive protest, FDA abolished the ban over saccharin, and the saccharin products were permitted to reach the market along with a warning label. But very soon, a study came up in favor of O-toluene sulphonamide contaminated saccharine in which it failed to promote bladder cancer [29]. Later on, many studies suggested the safety of saccharin as a sugar substitute; and saccharin was removed from the list of potential carcinogens, in 2000.

6.3.1.3   SUCRALOSE

Chemically, sucralose is 1,6-dichloro-1,6-dideoxy-β-D-fructofuranosyl-4-chloro-4-deoxy-α-D-galactopyranoside. The key precursor of sucralose synthesis is sucrose. Sucrose undergoes multiple steps of chlorination to get converted into sucralose, which is 600 times sweeter than sucrose. There are contradicting reports on the inert nature of sucralose in the human body. Sucralose injection was shown to increase plasma glucose level and insulin secretion rate [74]; but the same result was not reproduced in another study [22], where sucralose failed to affect the insulin level in humans.

After consumption, only 15–40% of sucralose is absorbed into the bloodstream, and it is excreted through urine and feces in an unaltered form. However, sucralose gets concentrated in the liver, kidney, and gastrointestinal tract (GIT). The sweetener was first approved to be used in food and beverages in Canada in 1991 and later by FDA in 1998–1999.

Widespread use of sucralose over other sweeteners in the food industry is attributed to its solubility in fat, water, and alcohol-based media. Splenda and Nevella are brand names of sucralose in the market. Sucralose is stable at high temperature and hence can be used in the baking industry. However, some other studies suggested decomposition of sucralose to water and hydrogen chloride upon heating [10], which was shown to further react with glycerol releasing a dangerous toxin, chloropropanol [76] that is genotoxic, carcinogenic, nephrotoxic, and hepatotoxic [123] molecule. Sucralose induced deoxyribonucleic acid (DNA) damage in mouse GIT and a higher incidence of pelvic mineralization and epithelial hyperplasia of the renal pelvis in female rats [48]. The approved daily acceptable intake of this sweetener is about 5 mg/kg body weight. In 2004, European Union approved the use of sucralose-based foods with adailyn acceptable daily intake of 15 mg/kg body weight. But, sucralose reduces the number of beneficial microorganisms in the gut and increases the expression of gut proteins like *P-glycoprotein* (P-gp), cytochrome P450 3A4 (CYP3A4), and cytochrome P450 2D1 (CYP2D1), which are involved in limiting the bioavailability of orally administered drugs [2]. Sucralose is also reported to be associated with weight gain [88].

6.3.1.4   ADVANTAME

Advantame (*N*-[*N*-[3-(3-hydroxy-4-methoxyphenyl) propyl]-α-L-aspartyl]-L-phenyl alanine1-methyl ester) was introduced by the Chinese company Ajinomoto. Advantame is synthesized from isovaline and aspartame. It was approved by FDA as a general-purpose sugar in 2014. It is preferred in the food industry because of its ultra-sweetening property. It is about 20,000 times sweeter than sucrose.

Advantame is water soluble; and since it withstands high and low temperature well, it can be used for baking and also in frozen food products. Advantame is commonly used in syrups, gelatins, jams, jellies, puddings, chewing gum, confectionaries, frozen desserts, processed fruit juices, and soft drinks. The approved daily intake level for advantame by FDA is about 32.8 mg/kg body weight. Advantame that reaches the body is incompletely absorbed and excreted via feces. It has been proved safe till date, but more research for evaluating the safety of advantame is being carried out.

6.3.1.5   NEOTAME

It is a combination of aspartame and 3, 3-dimethyl butyl; and is 13,000 times sweeter than table sugar. The brand name is Newtame. In 2006, it was approved by FDA for the general sugar category. It was approved as a flavor enhancer by European Union in 2010. It is not metabolized by oral bacteria and has a low glycemic index. It is used alone or in combination with other sweeteners.

It is non-mutagenic and is well tolerated even at high concentration. When digested, 92% of neotame is converted to de-esterified neotame [114], which is excreted and the rest is converted to methanol in the body. Neotame is also used as a substitute for molasses in cattle feed. Since it is produced from aspartame, it may have side effects, but no side effects have been reported so far.

6.3.1.6   ACESULFAME POTASSIUM

Once ingested, acesulfame-K is readily absorbed by the body and rapidly excreted unaltered in urine. The brand names Sunett and Sweet-One belongs to acesulfame-K. This sugar substitute is 200 times sweeter than sucrose but at high concentration shows a bitter after-taste. It is stable at high temperature and on long term storage.

Safety regarding acesulfame use still remains controversial. However, FDA approved its use in the food industry in 1993 and categorized it as a general purpose sweetener and flavoring agent in 2003. It is used in combination with sweeteners like aspartame and sucralose. Acceptable daily intake of this sweetener is about 15 mg/kg body weight. Meanwhile, acesulfame potassium was suggested to be a multipotential carcinogenic agent [38]. More studies have to be conducted to evaluate the toxicity of this compound.

6.3.2   MAJOR ARGUMENTS RELATED TO ARTIFICIAL SWEETENERS

Though artificial sweeteners have been approved by the major food control organizations, yet controversy still exists about the use of artificial sweeteners in human food because of serious health problems associated with the use of artificial sweeteners (Table 6.1).

TABLE 6.1   Artificial Sweeteners and Their General Side Effects

Artificial sugar	E- Number	Safety	Glycemic index	Brand name	General side effects encountered during consumption	Ref.
Acesulfame potassium $C_4H_4KNO_4S$	E950	Avoid	0	Sunett Sweet One Sweet & Safe	Lung tumor, breast tumor, leukemia, chronic respiratory diseases.	[133]
Advantame $C_{24}H_{30}N_2O_7$	E969	Avoid	0	No brand names	Gastrointestinal disturbances in animals, compromised immune system function.	[131]
Aspartame $C_{14}H_{18}N_2O_5$	E951	Avoid	0	Nutra sweet Equal Spoonful Equal-Measure Canderel Benevia AminoSweet NatraTaste	Attention deficit disorder, birth defects, depression, dizziness gastrointestinal problems, respiratory problems, headaches, skin complications, weight gain, carcinogenic effect, genotoxic effect, play a role in the pathogenesis of Alzheimer's disease and multiple sclerosis.	[124, 125, 126]
Neotame $C_{20}H_{30}N_2O_5$	E961	Safe	0	Newtame	Drying or chapping of the skin, coughing inhaled. Note: No scientific studies reporting health hazards till date	[132]
Saccharin $C_7H_5NO_3S$	E954	Avoid	0	Sweet'N Low Necta Sweet Cologran Heremesetas Sucaryl Sucron Sugar Twin Sweet 10	Bladder tumor in rat, diarrhea, difficulty in breathing headache, skin rash or hives, weight gain, allergic reactions, risk for developing diabetes.	[127, 128]
Sucralose $C_{12}H_{19}Cl_3O_8$	E955	Avoid	0	Splenda Nevella Zerocal Sukrana SucraPlus Candys Cukren	Acne, bloating, blood sugar increases, blurred vision, chest pain, dizziness, gastrointestinal problems, gum bleeding, headaches, migraines, rashes, seizures, tinnitus, weight gain, allergy, heart palpitations, shortness of breath, joint pain, nausea, bloating, tinnitus, dizziness, anxiety or depression.	[129, 130]

6.4 EMERGENCE OF NUTRACEUTICALS

Nutraceuticals can be defined as an 'innovation in food or food products that can either cure disease or can ameliorate the pathophysiological condition.' The concept existed earlier from the time of Hippocrates. In 1989, Stephen Defelice introduced the term nutraceuticals by combining "nutrition" and "pharmaceuticals." Nutraceuticals are considered safe, effective, and attractive.

Nutraceuticals include fortified nutraceuticals (foods with added nutrients), probiotics (live microbial supplements to improve health), recombinant nutraceuticals (nutraceutical foods produced through recombinant DNA technology) and enzymes. Phytochemicals, herbal products, and nutrients are also included in the nutraceutical classification based on their chemical constituents; modified from [77, 92].

Rather than intended to provide basic nutrition, the concept of nutraceuticals is focused more on enhancing health by preventing or curing disease. For nutraceutical formulation, isolated refined components or conventional and/or fortified foods with less toxicity and fewer side effects are used.

Nutraceuticals are classified (Figure 6.2) based on their chemical constituents [77] and availability of food in the market (traditional nutraceuticals and non-traditional nutraceuticals) [92]. The stability, efficacy, and bioavailability are key factors concerned with treatment regime, and hence, nanotechnology-based nutraceuticals formulations are also being experimented.

6.5 XYLITOL: A POTENT NUTRACEUTICAL

Xylitol, a sugar alcohol, is a white crystalline compound found naturally in some fruits (e.g., plum, strawberry, and banana), vegetables (e.g., cauliflower, spinach, and carrot), lichens, algae, mushrooms [61, 112], hardwood trees like birch, etc. It is also produced as a by-product during human metabolism and is estimated that about 5–15 g of xylitol is produced within the human body every day [66].

Since it can be synthesized artificially, it comes under the category of artificial sweeteners. During the World War II and around the 1960s, the sugar shortage in Finland forced the Finns to rediscover xylitol as a substitute for sugar.

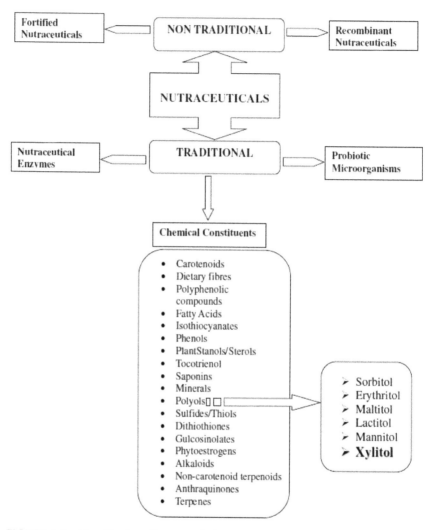

FIGURE 6.2   Classification of nutraceuticals.

Xylitol contains many functional hydroxyl groups (Figure 6.3), and hence it is placed under the category of polyol or sugar alcohol [140]. Bulk production of xylitol is usually made from wood or corn cobs rich in xylan. In the process, xylan is chemically hydrolyzed to xylulose and is finally hydrogenated to xylitol. Different microbial strains that are useful in the production of xylitol are also under study [17, 68, 78].

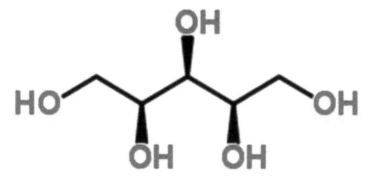

FIGURE 6.3 **(See color insert.)** Structure of xylitol.

Xylitol was permitted for special dietary use in 1963. Later in 1986, xylitol was approved by FDA as a safe sugar. Xylitol provides sweetness equivalent to that of sucrose. The pleasant taste, cooling effect and its increased solubility are the advantages of xylitol over sucrose. Since they are inert, they can be conjugated with other polyols, artificial sweeteners and even with amino acids. Gradually, xylitol made a place among the food additives and this triggered research on health benefits as well as side-effects of this sugar alcohol.

Xylitol in food is absorbed at a very low rate from intestine [23] *via* non-active transport mechanism [12]. Once absorbed, the major portion of xylitol is rapidly metabolized in the hepatic tissue, and the rest is metabolized in the extrahepatic tissues [120]. Intestinal microbial flora is also capable of fermenting small amount of xylitol to produce small fatty acids. In the liver, xylitol is metabolized to D-xylulose by a non-specific *nicotinamide adenine dinucleotide* (NAD) linked polyol dehydrogenase, which enters into the pentose phosphate pathway. A very small amount of xylitol is converted to glucose, but the plasma insulin and visceral fat accumulation remain unaltered [6].

Xylitol reduces absorption of glucose and water from small intestine [23]. It inhibits gluconeogenesis and reduces the production of glucose from pyruvate and other substrates. Thus, this low-calorie sugar substitute can be a good choice for diabetic patients. During the metabolism, the ratio of NADH to NAD is increased that leads to an increase in the redox status of the liver. This may lead to inhibition of the citric acid cycle while the mitochondrial electron transport chain utilizes the NADH for energy production [35]. Xylitol is converted to xylulose–5-phosphate and consumes adenosine triphosphate (ATP) at a very fast rate and consequently will result in a decrease in hepatic ATP level [116]. As xylitol is metabolized rapidly, xylitol concentration in blood remains the same in the range between 0.03–0.06 mg/100 ml blood [121]. However, the intake of very high doses of xylitol led to osmotic diarrhea in humans and experimental animals. But gradual increase in xylitol intake is well tolerated. An initial intake of smaller dose followed by a continuous and gradual increase in the intake of this sugar alcohol helps the intestine to get adapted to xylitol along with an enhanced intestinal absorption [12]. The amount of xylitol that can be consumed without any side effect is around 30–60 g, and after adaptation, the dose can be increased up to 400 g daily [46]. Being inert, it can be blended with any type of food material. All these properties make xylitol an excellent candidate to be a nutraceutical.

6.6   PHARMACOLOGICAL IMPORTANCE OF XYLITOL

Xylitol is a preferred candidate in food formulations, due to its ability to prevent and control diseases like diabetes, lipids metabolism disorders, renal, and parenteral lesions, dental caries, pulmonary infections, acute otitis media (AOM), and osteoporosis [121].

6.6.1   XYLITOL IN DIABETIC FOOD FORMULATION

The replacement of sugar from diet with alternatives like xylitol may add to a better health for diabetic patients. In contrast to glucose with a caloric value of 4 kcal/g, xylitol exhibits a low caloric value of 2.4 kcal/g and low glycemic index. This is a major advantage that is taken into account while formulating a diabetic diet.

Studies in humans [18, 89] and animal models [85] have shown that intake of xylitol delays gastric emptying and decreases the intake of food. The gut hormones namely Glucagon-like peptide 1 (GLP–1) and cholecystokinin (CCK) that retard gastric emptying, are increased upon the xylitol intake [115]. Unabsorbed xylitol acts like dietary fiber, helping to maintain healthy gut functions. Partial bacterial fermentation in the gut produces volatile short-chain fatty acids by insulin-dependent energy pathways. Lag in the absorption of xylitol and its metabolism in the liver results in an insignificant rise in blood glucose concentration.

Xylitol intake by healthy human subjects [30, 77] and rat models [6, 77] showed a minimal elevation in blood glucose level, and plasma insulin level remained unchanged. Moreover, diabetic patients also showed no increase in blood glucose level when xylitol was given orally [33, 55]. This is because glucose produced from xylitol is stored as glycogen [32, 54]. Xylitol also enhances muscle glucose uptake in an insulin-like manner. Carbohydrate digesting enzymes (alpha-amylase and alpha-glucosidase) were inhibited *in vitro* by xylitol in a dose-dependent manner [18]. Xylitol has been found to be an excellent source of energy in the liver of rats treated with Carbon tetrachloride (CCl_4) or alloxan [31]. The 45–60 g/day of xylitol intake in 20 diabetic patients showed no adverse effects on the metabolic parameters [56].

Xylitol significantly decreases total serum cholesterol and low-density lipoprotein-cholesterol. Though visceral fat accumulation and serum lipid concentration was less in xylitol-fed rat [3, 32], yet there was an increase in HDL profile [3]. Transcription factor that activates lipogenic enzyme genes (Carbohydrate-responsive element-binding protein (ChREBP) and lipogenic enzymes) and fatty acid oxidation-related genes were significantly increased in xylitol-fed rats [6]. All these studies stress the advantages of including xylitol in the diabetic diet.

6.6.2 ANTIBACTERIAL AND ANTIVIRAL ACTIVITIES OF XYLITOL

Inhibitory effect of xylitol against many bacteria has been well documented. Xylitol is internalized by the bacterial cell through fructose specific phosphotransferase system (PTS) [7, 75]. Inside the cell, xylitol is metabolized to xylitol-5-phosphate, which cannot be further metabolized by the bacterial cell. Thus, the overall metabolism of bacteria feeding on xylitol ends up in a "futile cycle." Xylitol inhibits bacterial multiplication by stably integrating within the bacterial cell wall of gram positive and negative bacteria, and this affects bacterial multiplication [69]. Cell degradation, autolysis, and vacuole formation have also been reported in *Mutans streptococci* treated with xylitol [109].

Xylitol reduces cell adherence and virulence properties of *E. coli*, *Streptococcus pneumonia*, *Pseudomonas aeruginosa*, *Staphylococcus aureus*, and *Haemophilus influenza* [20, 42, 91, 99]. Adherence of *Clostridium difficile* to human colon adenocarcinoma (Caco–2 cells) was decreased by xylitol in a dose-dependent manner [70]. Lipoteichoic acid was reduced in *streptococci* and *lactobacilli* upon xylitol treatment [8, 109]. Xylitol inhibited oral bio-film formation in an *in vitro* model composed of *Streptococcus mutans, Streptococcus sobrinus, Lactobacillus rhamnosus, Actinomyces viscosus, Porphyromonas gingivalis,* and *Fusobacterium nucleatum* [8]. Xylitol attacks the bio-film by reducing the bacterial proliferation and colonization [84]. Xylitol also showed inhibitory effect on the growth of *P. gingivalis* [27].

The studies related to the antiviral activity of xylitol are limited. Xylitol has a direct inhibitory effect on human respiratory syncytial virus (hRSV) [117]. Red ginseng (RG) is commonly used for the prevention of influenza viral infection. RG, when combined with xylitol fractions significantly reduced lung virus titers after infection and the effect, was increased with increasing doses of xylitol [119]. More studies have to be conducted to identify the prophylactic possibility of xylitol against different viral species.

6.6.3   XYLITOL VERSUS PREVENTION OF DENTAL CARIES

Dental caries is a multi-factorial disease. Among the bacteria involved in plaque formation, *Streptococcus mutans* is the prominent one. Inhibition of *S. mutans* by xylitol is well documented [49, 50, 87]. Adhesion of bacteria to the pellicle is a key event in the formation of bio-film in caries formation which is followed by fermentation of sugar to acid and caries formation. Salivary flow rate is another important factor that determines the occurrence of caries. Search for a remedy to prevent caries formation is an area that has been extensively researched. Anticariogenic property of xylitol was discovered in the 1970s, in Turku, Finland. A quality attributed to xylitol as an anti-caries agent is its ability to enhance the flow of saliva [82].

Xylitol may easily remove *S. mutans* from the plaque by decreasing the adherence property of bacteria [84, 93]. As the concentration of xylitol increases, the number of *S. mutans* in the saliva and dental plaque decreases. Similarly, chewing 55% and 100%, xylitol gum caused a reduction in the amount of *S. mutans* in the dental plaque, and significant decrease of *S. mutans* in saliva was observed in 100% xylitol gum [107].

Xylitol chewing gum decreased salivary levels of *S. mutans* in subjects with poor oral hygiene also [25]. The decrease in bacterial number may reduce the acid production [4, 9] and thus may reduce dental caries in xylitol treated group.

Xylitol is non-fermentable to oral pathogens including *S. mutans*. Phosphoenol pyruvate: fructose-PTS in *S. mutans* takes in xylitol and phosphorylates it but cannot be further metabolized by the organism [7, 75]. In a comparative study using xylitol and sorbitol chewing gums, xylitol showed a greater impact on plaque reduction [5, 19, 49, 50, 82] compared to sorbitol [14, 19]; and a dose of at least 6–10g/day for at least 3–4 times a day has been suggested for anti-caries effect and the lower doses was not seemed effective [14, 58, 65].

The incidence of transmission of *S. mutans* from mother to child was also found to be reduced by xylitol chewing [4, 34, 63, 94]. *S. mutans* biofilm formation in the presence of sucrose was compromised by multiple exposures to xylitol in combination with chlorhexidine [59]. In addition, xylitol helps in the re-mineralization of enamel [57, 98] and this may be attributed to the ability of xylitol to complex with calcium and to increase its absorption [45]. However, xylitol, when used in combination with fluoride, is more effective against dental plaque formation and re-mineralization [86, 103].

6.6.4   XYLITOL IN OSTEOPOROSIS TREATMENT

Xylitol can be used for the treatment of osteoporosis by enhancing calcium absorption in a vitamin D independent manner [26], and this could restore the bone biochemical properties. Xylitol increased bone volume and bone mineral density in aged Sprague-Dawley rats. A decrease in bone density, trabecular bone loss [52, 53], and amount of collagen and the number of its mature cross-links in rats subjected to ovariectomy was suppressed by xylitol treatment [101]. Moreover, the bone calcium, phosphorus, and citrate concentration were also increased by xylitol treatment [100]. Xylitol also confined the ovariectomy-induced increase of bone turnover in rats. A defective bone metabolism resulted from Collagen type II-induced arthritis could be restored by xylitol treatment [36].

6.6.5   XYLITOL IN THE PREVENTION OF ACUTE OTITIS MEDIA (AOM)

Viral and bacterial involvement in the pathogenesis of the AOM has been well documented. *Streptococcus pneumoniae, Haemophilus influenza*, and *Moraxella catarrhalis* are the reported microorganisms in AOM [51]. Treatment regimens against this nasopharyngeal flora may help in preventing disease. The striking ability of xylitol (1% and 5%) to reduce the growth of *S. pneumoniae* was first reported in 1995 [43]. Randomized trial using xylitol-containing chewing gum and syrup in children showed a decline in the incidence of AOM [110, 111]. However, xylitol failed to prevent AOM in the presence of respiratory infection [28, 105]. Xylitol reduces the adherence and virulence of *pneumococci* by altering the polysaccharide capsule and cell wall structures [106]. Adherence of *S. pneumoniae, H. influenzae*, and *M. catarrhalis* to epithelial cells is decreased in the presence of xylitol [42]. Fructose PTS is also involved in the prevention of xylitol mediated growth inhibition in *pneumococci* [104].

6.7   IMMUNE MODULATION BY XYLITOL

Studies have projected the ability of xylitol in influencing the cytokine and chemokine production *in vitro*. The recruitment of immune cells and proteins released by these cells modulate the immune response. However, when the pro-inflammatory status of the immune response is deranged, the ultimate result is chronic inflammation and tissue destruction. The expression of IL–8, a neutrophil recruiting cytokine is un-affected in human middle ear epithelial cell lines (HMEECs), A549 cells (lung carcinoma cell line) and RAW 264.7 (macrophage cell line) cells in the presence of xylitol [44, 108]. However, eotaxin (a chemokine for eosinophils) and monocyte chemotactic protein–1 and macrophage inflammatory protein–1 that recruit monocyte/macrophage are inhibited by xylitol in *P. gingivalis* stimulated THP–1-derived macrophage cell lines. Similarly, xylitol inhibited interferon γ–induced protein 10 in THP–1-derived macrophages stimulated by *P. gingivalis* [71]. Macrophages produce interleukin 12 (IL–12) that activate natural killer cells and T cells during infection. Over-expression of IL–12 is seen in diseases like Crohn's disease, and rheumatoid arthritis.

Xylitol is capable of inhibiting p40 subunit of IL–12 and regulating its pro-inflammatory effects [71]. Though Interleukin 6 (IL–6) forms a part of an exemplified inflammation, yet it is also involved in controlling the immune response. Xylitol showed no effect on the production of IL–6 in A549 cells and RAW 264.7 [108]. The transcription factor, nuclear factor kappa B (NF-kB) influences the expression of several genes involved in the inflammatory process. Lipopolysaccharide (LPS)-induced mobilization of NF-κB was also inhibited by pre-treatment with xylitol in a dose-dependent manner [27].

Tumor necrosis factor alpha (TNF-α), a pro-inflammatory cytokine, was found to influence the release of prostaglandin E2 (PGE2) by cyclooxygenase–2 (COX–2) enzyme in NF-κB dependent pathway [64]. PGE2 can also influence TNF-α production [80]. Upon xylitol treatment, in A549 cells and RAW 264.7, the LPS induced PGE2 production was decreased, and TNF-α was found to be increased [108]. However, TNF-α was unaffected by xylitol treatment in RAW macrophages and HMEECs [44]. Similarly, pre-treatment of RAW 264.7 with xylitol inhibited LPS-induced TNF-α and interleukin 1 (IL–1) gene expression and protein synthesis [27]. But, xylitol had no influence on COX–2 expression and mucin gene expression in HMEECs [44]. Though xylitol inhibits *sodium lauryl sulfate*-induced IL–1ß and myeloperoxidase, yet it has no effect on IL–1α [102]. Xylitol also inhibited *Aggregati bacter actinomycetemcomitans* induced IL–1β production and absent in Melanoma 2 (AIM2) inflammasome activation [40].

CD3+ and CD3+CD8+ lymphocytes recruitment was significantly reduced in xylitol treated animal model with human respiratory syncytial virus (hRSV) infection thus improving the immune status in these animals [117]. Xylitol in combination with lactobacilli protect against *Clostridium difficile* infection and reduce tissue destruction and pseudo-membrane formation in hamster model for infection [62].

Since xylitol has many pharmaceutical applications, the toxicity of xylitol has been studied extensively. The J744A.1 macrophage [21], THP–1 macrophage [72] A549, HMEECs, and RAW macrophages [108] were viable when treated with varying concentrations of xylitol *in vitro*. Oxidation of hepatocyte glutathione (GSH) by hydrogen peroxide (H_2O_2) can be increased by glyoxal, a sugar metabolite. The compromise in the cellular antioxidant enzyme system will result in increased cytotoxicity [90, 118]. Xylitol has also been shown to be cytoprotective during oxidative stress.

Xylitol can inhibit glyoxal-induced cytotoxicity and prevent the decrease in mitochondrial membrane potential by increasing glyoxal metabolism [90]. Apoptotic or necrotic mode of cell death was not induced by xylitol in HEI-OC1s and HMEECs [44].

However, xylitol was found to be cytotoxic to the following cancerous cell line; A549, Caki, NCI-H23, HCT–15, HL–60, K562 and SK MEL–2; but were found to keep the normal cells viable. Cell proliferation of these cancer cell lines were inhibited by xylitol. The morphological changes of confocal laser-scanning microscopic examination of A549 cells treated with xylitol for 72 h showed a dose-dependent decrease in cell number, cell shrinkage, and loss of cellular contact. Xylitol treatment markedly elevated levels of autophagolysosomes in A549 (lung cancer) cells [72]. Moreover, angiogenic events like cell migration, invasion, and tube formation were inhibited in human umbilical vein endothelial cells (HUVECs) by xylitol. A similar result was also reported *in vivo* in experimental mouse Matrigel plug assay. Furthermore, xylitol showed a decrease in the mRNA expression of some angiogenic proteins including vascular endothelial growth factor (VEGF), VEGFR-II (KDR), basic fibroblast growth factor (bFGF), bFGFR-II, matrix metalloproteinase–2 (MMP–2) and MMP–9 in HUVECs. These anti-angiogenic effects of xylitol are exerted through inhibition of NF-κB and Akt activation [119]. These results are promising for xylitol, and this molecule could be exploited for the cancer therapy.

Dietary xylitol may improve growth and inflammation in chickens treated with LPS plus Sephadex. Inflammatory markers like alpha–1-acid glycoprotein and interleukin–1-like activity in plasma was found to be reduced in them. Dietary xylitol enhanced activation and oxidative killing by neutrophils and prolonged the survival of rats suffering from sepsis caused by *S. pneumoniae* type 3 [79]. These results together suggest a strong immunomodulatory potential of this molecule. Multifaceted action of xylitol is summarized in Figure 6.4.

6.8 FUTURE PERSPECTIVES

Modern lifestyle and food culture characterized by high fat, high sugar, high calorie, and low nutrition accelerates the incidents of lifestyle diseases. Hence, researchers have started focusing on developing new molecules from natural source and to invent therapeutic strategies for preventing diseases or to reduce its pathophysiological effects. Research studies on this line came up with the suggestion that use of natural foods with antioxidant capacity and fiber content rather than processed food will be more suitable for human health. Consumers are also becoming aware of the health hazards caused by the processed and canned foods. This diet-based approach demands the consumption of less sugar, and this managed to bring in sugar substitutes. Of the several artificial sweeteners available, xylitol is a globally accepted sugar substitute with very small side effects. Xylitol have proven benefits to health and may come out as an alternative to current conventional sweeteners.

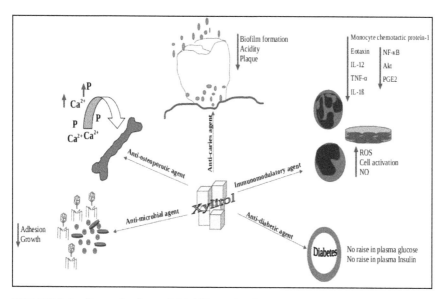

FIGURE 6.4 **(See color insert.)** Multifaceted action of xylitol.

Studies on the additional benefits of xylitol mainly revolve around its ability to prevent plaque formation and the development of dental caries. Inhibition of caries-related and other bacterial species by xylitol also gained attention. But, only a very few studies have focused on antiviral properties of xylitol. Therefore, use of xylitol as a prophylactic strategy against viral infections can be evaluated. Besides formulation containing additives which make the xylitol more active should be worked out. Influence of xylitol on inflammation can also be further pursued to unravel the exact mechanism of this response. Extensive research is needed to unleash the possible benefits of xylitol as a sugar substitute and a physiologically relevant health modulator.

6.9   SUMMARY

Efforts, for developing food and food ingredients for bringing about better health, have resulted in the discovery of artificial and natural sweeteners. Moreover, the pharmaceutical companies are trying to incorporate their products with food to formulate effective nutraceuticals. Xylitol is a naturally occurring polyol, which has been established as a sugar substitute. Xylitol has anti-diabetic, anti-cariogenic, anti-osteoporotic, and anti-bacterial effects. It is also a lead molecule with potent immunomodulatory activity. Healing properties of xylitol has to be researched further to bring it as a potent therapeutic agent in the future.

KEYWORDS

- **acute otitis media**
- **artificial sweeteners**
- **cytotoxicity**
- **nutraceuticals**
- **osteoporosis**
- **phosphotransferase system**
- **xylitol**

REFERENCES

1. Abdelfatah, A. A. M., Ghaly, I. S., & Hanafy, S. M., (2012). Cytotoxic effect of aspartame (diet sweet) on the histological and genetic structures of female albino rats and their offspring. *Pakistan J. Biol. Sci.*, *15*(19), 904–918.
2. Abou-Donia, M. B., El-Masry, E. M., Abdel-Rahman, A. A., McLendon, R. E., & Schiffman, S. S., (2008). Splenda alters gut microflora and increases intestinal p-glycoprotein and cytochrome p–450 in male rats. *J. Toxicol. Environ. Heal. A.*, *71*(21), 1415–1429.

3. Ahmad, N. Q., & Yusoff, N. M., (2015). The effects of xylitol on body weight loss management and lipid profile on diet-induced obesity mice. *J. Biosci. Med.*, *3*, 54–58.
4. Alamoudi, N. M., Hanno, A. G., Sabbagh, H. J., Masoud, M. I., Almushayt, A. S., El Derwi, D. A., (2012). Impact of maternal xylitol consumption on *mutans streptococci*, plaque and caries levels in children. *J. Clin. Pediatr. Dent.*, *37*(2), 163–166.
5. Alanen, P., Isokangas, P., & Gutmann, K., (2000). Xylitol candies in caries prevention: Results of a field study in Estonian children. *Community Dent. Oral Epidemiol.*, *28*(3), 218–224.
6. Amo, K., Arai, H., Uebanso, T., Fukaya, M., Koganei, M., Sasaki, H., et al., (2011). Effects of xylitol on metabolic parameters and visceral fat accumulation. *J. Clin. Biochem. Nutr.*, *49*(1), 1–7.
7. Assev, S., & Rölla, G., (1986). Further studies on the growth inhibition of *Streptococcus mutans* OMZ 176 by xylitol. *Acta Pathol. Microbiol. Immunol Scand. B.*, *94*(2), 97–102.
8. Badet, C., Furiga, A., & Thébaud, N., (2008). Effect of xylitol on an *in vitro* model of oral biofilm. *Oral Health Prev. Dent.*, *6*(4), 337–341.
9. Bahador, A., Lesan, S., & Kashi, N., (2012). Effect of xylitol on cariogenic and beneficial oral *streptococci*: A randomized, double-blind crossover trial. Iran. *J. Microbiol.*, *4*(2), 75–81.
10. Bannach, G., Almeida, R. R., Lacerda, L. G., Schnitzler, E., & Ionashiro, M., (2009). Thermal stability and thermal decomposition of sucralose. *Eclética. Química*, *34*(4), 21–26.
11. Bansal, V., Siddiqui, M. W., & Rahman, M. S., (2015). Minimally processed foods: Overview. In: Siddiqui, M. W., & Rahman, M. S., (eds.), *Minimally Processed Foods: Technologies for Safety, Quality and Convenience* (pp. 1–15). Springer International Publishing, New York.
12. Bässler, K. H., (1969). Adaptive processes concerned with absorption and metabolism of xylitol. In: Horecker, B. L., Lang, K., & Takagi, Y., (eds.), *International Symposium on Metabolism, Physiology, and Clinical Use of Pentoses and Pentitols* (pp. 190–196). Springer, Berlin.
13. Bryan, G. T., Ertürk, E., & Yoshida, O., (1970). Production of urinary bladder carcinomas in mice by sodium saccharin. *Science, 168*(3936), 1238–1240.
14. Burt, B. A., (2006). The use of sorbitol- and xylitol-sweetened chewing gum in caries control. *JADA, 137*(137), 190–196.
15. Butchko, H. H., & Stargel, W. W., (2001). Aspartame: Scientific evaluation in the postmarketing period. *Regul. Toxicol. Pharmacol.*, *34*(3), 221–233.
16. Butchko, H. H., Stargel, W. W., & Comer, C. P., (2002). Aspartame: Review of safety. *Regul. Toxicol. Pharmacol.*, *35*(2), S1–S93.
17. Chen, X., Jiang, Z. H., Chen, S., & Qin, W., (2010). Microbial and bioconversion production of D-xylitol and its detection and application. *Int. J. Biol. Sci.*, *6*(7), 834–844.
18. Chukwuma, C. I., & Islam, M. S., (2015). Effects of xylitol on carbohydrate digesting enzymes activity, intestinal glucose absorption and muscle glucose uptake: A multi-mode study. *Food Funct.*, *6*(3), 955–962.
19. Cronin, M., Gordon, J., Reardon, R., & Balbo, F., (1994). Three clinical trials comparing xylitol- and sorbitol-containing chewing gums for their effect on supragingival plaque accumulation. *J. Clin. Dent.*, *5*(4), 106–109.

20. Ferreira, A. S., Barbosa, N. R., Junior, D. R., & Silva, S. S., (2009). *In vitro* mechanism of xylitol action against *Staphylococcus aureus* ATCC 25923. In: Mendez-Vilas, A., (ed.), *Current Research Topic in Applied Microbiology and Microbial Biotechnology* (pp. 505–509). World Scientific Publishing Co. Pte. Ltd, Seville.

21. Ferreira, A. S., De Souza, M. A., Barbosa, N. R., Da Silva, & Silvério, S., (2008). *Leishmania amazonensis*: Xylitol as inhibitor of macrophage infection and stimulator of macrophage nitric oxide production. *Exp. Parasitol.*, *119*(1), 74–79.

22. Ford, H. E., Peters, V., Martin, N. M., Sleeth, M. L., Ghatei, M. A., Frost, G. S., & Bloom, S. R., (2011). Effects of oral ingestion of sucralose on gut hormone response and appetite in healthy normal-weight subjects. *Eur. J. Clin. Nutr.*, *65*(4), 508–513.

23. Frejnagel, S. S., Gomez-villalva, E., & Zduñczyk, Z., (2003). Intestinal absorption of xylitol and effect of its concentration on glucose and water absorption in the small intestine of rat. *Pol. J. Food. Nutr. Sci.*, *12*(53), 32–34.

24. Gallus, S., Scotti, L., & Negri, E., (2007). Artificial sweeteners and cancer risk in a network of case-control studies. *Ann. Oncol. Off. J. Eur. Soc. Med. Oncol.*, *18*(1), 40–44.

25. Haghgoo, R., Afshari, E., Ghanaat, T., & Aghazadeh, S., (2015). Comparing the efficacy of xylitol-containing and conventional chewing gums in reducing salivary counts of *Streptococcus mutans*: An *in vivo* study. *J. Int. Soc. Prev. Community Dent.*, *5*(2), S112–S117.

26. Hämäläinen, M. M., Mäkinen, K. K., Parviainen, M. T., & Koskinen, T., (1985). Peroral xylitol increases intestinal calcium absorption in the rat independently of vitamin D action. *Miner. Electrolyte Metab.*, *11*(3), 178–181.

27. Han, S. J., Jeong, S. Y., Nam, Y. J., Yang, K. H., Lim, H. S., & Chung, J., (2005). Xylitol inhibits inflammatory cytokine expression induced by lipopolysaccharide from Porphyromonasgingivalis. *Clin. Diagn. Lab. Immunol.*, *12*(11), 1285–1291.

28. Hautalahti, O., Renko, M., Tapiainen, T., Kontiokari, T., Pokka, T., & Uhari, M., (2007). Failure of xylitol given three times a day for preventing acute otitis media. *Pediatr. Infect. Dis. J.*, *26*(5), 423–427.

29. Hooson, J., Hicks, R. M., Grasso, P., & Chowaniec, J., (1980). Ortho-toluene sulphonamide and saccharin in the promotion of bladder cancer in the rat. *Br. J. Cancer.*, *42*(1), 129–147.

30. Huttunen, J. K., (1976). Serum lipids, uric acid and glucose during chronic consumption of fructose and xylitol in healthy human subjects. *Int. Z. Vitam. Ernahrungsforsch. Beih.*, *15*, 105–115.

31. Ishii, H., Takahashi, H., Mamori, H., Murai, S., & Kanno, T., (1969). Effects of xylitol on carbohydrate metabolism in rat liver treated with carbon tetrachloride or alloxan. *Keio J. Med.*, *18*, 109–114.

32. Islam, M. S., (2011). Effects of xylitol as a sugar substitute on diabetes-related parameters in nondiabetic rats. *J. Med. Food.*, *14*(5), 505–511.

33. Islam, M. S., & Indrajit, M., (2012). Effects of xylitol on blood glucose, glucose tolerance, serum insulin and lipid profile in a type 2 diabetes model of rats. *Ann. Nutr. Metab.*, *61*(1), 57–64.

34. Isokangas, P., Söderling, E., Pienihäkkinen, K., & Alanen, P., (2000). Occurrence of dental decay in children after maternal consumption of xylitol chewing gum, a follow-up from 0 to 5 years of age. *J. Dent. Res.*, *79*(11), 1885–1889.

35. Jakor, A., Williamson, J. R., & Toshio, A., (1971). Xylitol metabolism in perfused rat liver. *J. Biol. Chem.*, *246*(24), 7623–7631.

36. Kaivosoja, S. M., Mattila, P. T., & Knuuttila, M. L. E., (2008). Dietary xylitol protects against the imbalance in bone metabolism during the early phase of collagen type II-induced arthritis in dark agouti rats. *Metabolism, 57*(8), 1052–1055.

37. Kamath, S., Vijaynarayana, K., Shetty, D. P., & Shetty, P., (2010). Evaluation of genotoxic potential of aspartame. *Pharmacologyonline*, *1*, 753–769.

38. Karstadt, M. L., (2006). Testing needed for acesulfame potassium, an artificial sweetener. *Environ. Health Perspect.*, *114*(9), A516.

39. Khan, M. A., Schultz, S., Othman, A., Fleming, T., Lebrón-Galán, R., Rades, D., Clemente, D., Nawroth, P. P., & Schwaninger, M., (2016). Hyperglycemia in stroke impairs polarization of monocytes/macrophages to a protective non-inflammatory cell type. *J. Neurosci.*, *36*(36), 9313–9325.

40. Kim, S., Park, M. H., Song, Y. R., Na, H. S., & Chung, J., (2016). *Aggregatibacter actinomycetemcomitans*-induced aim2 inflammasome activation is suppressed by xylitol in differentiated THP–1 macrophages. *J. Periodontol.*, *87*(6), e116–e126.

41. Kirkland, D., & Gatehouse, D., (2015). Aspartame: A review of genotoxicity data. *Food Chem. Toxicol.*, *84*, 161–168.

42. Kontiokari, T., Uhari, M., & Koskela, M., (1998). Antiadhesive effects of xylitol on otopathogenic bacteria. *J. Antimicrob. Chemother.*, *41*(5), 563–565.

43. Kontiokari, T., Uhari, M., & Koskela, M., (1995). Effect of xylitol on growth of nasopharyngeal bacteria *in vitro. Antimicrob. Agents Chemother.*, *39*(8), 1820–1823.

44. Lee, B. D., & Park, K., (2014). Effects and safety of xylitol on middle ear epithelial cells. *Int. Adv. Otol.*, *10*(1), 19–24.

45. Lee, Y. E., Choi, Y. H., Jeong, S. H., Kim, H. S., Lee, S. H., & Song, K. B., (2009). Morphological changes in *Streptococcus mutans* after chewing gum containing xylitol for twelve months. *Curr. Microbiol.*, *58*(4), 332–337.

46. Lifholgerson, P., (2007). *Xylitol and its Effect on Oral Ecology: Clinical Studies in Children and Adolescents* (p. 165). PhD dissertation, Umeåuniversitet, Umeå.

47. Lim, U., Subar, A. F., Mouw, T., & Hartge, P., (2006). Consumption of aspartame-containing beverages and incidence of hematopoietic and brain malignancies. *Cancer Epidemiol. Biomarkers Prev.*, *15*(9), 1654–1659.

48. Lord, G. H., & Newberne, P. M., (1990). Renal mineralization--a ubiquitous lesion in chronic rat studies. *Food Chem. Toxicol.*, *28*(6), 449–455.

49. Mäkinen, K. K., (1998). Xylitol-based caries prevention: Is there enough evidence for the existence of a specific xylitol effect? *Oral Dis.*, *4*(4), 226–230.

50. Mäkinen, K. K., Mäkinen, P. L., & Pape, H. R., (1996). Conclusion and review of the Michigan xylitol program (1986–1995) for the prevention of dental caries. *Int. Dent. J.*, *46*(1), 22–34.

51. Massa, H. M., Cripps, A. W., & Lehmann, D., (2009). Otitis media: Viruses, bacteria, biofilms and vaccines. *Med. J. Aust.*, *191*(9), 44–49.

52. Mattila, P. T., Svanberg, M. J., & Knuuttila, M. L. E., (2001). Increased bone volume and bone mineral content in xylitol-fed aged rats. *Gerontology, 47*(6), 300–305.

53. Mattila, P., Knuuttila, M., & Svanberg, M., (1998). Dietary xylitol supplementation prevents osteoporotic changes in streptozotocin-diabetic rats. *Metabolism, 47*(5), 578–583.

54. Mccormickt, D. B., & Touster, O., (1957). The conversion *in vivo* of xylitol to glycogen via the pentose phosphate pathway. *J. Biol. Chem., 229*, 451–461.

55. Mehnert, H., (1976). Sugar substitutes in the diabetic diet. *Int. J. Vitam. Nutr. Res., 15*, 295–324.

56. Mellinghoff, C. H., (1961). The usefulness of xylitol as a sugar substitute in diabetics. *Klin. Wochenschr., 39*, 447.

57. Miake, Y., Saeki, Y., Takahashi, M., & Yanagisawa, T., (2003). Remineralization effects of xylitol on demineralized enamel. *J. Electron Microsc.* (Tokyo), *52*(5), 471–476.

58. Milgrom, P., Ly, K. A., Roberts, M. C., Rothen, M., Mueller, G., & Yamaguchi, D. K., (2006). *Mutans Streptococci* dose response to xylitol chewing gum. *J. Dent. Res., 85*(2), 177–181.

59. Modesto, A., & Drake, D. R., (2006). Multiple exposures to chlorhexidine and xylitol: Adhesion and biofilm formation by *Streptococcus mutans. Curr. Microbiol., 52*(6), 418–423.

60. Monteiro, C., Levy, R., Claro, R., De Castro, I., & Cannon, G., (2010). A new classification of foods based on the extent and purpose of their processing. *Cad. Saude Publica., 26*(11), 2039–2049.

61. Mussatto, S. I., (2012). Application of xylitol in food formulations and benefits for health. In: Da Silva, S. S., & Chandel, A. K., (eds.), *D-Xylitol: Fermentative Production, Application and Commercialization* (pp. 309–323). Springer, Berlin.

62. Naaber, P., Mikelsaar, R. H., Salminen, S., & Mikelsaar, M., (1998). Bacterial translocation, intestinal microflora and morphological changes of intestinal mucosa in experimental models of *Clostridium difficile* infection. *J. Med. Microbiol., 47*(7), 591–598.

63. Nakai, Y., Shinga-Ishihara, C., Kaji, M., Moriya, K., Murakami-Yamanaka, K., & Takimura, M., (2010). Xylitol gum and maternal transmission of *mutans streptococci. J. Dent. Res., 89*(1), 56–60.

64. Nakao, S., Ogtata, Y., Shimizu, E., Yamazaki, M., Furuyama, S., & Sugiya, H., (2002). Tumor necrosis factor-alpha (TNF-alpha)-induced prostaglandin E2 release is mediated by the activation of cyclooxygenase–2 (COX–2) transcription via NF-kappa-B in human gingival fibroblasts. *Mol. Cell. Biochem., 238*(1 & 2), 11–18.

65. Nayak, P. A., Nayak, U. A., & Khandelwal, V., (2014). The effect of xylitol on dental caries and oral flora. *Clin. Cosmet. Investig. Dent., 6*, 89–94.

66. Olinger, P. M., & Pepper, T., (2001). Xylitol. In: Nabors, L. O., (ed.), *Alternative Sweeteners* (3rd edn., pp. 335–365). Revised and Expanded, Marcel Dekker, Inc., New York.

67. Olney, J. W., Farber, N. B., Spitznagel, E., & Robins, L. N., (1996). Increasing brain tumor rates: Is there a link to aspartame? *J. Neuropathol. Exp. Neurol., 55*(11), 1115–1123.

68. Onishi, H., & Suzuki, T., (1969). Microbial production of xylitol from glucose. *Appl. Microbiol., 18*(6), 1031–1035.

69. Palchaudhuri, S., Rehse, S. J., Hamasha, K., Syed, T., Kurtovic, E., Kurtovic, E., & Stenger, J., (2011). Raman spectroscopy of xylitol uptake and metabolism in gram-positive and gram-negative bacteria. *Appl. Environ. Microbiol., 77*(1), 131–137.

70. Paredes-Sabja, D., & Sarker, M. R., (2012). Adherence of *Clostridium difficile* spores to Caco–2 cells in culture. *J. Med. Microbiol., 61*(9), 1208–1218.

71. Park, E., Na, H. S., Kim, S. M., Wallet, S., Cha, S., & Chung, J., (2014). Xylitol, an anticaries agent, exhibits potent inhibition of inflammatory responses in human THP–1-derived macrophages infected with Porphyromonasgingivalis. *J. Periodontol.*, *85*(6), e212–e223.

72. Park, E., Park, M. H., Na, H. S., & Chung, J., (2015). Xylitol induces cell death in lung cancer A549 cells by autophagy. *Biotechnol. Lett.*, *37*(5), 983–990.

73. Peldyak, J., & Mäkinen, K., (2002). Xylitol for caries prevention. *J. Dent. Hyg.*, *76*, 276–285.

74. Pepino, M. Y., Tiemann, C. D., Patterson, B. W., Wice, B. M., & Klein, S., (2013). Sucralose affects glycemic and hormonal responses to an oral glucose load. *Diabetes Care*, *36*(9), 2530–2535.

75. Pihlanto-Lepfala, A., Soderling, E., Kauko, K., & Soderling, E., (1990). Expulsion mechanism of xylitol 5-phosphate in *Streptococcus mutans*. *Scand. J. Dent. Res.*, *98*, 112–119.

76. Rahn, A., & Yaylayan, V. A., (2010). Thermal degradation of sucralose and its potential in generating chloropropanols in the presence of glycerol. *Food Chem.*, *118*(1), 56–61.

77. Rajat, S., Manisha, S., Robin, S., & Sunil, K., (2012). Nutraceuticals: A review. *Int. Res. J. Pharm.*, *3*(4), 95–99.

78. Ravella, S. R., Gallagher, J., Fish, S., & Prakasham, R. S., (2012). Overview on commercial production of xylitol, economic analysis and market trends. In: Da Silva, S. S., & Chandel, A. K., (eds.), *D-Xylitol: Fermentative Production, Application and Commercialization* (pp. 291–306). Springer, Berlin.

79. Renko, M., Valkonen, P., Tapiainen, T., & Kontiokari, T., (2008). Xylitol-supplemented nutrition enhances bacterial killing and prolongs survival of rats in experimental *Pneumococcal* Sepsis. *BMC Microbiol.*, *8*(1), S45.

80. Renz, H., Gong, J. H., Schmidt, A., Nain, M., & Gemsa, D., (1988). Release of tumor necrosis factor-alpha from macrophages enhancement and suppression are dose-dependently regulated by prostaglandin E2 and cyclic nucleotides. *J. Immunol.*, *141*(7), 2388–2393.

81. Reuber, M. D., (1978). Carcinogenicity of saccharin. *Environ. Health Perspect.*, *25*, 173–200.

82. Ribellesllop, M., Guinot-Jimeno, F., Mayné-Acién, R., & Bellet-Dalmau, L. J., (2010). Effects of xylitol chewing gum on salivary flow rate, pH, buffering capacity and presence of *Streptococcus mutans* in saliva. *Eur. J. Paediatr. Dent.*, *11*(1), 9–14.

83. Roncal-Jimenez, C. A., & Lanaspa, M. A., (2011). Sucrose induces fatty liver and pancreatic inflammation in male breeder rats independent of excess energy intake. *Metabolism*, *60*(9), 1259–1270.

84. Salli, K. M., Forssten, S. D., Lahtinen, S. J., & Ouwehand, A. C., (2016). Influence of sucrose and xylitol on an early *Streptococcus mutans* biofilm in a dental simulator. *Arch. Oral Biol.*, *70*, 39–46.

85. Salminen, S., Salminen, E., & Marks, V., (1982). The effects of xylitol on the secretion of insulin and gastric inhibitory polypeptide in man and rats. *Diabetologia*, *22*(6), 480–482.

86. Sano, H., Nakashima, S., Songpaisan, Y., & Phantumvanit, P., (2007). Effect of a xylitol and fluoride-containing toothpaste on the remineralization of human enamel *in vitro*. *J. Oral Sci.*, *49*(1), 67–73.

87. Scheinin, A., Mäkinen, K. K., Tammisalo, E., & Rekola, M., (1975). Sugar studies XVIII: Incidence of dental caries in relation to 1-year consumption of xylitol chewing gum. *Acta Odontol. Scand.*, *33*(5), 269–278.

88. Schiffman, S. S., & Rother, K. I., (2013). Sucralose, a synthetic organochlorine sweetener: Overview of biological issues. *J. Toxicol. Environ. Health. B. Crit. Rev.*, *16*(7), 399–451.

89. Shafer, R. B., Levine, A. S., Marlette, J. M., & Morley, J. E., (1987). Effects of xylitol on gastric emptying and food intake. *Am. J. Clin. Nutr.*, *45*(4), 744–747.

90. Shangari, N., & O'Brien, P. J., (2004). The cytotoxic mechanism of glyoxal involves oxidative stress. *Biochem. Pharmacol.*, *68*(7), 1433–1442.

91. Silva, A. F., Suzuki, É. Y., Ferreira, A. S., Oliveira, M. G., Silva, S. S., & Raposo, N. R. B., (2011). *In vitro* inhibition of adhesion of *Escherichia coli* strains by xylitol. *Brazilian Arch. Biol. Technol.*, *54*(2), 235–241.

92. Singh, J., & Sinha, S., (2012). Classification, regulatory acts and applications of nutraceuticals for health. *Int. J. Pharm. Biol. Sci.*, *2*(1), 177–187.

93. Söderling, E., Alaräisänen, L., Scheinin, A., & Mäkinen, K. K., (1987). Effect of xylitol and sorbitol on polysaccharide production by and adhesive properties of *Streptococcus mutans*. *Caries Res.*, *21*(2), 109–116.

94. Söderling, E., Isokangas, P., Pienihäkkinen, K., & Tenovuo, J., (2000). Influence of maternal xylitol consumption on acquisition of *mutans streptococci* by infants. *J. Dent. Res.*, *79*(3), 882–887.

95. Soffritti, M., Belpoggi, F., Esposti, D. D., & Lambertini, L., (2005). Aspartame induces lymphomas and leukaemias in rats. *Eur. J. Oncol.*, *10*, 107–116.

96. Soffritti, M., Belpoggi, F., DegliEsposti, D., Lambertini, L., Tibaldi, E., & Rigano, A., (2006). First experimental demonstration of the multipotential carcinogenic effects of aspartame administered in the feed to Sprague-Dawley rats. *Environ. Health Perspect.*, *114*(3), 379–385.

97. Soffritti, M., Belpoggi, F., Manservigi, M., Tibaldi, E., Lauriola, M., Falcioni, L., & Bua, L., (2010). Aspartame administered in feed, beginning prenatally through life span, induces cancers of the liver and lung in male Swiss Mice. *Am. J. Ind. Med.*, *53*(12), 1197–1206.

98. Soufyan, A., Alkatiri, F., Verisqa, F., Megantoro, A., Sumawinata, N., & Mangundjaja, S., (2010). Effect of xylitol on remineralization of demineralized dental enamel. *Int. J. Clin. Prev. Dent.*, *6*(2), 73–77.

99. Sousa, L. P., Silva, A. F., Calil, N. O., Oliveira, M. G., Silva, S. S., & Raposo, N. R. B., (2011). *In vitro* inhibition of *Pseudo-monasaeruginosa* adhesión by xylitol. *Brazilian Arch. Biol. Technol.*, *54*(5), 877–884.

100. Svanberg, M., & Knuuttila, M., (1994). Dietary xylitol prevents ovariectomy-induced changes of bone inorganic fraction in rats. *Bone Miner.*, *26*(1), 81–88.

101. Svanberg, M., Mattila, P., & Knuuttila, M., (1997). Dietary xylitol retards the ovariectomy-induced increase of bone turnover in rats. *Tissue Int.*, *60*(5), 462–466.

102. Szél, E., Polyánka, H., Szabó, K., & Hartmann, P., (2015). Anti-irritant and anti-inflammatory effects of glycerol and xylitol in sodium lauryl sulphate-induced acute irritation. *J. Eur. Acad. Dermatology Venereol.*, *29*(12), 2333–2341.

103. Tange, T., Sakurai, Y., Hirose, M., Noro, D., & Igarashi, S., (2004). The effect of xylitol and fluoride on remineralization for primary tooth enamel caries *in vitro*. *Pediatr. Dent. J.*, *14*(1), 55–59.

104. Tapiainen, T., Kontiokari, T., Sammalkivi, L., Ikaheimo, I., Koskela, M., & Uhari, M., (2001). Effect of xylitol on growth of *Streptococcus pneumoniae* in the presence of fructose and sorbitol. *Antimicrob. Agents Chemother.*, *45*(1), 166–169.

105. Tapiainen, T., Luotonen, L., Kontiokari, T., Renko, M., & Uhari, M., (2002). Xylitol administered only during respiratory infections failed to prevent acute otitis media. *Pediatrics*, *109*(2), E19.

106. Tapiainen, T., Sormunen, R., Kaijalainen, T., Kontiokari, T., Ikäheimo, I., & Uhari, M., (2004). Ultrastructure of *Streptococcus pneumoniae* after exposure to xylitol. *J. Antimicrob. Chemother.*, *54*, 225–228.

107. Thaweboon, S., (2004). Effect of xylitol on *mutans streptococci*. *Southeast Asian. J. Trop. Med. Public Health*, *35*(4), 1024–1027.

108. Thompson, M., (2013). *Anti-Inflammatory Properties of Xylitol in a Model of Chronic Sinus Disease* (p. 140). M. Phil. Dissertation, Griffith University.

109. Tuompo, H., Meurman, J. H., Lounatmaa, K., & Linkola, J., (1983). Effect of xylitol and other carbon sources on the cell wall of *Streptococcus mutans*. *Eur. J. Oral Sci.*, *91*(1), 17–25.

110. Uhari, M., Kontiokari, T., Koskela, M., & Niemelä, M., (1996). Xylitol chewing gum in prevention of acute otitis media: Double-blind randomized trial. *BMJ*, *313*(7066), 1180–1184.

111. Uhari, M., Kontiokari, T., & Niemelä, M., (1998). A novel use of xylitol sugar in preventing acute otitis media. *Paediatrics*, *102*(1), 879–884.

112. Ur-Rehman, S., Mushtaq, Z., Zahoor, T., Jamil, A., & Anjum, M. M., (2013). Xylitol, a review on bio-production, application, health benefits and related safety issues. *Crit. Rev. Food Sci. Nutr.*, *8398*, 37–41.

113. Weaver, C. M., Dwyer, J., Fulgoni, V. L., King, J. C., Leveille, G. A., McDonald, R. S., Ordovas, J., & Schnakenberg, D., (2014). Processed foods: Contributions to nutrition. *Am. J. Clin. Nutr.*, *99*(6), 1525–1542.

114. Whitehouse, C. R., (2008). The potential toxicity of artificial sweeteners. *Am. Assoc. Occup. Heal. Nurses*, *56*(6), 251–259.

115. Wölnerhanssen, B. K., Cajacob, L., Keller, N., Doody, A., Rehfeld, J. F., Drewe, J., Peterli, R., Beglinger, C., & Meyer-Gerspach, A. C., (2016). Gut hormone secretion, gastric emptying, and glycemic responses to erythritol and xylitol in lean and obese subjects. *Am. J. Physiol. Endocrinol. Metab.*, *310*(11), E1053–E1061.

116. Woods, H. F., & Krebs, H. A., (1973). Xylitol metabolism in the isolated perfused rat liver. *Biochem. J.*, *134*, 437–443.

117. Xu, M. L., Wi, G. R., Kim, H. J., & Kim, H. J., (2016). Ameliorating effect of dietary xylitol on human respiratory syncytial virus (hRSV) infection. *Biol. Pharm. Bull.*, *39*(4), 540–546.

118. Yang, K., (2011). *Formation and Metabolism of Sugar Metabolites, Glyoxal and Methylglyoxal, and Their Molecular Cytotoxic Mechanisms in Isolated Rat Hepatocytes* (p. 115). MSc dissertation, University of Toronto, Toronto.

119. Yin, S. Y., Kim, H. J., & Kim, H. J., (2014). Protective effect of dietary xylitol on influenza a virus infection. *PLoS One*, *9*(1), e84633.

120. http://www.inchem.org/documents/jecfa/jecmono/v12je22.htm (accessed on 2 December 2016).

121. https://www.globalsweet.com/merchant2/merchant.mvc?Screen=XYM1 (accessed on 2 December 2016).
122. https://www.reference.com/food/examples-unprocessed-foods–8be11607a5e15181? qo=cdpArticles (accessed on 2 December 2016).
123. http://www.inchem.org/documents/jecfa/jecmono/v32je17.htm (accessed on 2 December 2016).
124. http://www.sweetpoison.com/aspartame-side-effects.html (accessed on 2 December 2016).
125. http://www.fda.gov/ohrms/dockets/dailys/03/jan03/012203/02p–0317_emc–000199. txt (accessed on 2 December 2016).
126. http://www.sweetpoison.com/aspartame-sweeteners.html (accessed on 2 December 2016).
127. http://www.livestrong.com/article/470503-side-effects-of-saccharin-sodium/ (accessed on 2 December 2016).
128. https://www.cancer.gov/about-cancer/causes-prevention/risk/diet/artificial-sweeteners-fact-sheet (accessed on 2 December 2016).
129. http://articles.mercola.com/sites/articles/archive/2009/02/10/new-study-of-splenda-reveals-shocking-information-about-potential-harmful-effects.aspx (accessed on 2 December 2016).
130. http://hubpages.com/health/Splenda-is-Not-Splendid (accessed on 2 December 2016).
131. http://www.thealternativedaily.com/watch-advantame-newest-artificial-sweetener/ (accessed on December 2016).
132. http://food-additive.wikispaces.com/Side+effect+of+Neotame (accessed on 2 December 2016).
133. http://www.sweetpoison.com/aspartame-sweeteners.html (accessed on 2 December 2016).
134. https://pubchem.ncbi.nlm.nih.gov/compound/aspartame#section=Top (accessed on 2 December 2016).
135. https://pubchem.ncbi.nlm.nih.gov/compound/saccharin#section=Top (accessed on 2 December 2016).
136. https://pubchem.ncbi.nlm.nih.gov/compound/Sucralose (accessed on 2 December 2016).
137. https://pubchem.ncbi.nlm.nih.gov/compound/10389431#section=Top (accessed on 2 December 2016).
138. https://pubchem.ncbi.nlm.nih.gov/compound/Neotame (accessed on 2 December 2016).
139. https://pubchem.ncbi.nlm.nih.gov/compound/Acesulfame_K (accessed on 2 December 2016).
140. https://pubchem.ncbi.nlm.nih.gov/compound/xylitol (accessed on 2 December 2016).
141. https://www.sciencedaily.com/releases/2013/02/130201100149.htm (accessed on 2 December 2016).
142. http://drsircus.com/cancer/sugar-cancer-growth-research/ (accessed on 2 December 2016).
143. http://www.human-health-and-animal-ethics.com/health/dental-care/tooth-decay. php-(accessed on 2 December 2016).

CHAPTER 7

ALDOSE REDUCTASE INHIBITORS FOR MULTI-FACETED ANTI-DIABETIC ACTIONS

Y. ANIE,[*] V. N. HAZEENA, V. H. HARITHA, and V. S. BINCHU

[*]Corresponding author. E-mail: aniey@mgu.ac.in.

ABSTRACT

Pathophysiology of various complications in diabetes is attributed to the altered functioning of the polyol pathway, activated AGEs, PKC activation and hexosamine pathway. Altogether, these add up to the oxidative stress, glycative stress, reductive stress and osmotic stress paving the way for cell and tissue damage. Based on the findings that aldose reductase (AR) enzyme is implicated in the pathogenesis of different diabetic complications as well as of certain non-diabetic pathologies, ARI inhibition is being looked upon as a therapeutic approach in the treatment of such conditions. Moreover, many research groups established the efficiency of aldose reductase inhibitors (ARIs) in the treatment of different compromised health conditions – both diabetic-related and non-diabetic-related.

7.1 INTRODUCTION

Present-day health scenario represents the overall effect of an unhealthy diet and a polluted environment with emerging and increased lifestyle diseases. Diabetes is one such lifestyle disease, which is finding its way to every household these days, making it difficult to come across homes without a diabetic patient. It is among the leading chronic diseases that cause death across the world. Both developing and developed countries are victims of this disease that is progressing at an alarming rate.

Modern lifestyle and physical inactivity have led to obesity and lifestyle diseases like diabetes. Different pathological mechanisms lead to diabetic progression. First one is the destruction of β cells of the pancreas, the cells responsible for the production of insulin, causing insulin insufficiency that eventually leads to insulin resistance. Metabolic abnormalities in diabetes result from defective insulin production and inappropriate responses to insulin by its hormonal receptors in the metabolic pathways. Sometimes, a combination of both defects, i.e., impairment in the insulin production as well as in its action is seen in the same patient, making it difficult to trace the etiology or the primary cause of hyperglycemia.

This chapter discusses the involvement of aldose reductase (AR) enzyme in diabetes mellitus (DM) and the anti-diabetic role played by aldose reductase inhibitors (ARIs) in combating glycation, hyperglycemia, immune functions, oxidative stress, ROS production, extracellular trap formation, etc., with reference to outcomes from different experimental studies.

7.2   DIABETES MELLITUS (DM)

DM can be termed as related metabolic disorders of multiple etiologies characterized by chronic hyperglycemia. The disease expresses altered carbohydrate, lipid, and protein metabolisms due to insulin insufficiency resulting from defects in insulin production, insulin action or both. Polyuria, polydipsia, weight loss, polyphagia, and blurred vision are characteristic symptoms of diabetes. This disorder affects over 177 million people worldwide. About 1.1 million people die from diabetes annually. In 2010, 285 million adults were affected with DM, which is expected to reach 439 million in 2030. DM is a common cause of death in developing and underdeveloped countries, where high medical expense needed to treat diabetes, and its complications become a burden for middle/low-income people [126]. People live with diabetes for years, and in them, heart diseases, kidney failure, and other causes are reported as the cause of death. Limited diabetic awareness and access to the required health care services lead to diabetic complications.

7.2.1 TYPE OF DIABETES MELLITUS

DM is categorized into different types: Type 1 [106, 140], type 2 [54, 61], gestational [122. 123, 133] and other types [39, 76]. The other specific types are caused by different specific causes.

7.2.2 SECONDARY DIABETIC COMPLICATIONS

Uncontrolled chronic hyperglycemia observed in DM patients leads to the development of secondary diabetic complications as persistent hypergly-cemia causes damages to tissues like lens, retina, kidney, peripheral nerves and blood vessels. These complications include diabetic neuropathy, diabetic nephropathy, ischemic damage in the feet and ulcers, diabetic retinopathy and blindness. Impaired immune function along with frequent episodes of infection and delayed wound healing are also reported in these patients. These patients also show an increased tendency to develop atherosclerosis and coronary artery diseases. Another acute life-threatening consequence of uncontrolled diabetes is keto-acidosis or the nonketotic hyperosmolar syndrome. Therefore, a major share of expenditure for the management of DM is contributed for treating long-term secondary complications caused by the disease.

Acute complications are Diabetic ketoacidosis (DKA) [20, 34]; hyper-glycaemic hyperosmolar state (HHS) [20, 70, 96]. In general, the deleterious effects of chronic hyperglycemia are broadly classified into microvascular complications (diabetic neuropathy [12, 37, 50], retinopathy [37, 99] and nephropathy [2, 19, 79, 129] and macrovascular complications (peripheral arterial disease, coronary artery disease and stroke) [36, 37, 52].

7.2.3 PATHOPHYSIOLOGY OF SECONDARY DIABETIC COMPLICATIONS

7.2.3.1 THE POLYOL PATHWAY

This is a minor alternate pathway for glucose metabolism and channels only a very small portion of glucose through it at normal blood glucose levels. However, during hyperglycaemic conditions, situations take a turn where up to 30% of blood glucose enters the polyol or the sorbitol pathway. This pathway involves the conversion of glucose to sorbitol by AR enzyme and subsequent generation of fructose from sorbitol by the enzyme sorbitol dehydrogenase [85, 125].

Polyol pathway (Figure 7.1) presents itself as one of the prime mechanisms for tissue damage and the resulting long-term diabetic complications in diabetics. Even though the rate-limiting enzyme AR posses only a low substrate affinity under normoglycemia, high blood glucose levels activate the AR enzyme and convert glucose to sorbitol. This results in intracellular sorbitol accumulation with subsequent depletion in the NADH levels. This, in turn, will affect another NADPH requiring enzyme glutathione reductase (GR), involved in the reduction of oxidized glutathione to reduced glutathione.

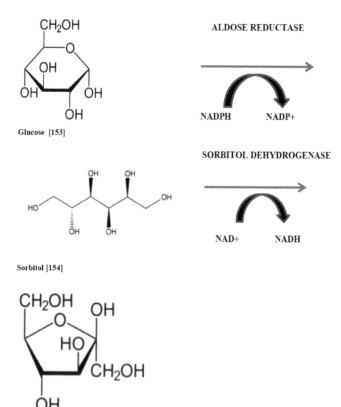

FIGURE 7.1   The polyol pathway.

The depletion in the vital antioxidant of the cellular system and the activation of NADH oxidase due to increased NADH levels from sorbitol dehydrogenase function augments ROS production which triggers cellular damage. Apart from these, accumulation of sorbitol, the product formed in the polyol pathway, due to its inability to cross cell-membrane and leave the cell may facilitate osmotic changes that result in tissue damage [36, 85].

7.2.3.2   OXIDATIVE STRESS

Presence of high amounts of oxidized proteins, lipids, and DNA in DM patients signifies the crucial role of oxidative stress in the pathogenesis of diabetic complications. Multiple mechanisms have been proposed for this diabetes-induced oxidative stress. Glyceraldehyde–3-phosphate dehydrogenase inhibition by hyperglycemia-induced action of superoxides increases the flux of glucose and its intermediates to various pathways like polyol pathway, protein kinase C (PKC) activation pathway, hexosamine pathway, AGEs pathway, etc. This causes augmented NO generation that favors the peroxynitrite formation. Peroxynitrite affects signal transduction by oxidizing sulfhydryl groups in proteins, nitrates, amino acids and initiates lipid peroxidation [97, 145].

Chronic hyperglycemia-induced up-regulation of polyol pathway depletes NADPH levels resulting in increased redox stress (Figures 7.2 and 7.3). Depletion of NADPH affects the regeneration of reduced glutathione (GSH), which plays a vital role in ROS scavenging. Enhanced levels of NADH formed during sorbitol dehydrogenase reaction provide an extra substrate for NADH oxidase, which is known for its superoxide production, further augmenting the oxidative stress levels. The cumulative effect of all the deranged systems would be the compromised anti-oxidative capacity, making the cells susceptible to the action of free radicals and consequential damages [36, 44]. Superoxides are directly involved in the inactivation of two anti-atherosclerotic enzymes; namely, prostacyclin synthase and endo-thelial nitric oxide (NO) synthase raising the chances for atherosclerosis.

7.2.3.3   GLYCATIVE STRESS

Advanced glycation end products (AGEs) are heterogeneous molecules produced by non-enzymatic glycosylation of proteins, lipids, and nucleic acids due to prolonged exposure to aldose sugars [21]. Increased polyol

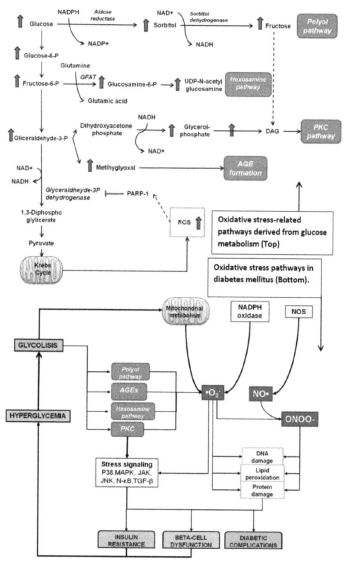

FIGURE 7.2   (See color insert.) Relationship between oxidative stress and diabetes mellitus. **Note:** Consequences of up-regulation of polyol pathway are: Increased AR activity leads to increased sorbitol levels causing osmotic stress. Depletion of NADPH reduces glutathione reductase and catalase activity and increased the action of sorbitol dehydrogenase results in the increased supply of NADH to NADH oxidase; both results in increased ROS generation. Osmotic and oxidative stress leads to secondary diabetic complications. [https://www.intechopen. com/books/oxidative-stress-and-chronic-degenerative-diseases-a-role-for-antioxidants/ oxidative-stress-in-diabetes-mellitus-and-the-role-of-vitamins-with-antioxidant-actions].

pathway generates higher levels of fructose, which is metabolized to fructose-3-phosphate and 3-deoxyglucosone that are non-enzymatic glycation agents that can ultimately lead to ROS generation [128]. Hyperglycemia also leads to the formation of the glycating agent methylglyoxal (MG) from glyceraldehyde–3-phosphate. Glycated hemoglobin (which is used as an index to measure the status of diabetes) is a major product of MG glycation. Depletion of cellular antioxidant systems (e.g., Glutathione peroxidase) and activation of NAD(P)H oxidase through the receptors of advanced glycation end product (RAGE) leads to the enhanced free radical formation and is yet another consequence of AGEs [48]. They also have a role in the release of pro-inflammatory cytokines (like IL-6, IL-β, and TNF-α), apart from stimulating the activities of various growth factors and adhesion molecules (like VEGF and CTGF, TGF-β1, IGF-1, and PDGF) [125]. Accumulated AGEs causes a proportional increase in oxidative stress and this affects the function of related enzymes or their receptors and subsequently leads to microvascular and macrovascular complications of DM [48].

7.2.3.4   REDUCTIVE STRESS

Under chronic hyperglycemia, an elevation in the levels of NADH and reduction in the NAD^+ levels will lead to an increase in redox potential (increased $NADH:NAD^+$ ratio). Production of fructose from sorbitol by sorbitol dehydrogenase involves utilization of NAD^+ reserves lowering the cytosolic availability of NAD^+. This affects the signal transduction, metabolism, and stress response adversely leading to a pseudo-hypoxic stress condition. The ample supply of NADH will activate NADH oxidase, which when accompanied by pseudohypoxia, will concomitantly add to the ROS production in the cell [82, 99, 145].

7.2.3.5   ACTIVATION OF PROTEIN KINASE C (PKC) PATHWAY

PKC isoforms that participate in the signaling pathways are activated by calcium, diacylglycerol (DAG) and phosphatidylserine. Chronic hyperglycemia leads to increased levels of dihydroxyacetone phosphate, which is then converted to glycerol–3 phosphates. Fatty acid residues are then attached to glycerol–3-phosphate to form DAG. PKC isoforms activated under hyperglycemic conditions alters the expression of growth factors. One mechanism of activation of the DAG-PKC pathway is due to increased DAG levels from

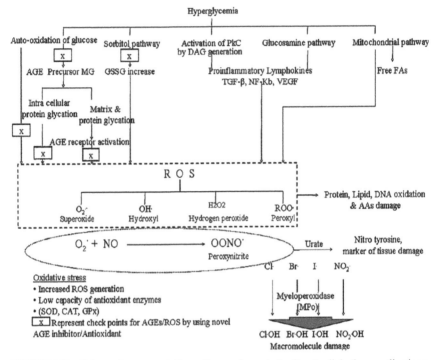

FIGURE 7.3   Schematic representation of hyperglycemia leading to diabetic complications: [https://www.researchgate.net/figure/263292979_fig3_Fig-3-Schematic-representation-of-the formation-of-ROS-and-RNS-under-hyperglycemic; and *Glycobiology* 2014, 1–12].

impaired glycolytic pathway [125, 144]. Enhanced PKC activation can also result from AGE- receptor interactions. PKC isoform activation is responsible for the majority of the vascular tissue related diabetic complications, as they are promoting the occurrence of leaky vasculature and increasing both mRNA synthesis of vascular endothelial growth factor (VEGF) and endothelium-leukocyte interaction [36]. Experimental studies on diabetes have proved that PKC activation leads to changes in renal blood flow by decreased production of NO, increases GFR, albuminuria, mesangial expansion and increases pro-inflammatory gene expression [125].

7.2.3.6   HEXOSAMINE PATHWAY

Hexosamine pathway involves the conversion of fructose-6-phosphate to glucosamine–6-phosphate by the enzyme glutamine: fructose-6-phosphate amidotransferase (GFAT), the rate determining enzyme [132]. Surfeit intracellular glucose is diverted to hexosamine pathway and the product UDP- N acetylglucosamine formed from glucosamine–6-phosphate facilitate glycosylation of lipids and proteins. Specific enzymes make use of N acetylglucosamine for post-translational modification of cytoplasmic and nuclear proteins [36]. GFAT promoter is activated by increased glucose levels angiotensin II in the mesangial cells, and this enhances the flux through the hexosamine pathway. Inhibition of GFAT by azaserine was shown to decrease the hyperglycemia-induced production of TGF-β and ECM, indicating that it is partly mediated by hexokinase. Increased hexosamine pathway is also related to PKC activation and increased ECM production [125, 132].

7.2.3.7 ALTERED INNATE IMMUNITY

Type 2 diabetes and insulin resistance have been linked to altered function and production of proinflammatory cytokines, circulating innate immune proteins and cellular pattern recognition receptors. Increased numbers of white blood cells and adipose tissue macrophages are associated with obesity and insulin resistance. Inflammatory acute phase markers such as IL-6 and CRP are increased in newly diagnosed and established type 2 diabetic patients. In middle-aged population, certain inflammatory markers like low serum albumin, fibrinogen, α1-acid glycoprotein, white blood cell count and sialic acid have been used for predicting the development of type 2 diabetes. Lipopolysaccharide-stimulated PBMCs and monocytes of diabetic patients showed a decrease in IL-1and IL-6 productions indicating induced tolerance to stimulation [42, 112].

a. **Neutrophil Dysfunctions:** Neutrophils are the first cells to arrive at the site of inflammation; and so, reduction in their functional activity contributes to the high susceptibility and severity of infection in DM [7]. In diabetes, hyperglycemia induces defects in neutrophil function such as chemotaxis, phagocytosis, increased extracellular superoxide generation and bactericidal killing. [15, 43, 45, 47, 63, 102, 136]. In them, the extent of neutrophil respiratory burst is correlated to high blood glucose levels as well as increased NADPH oxidase and PKC activities [47, 83]. DM patients also display elevated resting levels of IL-6, IL-8 and

TNF-α [94, 109, 152]. Hyperglycemia up-regulates the expression of several proinflammatory genes including those encoding for IL-β and TNF-α under *in vitro* conditions [124]. Moreover, some other properties of neutrophil functions such as leukotriene (LT) release, lysosomal enzyme secretion and the levels of superoxide and intracellular calcium are also altered in diabetes [86].

b. **Monocyte Dysfunctions**: Monocytes, the central cells of the inflammatory response, are affected by DM [88]. Monocyte plays a major part in providing protection against fungi and bacteria, and it also modulates the host immune response. Information on monocyte function in diabetic patients is not available as that of neutrophil functions. However, a few studies have reported that monocyte functions like adherence, phagocytosis, and chemotaxis are impaired in DM [27, 43, 69]. Monocyte linked intracellular bactericidal function was considerably less in patients with poorly controlled type 2 DM [26]. Also, reduced PBMC functions indicating an increased risk of infection has been reported in diabetic patients [42]. Monocytes from diabetic patients are seen to have increased hexose monophosphate shunt activity, superoxide generation, and chemiluminescence production, all pointing towards an enhanced metabolic activity [71]. This hyperactivity may cause autooxidative membrane damage to monocyte affecting its locomotory ability [53] and also, its phagocytic activity in type 2 DM patients [77].

When compared to neutrophils, monocytes have greater capability to produce pro-inflammatory and anti-inflammatory cytokines and growth factors which modulates inflammatory responses, of which TNF-α, IL-β, IL-6, C-reactive protein and monocyte chemoattractant protein–1 (MCP–1) common markers of inflammation are shown to be increased in DM. This may be the reason behind the increased inflammation in DM patients [24, 31, 38, 59, 67, 141]. Among the pro-inflammatory cytokines, a key role is played by IL-6 in DM. IL-6 promotes insulin resistance and is responsible for inducing acute phase responses such as serum amyloid A and C-reactive protein release [59, 152]. Under hyper-glucose conditions, enhanced release of IL-6 from mono-cytes occurs via reactive oxygen species, PKC, MAPK (Mitogen-activated protein kinase) and NF-κβ pathways and these pathways are incriminated in diabetic vasculopathies. Oxidative stress plays a major role in the pathogenesis of diabetic complications [82]. Superoxides liberation from monocytes during high glucose conditions is mediated through activation of PKC [83]. In rat smooth muscle cells, the synthesis of IL-6 and TNF- α is controlled by the p38 MAPK pathway [77]. PKC activation can also stimulate the MAPK pathway directly. Hyperglucose have also been revealed to induce TNF through p38MAPK activity in monocytes, oxidant stress and increased NF-κβ activity [83, 124].

c. **Augmented NETosis**: Neutrophils perform different defense strategies like phagocytosis, oxidative burst, degranulation, and NETosis. NETosis is a recently discovered mode of cell death in which PAD4 (peptidyl arginine deiminase 4 enzyme) activation and subsequent hypercitrullination of histones lead to the ejection of the chromosomal material from the cell into the external environment in the form of a trap entangled with antimicrobial peptides. ROS and some cytokines such as IL-6 and TNF α are strong stimulators of NETosis. These cytokine levels are increased in DM, and this may be the reason for increased NETosis in type 2 DM.

In chronic disease conditions like diabetes, the neutrophil function including NETosis is seemed to be altered. High level of circulating free DNA (De-oxy ribonucleic acid) and augmented NETosis is reported in type 2 diabetes. Increase in circulating nucleosome shows a positive correlation with glucose level and HbA1c level. However, the neutrophil number had no influence on the increase in circulating nucleosome. Circulating MPO, NE, and PR 3, the prominent markers of NETosis, are also increased in diabetes. More superoxide production in diabetic patients also elevates NETosis. Neutrophils with an increase in expression of PAD4 and citrullinated histone were reported in wounds of diabetic patients. Accelerated NETosis and NET protein components in the wound prevent it from healing, and this state can be reverted by DNAse treatment [60, 89, 118].

Different hyperglycemia-induced pathogenic mechanisms, AGE formation, RAGE ligand binding, Hexosamine pathway- all arise from overproduction of superoxides. Increased ROS generation leads to NETosis, proinflammatory cytokines production, and altered immune response. All these factors together contribute in inducing tissue damage and end-organ complications. The microvascular complication in diabetes is also seen as a result of augmented NETosis. Reducing oxidative stress and NET generation would thus be an additional approach in the treatment of secondary diabetic complications.

d. **Aldose Reductase Enzyme**: AR enzyme is the rate determining enzyme of the polyol pathway that catalyzes the first reaction, i.e., the conversion of glucose to sorbitol using NADPH as the cofactor. The sorbitol is then converted to fructose by a second enzyme sorbitol dehydrogenase that requires NAD^+ as a cofactor (Figure 7.1). Apart from its role in reducing glucose and related monosaccharides, it also has its role in the reduction of aldehydes and their conjugates derived from various lipid peroxidation. AR (EC 1.1.1.21) is a small monomeric protein (36kDa) of 315 amino acid residues [181]. It forms a part of the aldo-keto reductase superfamily. AR is a cytosolic enzyme, which is not in uniform distribution among the tissues but is seen in the majority of the mammalian cells.

Crystalline structures of porcineALR2 and human placental ALR2 are similar, and both have of α/β barrel motif. The core is made of 8 β strands and 8 α helices - β strands are arranged parallelly to each other, and they are joined at the periphery by 8 α helices arranged antiparallel to the eight β strands. At the center of the barrel, NADPH, the coenzyme of AR is bound to the C terminal of the β strand in an extended form. Thus, the nicotinamide ring of NADPH occupies the core of the large and hydrophobic active site. The active site of AR is lined by aromatic amino acids (Tyr 48, Trp 20, Trp111, Phe121, Trp79, and Trp219), polar amino acids (Cys298, His110, and Gln49) and nonpolar amino acids (Pro218, Val147, Leu301, and Leu300). His110, Tyr48, and Cys298 were identified as the three proton donors [18, 117, 142].

7.3   PHYSIOLOGICAL ROLE OF ALDOSE REDUCTASE (AR)

7.3.1   ROLE IN DETOXIFICATION OF TOXIC ALDEHYDES

AR enzyme plays a cytoprotective role with its broad range of substrate specificity by eliminating toxic reactive aldehydes and their GSH conjugates generated endogenously [137, 143].

7.3.2 ROLE IN SPERM PHYSIOLOGY AND MATURATION

Fructose is the main source of energy for sperms, and adequate amount of fructose is formed by AR activity. Interestingly, the enzyme was first identified in seminal vesicles. ALR2 and SDH, are synthesized in the bovine and human epididymal epithelium for the production of fructose. In humans, ALR2 and SDH are detected as components of epididymosomes and prostasomes, vesicles involved in the functioning and maintenance of spermatozoa. ALR2 is thought to be important for epididymal sperm motility, maturation, and metabolism [137, 143].

7.3.3 OTHER ROLES

Another part played by AR enzyme is in osmoregulation, as sorbitol is highly osmolar and also it has a part in catecholamine and steroid metabolism (Figure 7.4).

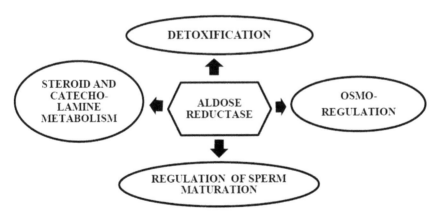

FIGURE 7.4   Physiological roles of aldose reductase (AR).

7.3.4 PATHOLOGICAL SIGNIFICANCE OF ALDOSE REDUCTASE

AR is also associated with secondary diabetic complications, inflammation, mood disorders, cardiovascular disorders, ovarian abnormalities, renal insufficiency and cancers [8].

7.3.5 DIABETIC COMPLICATIONS

AR has a major part in the development of secondary diabetic complications such as retinopathy, cataract, nephropathy, and neuropathy in cases of prolonged diabetes. Diabetic complications are believed to arise as a result of sorbitol accumulation in cells due to hyperactivity of AR. This sorbitol brings about osmotic stress in the cell and thus causes damage to tissues like lens, kidney, retina, and nerves. Moreover, upregulation of polyol pathway also activates and affects different pathways related to the cofactors like NADPH, NADH, and intermediates involved in it as described earlier. This leads to the building up of different types of stresses such as: osmotic stress, oxidative stress, glycative stress, and reductive stress, thus disturbing the normal functioning of the cells thereby resulting in cellular damages [8, 21].

7.3.6 CARDIOVASCULAR DISORDERS

Aberrant AR activity is also reported in cardiovascular diseases. An increase in the activity of ALR2 by activation by endogenous factors like NO production was found during ischemia. Inhibition of the enzyme seems to reduce injury during ischemia, retinosis, and atherosclerosis [52, 112].

7.3.7 INFLAMMATION

AR plays a mediator role in inflammation reaction. Signaling induced by stimuli like high glucose, growth factors, cytokines or bacterial endo-toxins leads to an increase in ROS. Membrane lipids are then oxidized by ROS into lipid aldehydes like 4-hydroxynonemal (HNE). HNE reacts with glutathione and gets converted into GSH-HNE conjugate. GSH-HNE conjugate is reduced to glutathionyl 1,4 dihydroxynonene (GS-DHN) by AR. GS-DHN activates phospholipase C (PLC) and PKC and facilitate the activation of NFkβ and activator protein 1 (AP-1). These transcription factors enter the nucleus and transcribe various inflammation marker genes, which cause inflammation. Diabetes is linked to increased levels of IL-8, IL-6, TNF-α and CRP. However, under sepsis condition, no differences are observed in circulating cell surface markers or coagulation markers in patients with or without diabetes [8, 44, 131, 132].

7.3.8   MOOD DISORDERS

Increased flux of the polyol pathway in the brain was reported in patients with unipolar and bipolar mood disorders. A higher level of sorbitol was observed in the cerebrospinal fluid (CSF) of these patients [8].

7.3.9   CANCER

Overexpression and increased activity of AR can also be seen in different tumors of the breast, ovarian, liver, cervical, and rectal tissues. The exact function of ALR in cancerous tissue is unknown, but it is related to the resistance of cells to chemotherapy [8].

7.4   ALDOSE REDUCTASE (AR) INHIBITION: A THERAPEUTIC APPROACH

Increased AR activity associated with persistent hyperglycemia is one of the major causes for the development of diabetic complications. Along with that, studies have established its involvement in many other pathophysiological conditions like atherosclerosis, asthma, different human cancers, sepsis, uveitis, and other inflammatory conditions also. Therefore, blocking the action of this enzyme is suggested as a promising strategy for the management of these diseases. This suggestion was further supported by experimental proofs from studies using ARIs. Based on these studies, a number of ARIs have been developed and analyzed *in vitro* and *in vivo* for their efficacy by different research groups. Potential inhibitors identified in these experiments were further validated for its potency as a drug, and some of them reached up to phase-III clinical trials. Considering the myriad role played by AR in human physiology, ARIs can affect the physiology in more than one way. Also, the different role of AR in physiological processes can predict the multiple effects an ARI molecule could possibly have.

Studies provide evidences for the multifaceted effects of ARI molecules. Apart from their role in dealing with diabetic complications, their safe use as anti-inflammatory drugs, potent growth inhibitor in proliferating cells and many more that have been reported are discussed below. Therefore, key inflammatory pathologies like cancer and cardiovascular diseases may use ARIs in down-regulating and relieving a few of the main health concerns worldwide [8, 143].

7.4.1 ALDOSE REDUCTASE INHIBITORS (ARIS) IN DIABETIC RETINOPATHY

High level of sorbitol within retina causes osmotic stress, resulting in excessive hydration and gain in Na^+ and loss in K^+ ions. Fructose-3-phosphate and 3-deoxyglucosone, by-products of the polyol pathway are great glycosylating agents which aid in the production of AGEs. AGEs are well-known factors that contribute to the pathogenicity of diabetic retinopathy [10, 81]. ALR2 is associated with diabetic macular pucker *via* activation of proteins like $NF\kappa\beta$ and monocyte chemotactic protein–1 (MCP–1) [98, 119]. Elevated activity of ALR2 leads to oxidative stress, up-regulates retinal VEGF and activates poly (ADP-ribose) polymerase (PARP), which may lead to cataract formation and diabetic retinopathy [99]. Involvement of AR in the pathogenesis diabetic retinopathy was validated by genetic level studies as well. Single nucleotide polymorphism in the promoter region of AR was shown to increase the tendency to bring about diabetic retinopathy in Japanese DM patients. [22]. Restriction fragment length polymorphism (RFLP) analyses on Japanese and Chinese type 2 DM patients suggested the involvement of the AR promoter region (the microsatellite DNA 59) in the early development of diabetic retinopathy. However, the study did not find any connection between the AR promoter and diabetic nephropathy [22, 81].

In many animal models, ARIs prevented early complement activation in the retinal vessel walls, apoptosis of vascular pericytes or endothelial cells and the development of a cellular capillaries [29]. A well-known ARI-Fidarestat prevents retinal oxidative-nitrosative stress and PARP formation in DM [32]. The VEGF levels in the ocular fluid and extensive damages in the region of the optic disc were reduced by Fidarestat administration in Spontaneously Diabetic Torii (SDT) rats. Ranirestat (another ARI) was seen to suppress accumulation of VEGF and Nε-(carboxymethyl) lysine in the retina of SDT [68]. In SDT rats, Ranirestat was also seen to reduce the retinal thickness and the area of stained glial fibrillary acidic protein (GFAP). Thus, potent ARIs were shown to prevent the development of diabetic retinopathy [32, 81, 89, 134] efficiently.

7.4.2 ALDOSE REDUCTASE INHIBITORS IN DIABETIC NEPHROPATHY

Physiological roles of AR involve regulation of osmotic pressure in the renal medulla, and detoxification of aldehyde compounds [100, 119]. Interaction of TGF-β expression with RAS and PKC activity has its role in the development of diabetic nephropathy [100, 119]. Hyperglycemia and the renin-angiotensin-aldosterone system (RAAS) may be the primary metabolic and hemodynamic drivers, respectively [99]. Diabetes-driven mesangial ECM accumulation leads to diabetic nephropathy. Therefore, data suggesting hyperglysolia-mediated mesangial expansion is inhibited by ARIs or that excessive glomerular metabolism via AR is profibrotic [97, 98]. Treatment of moderately diabetic rat with ARI averted mesangial expansion [21].

Glomerular morphometric analysis after the induction of diabetes established that either ARI or insulin treatment reduced 4.7% relative increase in the mesangial matrix by 80 to 90%. A general explanation for the mechanism by which ARI prevents mesangial expansion is that metabolism and gene transcription is activated by hyperglysolia and that autocrine signaling by TGF-β plays a key role to increase ECM synthesis and decrease ECM degradation [97]. ARIs inhibit glucose and AGE-induced activation of TGF-β in human and rodent mesangial cells. In addition, zopolrestat (an ARI) is seen to inhibit TGF-β upregulation by AGE in murine glomeruli cells [151]. Studies have analyzed many different ARIs like sorbinil, tolrestat, and ponalrestat in diabetic animals. In galactosemic animals, tolrestat given orally have been seen to reduce capillary lesions, pericyte degeneration, capillary dilation, endothelial cell proliferation, microaneurysm formation, and acellularity.

Albuminuria development and glomerular basement membrane thickening (GBMT) in diabetic rats were seen to be prevented by the administration of sorbinil and ponalrestat. Short and long duration clinical studies using ARIs in diabetic complications have reported the potential of ARIs in delaying the progression of the disease and also reduced urinary albumin excretion rate implicating the usefulness of ARIs in managing diabetic nephropathy [66]. ARIs capable of suppressing biomarkers of diabetes-induced oxidative stress and renal fibrosis would definitely be a welcome approach in dealing with diabetic nephropathic complications.

7.4.3   ALDOSE REDUCTASE INHIBITORS IN DIABETIC NEUROPATHY

Excessive accumulation of sorbitol causes the decrease of *myo*-inositol content in the peripheral nerves, which leads to changes in the levels of K^+ and Na^+, and ATPase pump activity present on the axonal membrane [17]. Functional impairment resulting from intra axonal Na^+ accumulation and reduction in trans-membrane Na^+ conduction alters the membrane potential that eventually leads to axonal death. Hyperactivity of the polyol pathway, advanced glycation end products (AGE) formation, PKC activation and uncoupling of the mitochondrial respiratory chain augment oxidative stress. Together, these factors lead to the development of diabetic neuropathy [23].

In peripheral nerves, AGEs interfere with axonal transport. Tissue damage by AGEs could be due to their reactive nature and protein cross-linking ability. Mesangial cells, endothelial cells, fibroblasts, and macrophages express receptors for AGE (RAGE) and are prone to AGE-mediated changes. The uptake of AGE-modified proteins by a macrophage/monocyte RAGE triggers the release of IL-1, IGF-1, TNF-α and platelet-derived growth factor and bring about cytokine-mediated changes. Endothelial RAGEs internalize AGE and take them to the sub-epithelium. This increases their permeability and endothelium-dependent coagulant activity. AGE also causes oxidative and other changes in lipo-proteins and RBCs [17].

The results of different clinical trials carried out in the past 15–20 years analyzing the effect of some selected ARIs in managing diabetic neurop-athy are ambiguous. Both satisfactory and disappointing results have been reported from these studies. Even for the ARIs that provided satisfac-tory results, a proper explanation about mechanisms involved is scanty. Therefore, this line of investigation is going on for obtaining adequate information in order to get the approval for conducting clinical trials. Even then, some of the studies provide some insight into the preventive potential of ARIs on the development of diabetic neuropathy. The action of ARIs was shown to improve blood flow to nerves and reduce sorbitol levels. Zenarestat (a potent ARI) considerably decreased the rat sciatic nerve sorbitol levels. Another ARI epalrestat prevented the development of diabetic cardiovascular autonomic neuropathy [56]. Fidarestat was seen to reverse the accumulation of erythrocyte sorbitol content. Fidarestat also showed desirable effects on the neuronal pathways that are considered to be affected in diabetic neuropathy. Moreover, it improved the adverse effects of neuropathy [13, 32].

A study using the ARI ranirestat concluded that this ARI can enter the sural nerve and prevent sorbitol accumulation; bringing relief to patients suffering from diabetic sensorimotor polyneuropathy [119]. Enthused by these positive benefits, Ranirestat is currently taken for phase III clinical trials in the United States. But when the available data on experimental evidence on the positive benefits are analyzed, among all studied ARIs, Epalrestat is the one that seems to be very effective and well suitable for long-term treatment. Though it has its own drawbacks, studies suggest that medication with Epalrestat reduces the severity of diabetic complications. High glucose-induced VSMCs, when treated with Epalrestat, showed promising results implicating its antiproliferative and antihypertrophic effects. Besides, it reduced both disease-associated elevations in intracellular NADH/NAD+ ratio and membrane-bound PKC activation [58, 65, 119].

7.4.4 ALDOSE REDUCTASE INHIBITORS IN NON-DIABETIC PATHOLOGIES

AR inhibition is considered a new treatment modality for diverse types of diseases associated with inflammation [112]. A number of cardiac disorders (like ischemic heart disease, heart failure, reperfusion injury, ventricular hypertrophy, and heart muscle disease) have been associated with increased ROS production and lipid peroxidation in the myocardium. Number of studies has been conducted in order to establish the part played by ALR2 in heart diseases, myocardial metabolism, inflammation, renal failure, ovarian abnormalities, CNS disorders and human cancers like breast, cervical, liver, ovarian, and rectal cancers [16, 33, 47, 59, 97, 107, 112, 119, 125, 130]. ARIs can be used in for treating inflammatory airway diseases, endotoxin-mediated inflammatory diseases and ocular inflammatory response such as uveitis [112]. In studies with transgenic mice expressing human AR (ARTg), AR was shown to have a major role during ischemia-reperfusion injury.

ARIs, by inhibiting AR, aid in reducing ischemic injury and decreasing the cytosolic NADH/NAD ratio. Cardiac ALR2 activity is correlated to ischemia and inhibition of ALR2 can reduce the elevated NADH/NAD ratio leading to increased glycolysis and ATP production. This normalizes myocardial glucose metabolism. ARIs have been shown to be beneficial in improving cardiac functioning and preventing myocyte damage in myocardial infarction if treated during the pre-ischemic stage, whereas, during the reperfusion phase, it could limit the myocyte injury considerably by decreasing the ROS production [57].

Proliferating cells of vascular smooth muscle cells (VSMC) and neointima expressed high levels of AR compared to inactive cells, indicating the upregulation of AR during growth. Cell growth regulatory function of AR was established in VSMC, based on the finding that ARIs attenuated the proliferation of VSMCs by preventing the TNF induced activation of NFκβ. AR-mediated generation of CML-AGEs and MG, a key precursor of AGEs, could affect endothelial-dependent relaxation (EDR) via RAGE. Impaired EDR in response to acetylcholine and increased MG levels in the aortas were seen in older rats compared to the young ones. Inhibition of AR and RAGE improves EDR in aged rats. Zopolrestat treatment significantly reduced MG levels and improved EDR in aged rats [48, 111, 146].

Role of ALR2 in the progression of colon cancer is linked to the growth factor-induced expression of COX–2. Increased COX–2 can cause altered prostaglandin synthesis and can lead to a proliferation of colon epithelial cells [115, 135]. Also, oxidative stress and ROS generation play roles in the activation of NFκβ by means of growth factors. ARI inhibition may be a good method for combating colon cancer as ALR2 was found to mediate the altered inflammatory brought about by the growth factors and inflammatory cytokine-induced cytotoxicity in colon cancer [135].

Considering the involvement of AR in etiology of various diseases prevalent in the 21st century, efforts are being taken to identify new efficient ARI molecules of well-defined structures that could be suited to use for the management of diverse diseases.

7.5 ALDOSE REDUCTASE INHIBITORS (ARIS)

Because of the reasons already mentioned, AR enzyme became a target for inhibition and subsequently, researchers focused more on the development of potential inhibitor molecules against the enzyme. As studies demonstrated the involvement of AR enzyme in diabetes pathology and improvement in diabetic subjects on AR inhibitor therapy, development of a potential inhibitor for AR is proposed as a pharmacological approach to treat the complications associated with hyperglycemia. As a result, a large variety of ARIs have been identified as potent in inhibiting the enzyme [30, 65].

7.5.1 CLASSIFICATION OF ARIS

7.5.1.1 CYCLIC IMIDES

Sorbinil is a cyclic imidine (spirohydantoin) compound which was studied comprehensively for AR inhibition. Experiments proved it to be an acceptable ARI in both *in vitro* and *in vivo* analyses. But during clinical trials, upon sorbinil administration, nearly 10% of patients developed hypersensitivity reactions in the form of skin rashes, myalgia, and fever. This was suggested to be due to the production of a toxic metabolic intermediate of the ARI. Sorbinil also showed more affinity towards aldehyde reductase compared to AR, and this made things further worse. While, the barriers of side effects and specificity hampered the development of this drug, sorbinil has made its identity as a notable leading inhibitor or a reference standard in the AR and polyol pathway research particularly for the development of new ARIs [150].

Hydantoin moiety was blamed for the severe hypersensitivity reaction observed on sorbinil administration. Hydantoins have some side effects like skin rash and hypersensitivity or liver toxicity which make it an undesirable moiety [46]. In order to resolve the hypersensitivity problems associated with hydantoins, hydantoin bioisosteric moiety such as thiazolidinedione was introduced. They caused lesser hypersensitivity when compared to hydantoins and additionally, possessed promising anti-hyperglycemic efficacy. So, compounds having thiazolidinedione (bioisoster of hydantoin moiety) were preferred as ARIs.

Derivatives of sorbinil such as methosorbinil, fidarestat (amide analog of sorbinil), and minalrestat have also been reported. Among this, a recently introduced improved derivative of sorbinil, fidarestat was 10 times more potent than sorbinil. In patients with diabetic neuropathy, it normalized sorbitol levels in erythrocytes without exhibiting observable side effects after continuous administration for one year. However, it did not attain worldwide recognition as a drug for preventing or delaying secondary diabetic complications.

Despite a promising ARI potential, Imirestat, another related ARI was withdrawn because of toxicity problems during clinical trials. However, cyclic imides ranirestat and fidarestat are gaining an identity as beneficial ARIs in the treatment of diabetic complications [37, 119, 150].

7.5.1.2   CARBOXYLIC ACID DERIVATIVES

Flexible carboxylate moieties present in a molecule along with other characters -that allow it to bind favorably to AR enzyme- make these molecules excellent ARIs. Epalrestat, a carboxylic acid moiety containing compound is the only ARI that has been given marketing approval to be used as a therapeutic drug, particularly in the clinical treatment of diabetic neuropathy. Another strong ARI, tolrestat was approved in a few markets, but severe liver toxicity and death risks led to the withdrawal of the compound.

The other carboxylic acid derivatives include: zenarestat, ponalrestat, and zopolrestat. Zenarestat is a beneficial drug used in the management of cataracts, diabetic retinopathy, and neuropathy. Though all the three drugs-zenarestat, ponalrestat, and zopolrestat- entered clinical trials because of their high ARI potential, but ponalrestat was taken off from clinical trials due to certain deficiencies like poor tissue penetration and low potency *in vivo*. Generally, carboxylic acid derivatives are considered less effective *in vivo* [8, 17, 21, 56, 98, 114].

7.5.1.3   OTHER CLASSES

Tetrahydropyrrolo [1, 2-a] pyrazine–1,3-dione derivatives, 5-arylidene–2,4-thiazolidinedione derivatives, N-nitromethyl sulfonanilide derivatives, sulfonylpyridazinone derivatives, N-(3,5- diflurophenol–4-hydroxyphenyl) derivatives, and carboxymethylated pyridoindole derivatives represent some other chemical classes, which can bind and inhibit AR. 2-Thioxo-4-thiazolidinone; and its N-unsubstituted analogs possess AR inhibitory effects at low micromolar doses. Addition of acetate side chain on N–3 of thiozolidinone backbone produced an analog with a very low IC_{50} value comparable to that of epalrestat.

Other examples of effective ARIs include ranirestat that entered the phase II clinical trial [84, 119]. A new ARI belonging to a new class of pyridazizones -ARI–809 was also found to be successful in preventing the development of retinal abnormalities associated with diabetic retinopathy.

7.5.1.4   ALDOSE REDUCTASE INHIBITORS FROM PLANT EXTRACTS

ARI molecules were also identified from natural sources like plant metabolites and compounds of fungal or bacterial origin; these molecules are believed to be less toxic and cost-effective. Plants provide diverse molecules having special structures as potential ARIs. Many plant extracts have been reported to have AR inhibitory activity. Mostly, plant-derived flavonoids and related components are studied for their ARI activity. For example, flavone constituents 3,'4,' dihydroxyflavone, 3,'4,'7-trihydroxyflavone, luteolin, and luteolin 7-O-β-D-glucopyranoside potently inhibited AR enzyme activity with IC_{50} values of 0.37, 0.30, 0.45, 0.99 μM respectively [75]. Many plant-derived compounds have been identified as potent ARIs and research provide proof for their efficacy in treating secondary diabetic complications. One example is β-glucogallin isolated from *Emblica officinalis*. It is similar to other ALR2 active site inhibitors in their mechanism of action. It inhibits AR in an uncompetitive or noncompetitive fashion. Favorable binding of β-glucogallin at the active site of AKR1B1 was demonstrated by molecular docking studies. [110].

Experimental studies have revealed the AR inhibitory potential of many commonly used dietary sources like fennel, lemon, spinach, cumin, black pepper and basil [30, 120]. Apart from that, ether, and methanolic extracts of *Gentiana lutea* roots were also shown to reduce sorbitol accumulation dose-dependently in human erythrocytes under hyperglycemic conditions [5]. Ethyl acetate fraction of *Hybanthus enneaspermus* Linn F. Muell with a very high flavonoid and phenolic content also exhibited excellent ARI potency [108]. *Houttuynia cordata* Thunb, *Colocasia esculenta* (L.), *Butea monosperma*, *Psidium guajava*, *Balanites aegyptiaca*, and *Vernonia anthelmintica* are some of the recently reported ARI potent plants [1, 11, 41, 51, 74, 78].

Microorganisms are also great sources for diverse biologically active molecules. More than 20 kinds of microbial ARIs have been reported. Many of these ARIs are proved to be effective, but they exhibited toxicity, less selectivity and this limited their clinical use. Examples are: WF–3681; fungal metabolite from *Chaetomella raphigera;* benzothiazole derivative from *Actinosynnema spp.* and *Paecilomyces lilacinus*; Aldostatin, a fungal metabolite from *Pseudeurotium zonatum;* YUA001 from the supernatant of cultured broth of *Corynebacterium sp.*; and sclerotiorin from *Penicillium frequentans* CFTRI A–24, etc. [30, 121, 147]. Extracts from some of the sponges (like *Dyidea sp.*, *Ircinia ramosa*, *Dactylospongia metachromia*, and *Hippospongia* sp. and red algae like *Asparagopsis taxiformis*) are also reported to have ARI activity [30].

7.5.2 ARI MOLECULES WITH MULTIPLE ANTI-DIABETIC EFFECTS

Diabetes is a multifactorial disease, and hence, drug molecule that can act in more than one way to ameliorate the symptoms would be appreciable. Some of the ARI molecules reported so far exhibit other beneficial properties like anti-hyperglycemic, antioxidant, and immune modulation activities apart from AR inhibition. Reported ARIs with a multitude of beneficial actions in diabetes are shown in Table 7.1.

There are many potent ARIs reported till date. But they all have one or other side effects, less selectivity and slower tissue penetration causing ineffective action. Carboxylic acid derived ARIs possess lower activity and lack the ability to penetrate physiological membranes while spirohydantoin derivatives show higher activity *in vivo* but is limited by hypersensitivity reactions. The only marketed ARI, Epalrestat, is also reported to have adverse effects like the elevation of liver enzymes, nausea, vomiting, hepatic dysfunction, hepatic failure, and jaundice. Experimental studies to rectify the limitations in the molecular interactions of different structures in binding to AR are also yielding promising results. Modifications such as addition or deletion of certain groups could alter the interaction of the ARI with AR in a positive manner. For instance, nitrosothiols cause multiple changes in the structure and function of AR.

Molecular docking analysis of a number of molecules suggested that besides carboxylic acid derivatives, other members of families such as sulfonic acids, nitro-derivatives, sulphonamides, and carboxyl derivatives are also putative inhibitors of AR [72, 113]. However, even though several therapeutic leads have been identified as having AR inhibitor activity, no molecules from natural sources have been made available as a drug in the market as such. Different reviews have analyzed various reports and included information regarding a number of ARI potential molecules. Both synthetic and natural molecules are classified and tabulated based on their chemical nature in many previous reviews [30, 119, 138].

TABLE 7.1 Some Reported ARIs with Anti-Diabetic Effects Other Than AR Inhibition

ARI Molecule	Source	Anti-diabetic effects other than AR inhibition	Reference
Apigenin	Plant	Facilitates GLUT4 translocation.	[55, 139, 164]
		Possess anti-oxidant and anti-hyperglycemic effects.	
Baicalein	Plant	Improves glucose tolerance and survival of islet β-cell.	[3, 40, 138, 139, 167]
		Possess anti-inflammatory, anti-oxidant, anti-hyperglycemic, and nephroprotective effects. Normalizes the liver function enzymes and the levels of serum pro-inflammatory cytokines.	
		Reduce AGEs and TNF-α level. Improves insulin resistance, counteracts the phosphorylation of protein kinase triggered by DM-mediated activation of p38 mitogen, oxidative, and nitrosative stress and activate lipoxygenase.	
Berberine	Plant	Alleviate hyperglycemia and ameliorate insulin resistance.	[80, 168]
Chlorogenic acid	Plant	Antihyperalgesic anti-oxidant and anti-inflammatory effects. Improves glucose tolerance and insulin sensitivity	[87, 103, 180]
Curcumin	Plant	Antihyperalgesic, antioxidant, and anti-inflammatory	[93, 116, 176]
Ellagic acid	Plant	Neuroprotective effects by the antioxidant property. Inhibits α- amylase	[75, 138, 173]
Epalrestat	Synthetic	Reduce platelet aggregation, antiproliferative, and anti-hypertrophic effects on VSMCs.	[58, 68, 157]
Epigallocatechin-gallate	Plant	Inhibits oxidative stress.	[75, 105, 178]
Eugenol	Plant	Improves vascular and nerve function. Anti-oxidant.	[105, 177]
Ferulic acid	Plant	Reduce leukocyte infiltration and inhibit endothelial pyknosis and ROS formation. Alleviate hypertension associated with diabetes	[14, 75, 174]
Fidarestat	Synthetic	Prevents activation of MAPK, improves contractile dysfunction and normalizes Ca^{2+} signaling.	[13, 68, 99, 156]

TABLE 7.1 (Continued)

ARI Molecule	Source	Anti-diabetic effects other than AR inhibition	Reference
Kaempferol	Plant	Ameliorate glycoprotein abnormalities	[9, 113, 179]
Minalrestat	Synthetic	Corrects the impaired responses to inflammatory mediators.	[4, 163]
Naringenin	Plant	Inhibits intestinal α-glucosidase. Reduces oxidative damage.	[35, 49, 92, 104, 139, 169]
		Decreases cholesterol and cholesterol ester synthesis.	
		Improves insulin sensitivity and glucose tolerance.	
		Decrease hyperglycemia and increase antioxidant enzyme (SOD). Suppress carbohydrate absorption from the intestine, thereby reducing blood GLU levels.	
		Neuroprotective.	
Puerarin	Plant	Dilates blood vessels and improves microcirculation.	[108, 149, 171]
		Decreases blood thickness leading to increased nerve conductivity.	
Quercetin	Plant	Regenerates the pancreatic islets and probably increases insulin release.	[6, 91, 138, 175]
		Inhibits NFκβ and caspase 3 expressions.	
		Ameliorates hyperglycemia and oxidative stress.	
		Prevents β-cell death by exerting its anti-inflammatory, antioxidant, and anti-apoptotic effects.	
		Accelerates the function of glucose transporter 4 (GLUT 4) and insulin receptor	
Ranirestat	Synthetic	Inhibits NET formation	[89, 161]
Resveratrol	Plant	Scavenging activity on ROS and vasorelaxant.	[28, 73, 144, 172]
		Inhibits NF-κB.	
Rosmarinic acid	Plant	Anti-oxidant, anti-inflammatory, and neuroprotective effects. Inhibits NF-κB activation.	[72, 165]

TABLE 7.1 *(Continued)*

ARI Molecule	Source	Anti-diabetic effects other than AR inhibition	Reference
Rutin	Plant	Inhibits inflammatory cytokines. Improves anti-oxidant and lipid profiles. Decreases glucose, caspase 3 and TBARS. Increases insulin and Bcl-2 protein. Decreases oxidative stress Decreases MDA levels. Increases SOD and CAT.	[62, 95, 114, 170]
Sorbinil	Synthetic	Prevents platelet-derived growth factor and fibroblast growth factor-mediated increase in the synthesis of prostaglandin E2 (PGE2). Restores normal levels of glutathione and glycerol 3-phosphate.	[66, 99, 135, 158]
Tolrestat	Synthetic	Reverts capillary lesions, pericyte degeneration, endothelial cell proliferation, capillary dilation, acellularity, and microaneurysm formation in galactosemic animals.	[107, 138, 162]
Trans-cinnamaldehyde	Plant	Antihyperglycemic effects, increase serum insulin levels, increase insulin-receptor signaling	[108, 148, 166]
Zenarestat	Synthetic	Alleviate nerve dysfunctions	[21, 56, 127, 160]
Zopolrestat	Synthetic	Reduce leukocyte infiltration and inhibits endothelial pyknosis and ROS formation. Alleviate hypertension associated with diabetes	[14, 159]

7.6   FUTURE PROSPECTS

The use of ARIs is a beneficial approach for preventing or delaying the development of secondary diabetic complications. Future studies on ARI should evaluate the AR inhibitory potential molecules for antidiabetic effects other than AR inhibition (e.g., hypoglycemic action, insulinogenic action, etc.) as well. This helps the patient to manage DM more effectively by reducing the amount of drug needed for alleviating different symptoms. Furthermore, current research needs to focus on the potential of these molecules in alleviating other diabetes-related complications (like oxidative stress, CVDs, AGE formation, etc.). Moreover, since AR is having its role in different non-diabetic pathologies as well, further analysis of ARIs for their different possible actions may unravel its potential in novel therapeutic options that are more efficient in combating diabetic complications.

7.7   SUMMARY

Secondary diabetic complications have been linked to the biochemical imbalance caused by AR enzyme. Apart from hyperglycemia, diabetes is also complicated with osmotic stress due to elevated AR activity, oxidative stress, increased glycation end product formation, activation of PKC pathway, activation of hexosamine pathway and altered innate immune function, which includes neutrophil dysfunctions, monocyte dysfunctions, altered cytokine release and augmented NETosis. AR inhibition proves to be a welcome strategy for dealing with all of these diabetic complications. Inhibitors of AR have been effective in delaying and even preventing several diabetic pathologies, including diabetic retinopathy, neuropathy, and nephropathy. The role of ARIs in preventing hyperglycemia-mediated glycation, oxidative stress and immune functions, ROS production, extracellular trap formation, etc., are discussed here with reference to outcomes from different experimental studies.

ACKNOWLEDGMENT

We gratefully acknowledge the financial assistance of DBT under DBT-MSUB IPLSARE program (No.BT/P44800/INF/22/152/2012; dated 23/03/2012).

KEYWORDS

- **aldose reductase inhibition**
- **LADA**
- **neutrophils dysfunction**
- **protein kinase C pathway**
- **sorbitol**
- **sorbitol dehydrogenase**

REFERENCES

1. Abdel, M. A., El-Askary, H., Crockett, S., Kunert, O., Sakr, B., Shaker, S., et al., (2015). Aldose reductase inhibition of a saponin-rich fraction and new furostanol saponin derivatives from *Balanites aegyptiaca*. *Phytomedicine, 22*(9), 829–836.
2. Adler, A. I., Stevens, R. J., Manley, S. E., Bilous, R. W., Cull, C. A., & Holman, R. R., (2003). Development and progression of nephropathy in type 2 diabetes: The United Kingdom Prospective Diabetes Study (UKPDS 64). *Kidney Int., 63*(1), 225–232.
3. Ahad, A., Mujeeb, M., Ahsan, H., & Siddiqui, W. A., (2014). Prophylactic effect of baicalein against renal dysfunction in type 2 diabetic rats. *Biochimie., 106*, 101–110.
4. Akamine, E. H., Hohman, T. C., Nigro, D., Carvalho, M. H., Tostes, R. C., & Fortes, Z. B., (2003). Minalrestat, an aldose reductase inhibitor, corrects the impaired microvascular reactivity in diabetes. *J. Pharmacol. Exp. Ther., 304*(3), 1236–1242.
5. Akileshwari, C., Muthenna, P., Nastasijevi, B., Joksi, G., Petrash, J. M., & Reddy, G. B., (2012). Inhibition of aldose reductase by *Gentiana lutea* extracts. *Exp. Diabetes Res.,* 1–8.
6. Alam, M. M., Meerza, D., & Naseem, I., (2014). Protective effect of quercetin on hyperglycemia, oxidative stress and DNA damage in alloxan induced type 2 diabetic mice. *Life Sci., 109*(1), 8–14.
7. Alba-Loureiro, T., Munhoz, C. D., Martins, J. O., Cerchiaro, G., Scavone, C., & Curi, R., (2007). Neutrophil function and metabolism in individuals with diabetes mellitus. *Brazilian J. Med. Biol. Res., 40*(8), 1037–1044.
8. Alexiou, P., Pegklidou, K., Chatzopoulou, M., Nicolaou, I., & Vassilis, J., (2009). Aldose reductase enzyme and its implication to major health problems of the 21st century. *Curr. Med. Chem., 16*(6), 734–752.
9. Alkhalidy, H., Moore, W., & Zhang, Y., (2015). Small molecule kaempferol promotes insulin sensitivity and preserved pancreatic β cell mass in middle-aged obese diabetic mice. *J. Diabetes Res.*, e–53298.
10. American Diabetes Association, (2013). Diagnosis and classification of diabetes mellitus. *Diabetes Care, 36*(1), 67–74.

11. Anand, S., Arasakumari, M., Prabu, P., & Amalraj, A. J., (2016). Anti-diabetic and aldose reductase inhibitory potential of *Psidium guajava* by in vitro analysis. *Int. J. Pharm. Pharm. Sci.*, *8*(9), 271–276.

12. Andrew, J. M. B., Malik R., Arezzo J. C., & Sosenko, J. M., (2004). Diabetic somatic neuropathies. *Diabetes Care*, *27*(6), 1458–1486.

13. Asano, T., Saito, Y., & Kawakami, M., (2002). Fidarestat (SNK–860): A potent aldose reductase inhibitor, normalizes the elevated sorbitol accumulation in erythrocytes of diabetic patients. *J. Diabetes Complications*, *16*, 133–138.

14. Badawy, D., El-Bassossy, H. M., Fahmy, A., & Azhar, A., (2013). Aldose reductase inhibitors zopolrestat and ferulic acid alleviate hypertension associated with diabetes: Effect on vascular reactivity. *Can. J. Physiol. Pharmacol.*, *91*(2), 101–107.

15. Bagdade, J. D., & Nielson, K. L., (1972). Reversible abnormalities in phagocytic function in poorly controlled diabetic patients. *Am. J. Med. Sci.*, *263*(6), 4403194.

16. Berry, G. T., (1995). The role of polyols in the pathophysiology of hypergalactosemia. *Eur. J. Pediatr.*, *154*(7 Suppl. 2), S53–S64.

17. Bhadada, S., Sahay, R., Jyotsna, V., & Agrawal, J., (2001). Diabetic neuropathy : Current concepts. *Indian Acad. Clin. Med.*, *2*(4), 305–319.

18. Borhani, D. W., Harter, T. M., & Petrash, J. M., (1992). The crystal structure of the aldose reductase. NADPH binary complex. *J. Biol. Chem.*, *267*(34), 2–3.

19. Božidar, V., Tamara, T., ŽC, O., & Rački, G. Đ., (2008). Diabetic nephropathy: Pathophysiology and complications of diabetes mellitus stage. *Diabetes Care*, *31*(4), 823–827.

20. Brenner, Z. R., (2006). Management of hyperglycemic emergencies. *AACN Clin. Issues*, *17*(1), 56–65.

21. Brownlee, M., (2001). Biology of diabetic complications. *Nature*, *414*, 813–820.

22. Busik, J. V., & Grant, M. B., (2014). Aldose reductase meets histone acetylation: A new role for an old player. *Diabetes*, *63*(2), 402–404.

23. Cameron, N. E., & Cotter, M. A., (1993). Potential therapeutic approaches to the treatment or prevention of diabetic neuropathy: Evidence from experimental studies. *Diabet. Med.*, *10*(0742–3071, SB-M), 593–605.

24. Cassatella, M. A., (1999). Neutrophil-derived proteins: Selling cytokines by the pound. *Adv. Immunol.*, *73*, 369–509.

25. Chandramohan, G., Al-Numair, K. S., Alsaif, M. A., & Veeramani, C., (2014). Antidiabetic effect of kaempferol a flavonoid compound, on streptozotocin-induced diabetic rats with special reference to glycoprotein components. *Prog. Nutr.*, *17*(1), 50–57.

26. Chang, F. Y., & Shaio, M. F., (1995). Respiratory burst activity of monocytes from patients with non-insulin-dependent diabetes mellitus. *Dia. Res. Clin. Pract.*, *29*(2), 121–127.

27. Chou, M. Y., Shian, L. R., Chang, F. Y., & Shaio, M. F., (1989). Opsonophagocytosis of *Staphylococcus aureus* by diabetics. *Taiwan Yi Xue Hui Za Zhi*, *88*(4), 1–2.

28. Ciddi, V., & Dodda, D., (2014). Therapeutic potential of resveratrol in diabetic complications: *In vitro* and *in vivo* studies. *Pharmacol. Reports*, *66*(5), 799–803.

29. Dagher, Z., Park, Y. S., Asnaghi, V., Hoehn, T., Gerhardinger, C., & Lorenzi, M., (2004). Studies of rat and human retinas predict a role for the polyol pathway in human diabetic retinopathy. *Diabetes*, *53*(9), 2404–2411.

30. De la Fuente, J. A., & Manzanaro, S., (2003). Aldose reductase inhibitors from natural sources. *Nat. Prod. Rep.*, *20*, 243–251.

31. Devaraj, S., & Jialal, I., (2000). Low density lipoprotein postsecretory modification, monocyte function, and circulating adhesion molecules in type-2 diabetic patients with and without macrovascular complications: The Effect of tocopherol supplementation. *Circulation*, *102*, 191–196.

32. Drel, V. R., Pacher, P., Ali, T. K., Shin, J., Julius, U., & Obrosova, I. G., (2008). Aldose reductase inhibitor fidarestat counteracts diabetes- associated cataract formation, retinal oxidative-nitrosative stress, glial activation, and apoptosis. *Int. J. Mol. Med.*, *21*(6), 667–676.

33. Dunlop, M., (2000). Aldose reductase and the role of the polyol pathway in diabetic nephropathy. *Kidney Int. Suppl.*, *58*(77), S3–S12.

34. Eledrisi, M. S., Alshanti, M. S., Shah, M. F., Brolosy, B., & Jaha, N., (2006). Overview of the diagnosis and management of diabetic ketoacidosis. *Am. J. Med. Sci.*, *331*(5), 243–251.

35. Fallahi, F., Roghani, M., & Moghadami, S., (2012). Citrus flavonoid naringenin improves aortic reactivity in streptozotocin-diabetic rats. *Indian J. Pharmacol.*, *44*(3), 382–386.

36. Fernández, M. C., & Mary, L. L. V. M., (2013). Oxidative stress in diabetes mellitus and the role of vitamins with antioxidant actions. *Oxidative Stress and Chronic Degenerative Diseases- A Role for Antioxidants*, 209–232.

37. Fowler, M. J., (2008). Microvascular and macrovascular complications of diabetes. *Clinical Diabetes*, *26*(2), 77–82.

38. Francesco, C., Francesco, C., Angelika, M., Matteo, M., Annalisa, I., Maria, F., et al., (2002). Circulating monocyte chemoattractant protein–1 and early development of nephropathy in type 1 diabetes. *Diabetes Care*, *25*(10), 1829–1834.

39. Froguel, P., Velho, G., & Vionnet, N., (1994). Genetics and diabetes. *Sang. Thromb. Vaiss.*, *6*(5), 39–46.

40. Fu, Y., Luo, J., Jia, Z., et al., (2014). Baicalein protects against type 2 diabetes via promoting islet beta-cell function in obese diabetic mice. *Int. J. Endocrinol.*, *84*, 6742.

41. Garapelli, R., Rao, A. R., & Veeresham, C., (2015). Aldose reductase inhibitory activity of *Butea monosperma* for the management of diabetic complications. *Pharmacologia*, *6*(8), 355–359.

42. Geerlings, S. C., & Hopelman, A. I., (1999). Immune dysfunction in patients with diabetes mellitus (DM). *FEMS Immunol. Med. Microbiol.*, *26*, 259–265.

43. Geisler, C., Almdal, T., Bennedsen, J., Rhodes, J. M., & Kølendorf, K., (1982). Monocyte functions in diabetes mellitus. *Acta Pathol. Microbiol. Immunol. Scand. C.*, *90*(1), 33–37.

44. Giacco, F., (2011). Oxidative stress and diabetic complications. *Circ. Res.*, *107*(9), 1058–1070.

45. Golub, L. M., Nicoll, G. A., Iacono, V. J., & Ramamurthy, N. S., (1982). *In vivo* crevicular leukocyte response to a chemotactic challenge: Inhibition by experimental diabetes. *Infect Immun.*, *37*(3), 1013–1020.

46. Graham, A., Brown, L., Hedge, P. J., Gammack, A. J., & Markham, A. F., (1991). Structure of the human aldose reductase gene. *J. Biol. Chem.*, *266*(11), 6872–6877.

47. Gyurko, R., Siqueira, C. C., Caldon, N., Gao L., Kantarki, A., & Dyke, V. E. T., (2006). Chronic hyperglycemia predisposes to exaggerated inflammatory response and leukocyte dysfunction in Akita mice. *J. Immunol.*, *177*(10), 7250–7256.

48. Hallam, K. M. C., Li, Q., & Ananthakrishnan, R., (2010). Aldose reductase and AGE-RAGE pathways: central roles in the pathogenesis of vascular dysfunction in aging rats. *Aging Cell*, *9*(5), 776–784.

49. Hasanein, P., & Fazeli, F., (2014). Role of naringenin in protection against diabetic hyperalgesia and tactile allodynia in male Wistar rats. *J. Physiol. Biochem.*, 70(4), 997–1006.

50. Haslbeck, A. M., Luft, D., Neundörfer, B., & Ziegler, D., (2004). *Diagnosis, Treatment and Follow-up of Diabetic Neuropathy* (pp. 1–78). *Updated and summarized version of the guidelines Diagnostics, therapy and follow-up of autonomic and sensorimotor diabetic neuropathy on the websites of the DDG,* http://www. deutsche-diabetes-Gesellschaft. de/EvidenzbasierteLeitlinien/Neuropathie.

51. Hazeena, V. N., Sruthi, C. R., Soumiya, C. K., Haritha, V. H., Jayachandran, K., & Anie, Y., (2016). *Vernonia anthelmintica* (L.) prevents sorbitol accumulation through aldose reductase inhibition. *Sch. Acad. J. Biosci.*, *4*, 787–795.

52. He, Z., Naruse, K., & King, G. L., (2005). Effects of diabetes and insulin resistance on endothelial functions. *Contemporary Cardiology: Diabetes and Cardiovascular Disease*, *5*, 25–46.

53. Hill, H. R., Augustine, N. H., Rallison, M. L., & Santos, J. I., (1983). Defective monocyte chemotactic responses in diabetes mellitus. *J. Clin. Immunol.*, *3*(1), 70–77.

54. Holt, R. I. G., (2004). Diagnosis, epidemiology, and pathogenesis of diabetes mellitus: An update for psychiatrists service. *Br. J. Psychiatry*, *4*, 55–63.

55. Hossain, C. M., Ghosh, M. K., Satapathy, B. S., Dey, N. S., & Mukherjee, B., (2014). Apigenin causes biochemical modulation, GLUT4, and CD38 alterations to improve diabetes and to protect damages of some vital organs in experimental diabetes. *Am. J. Pharmacol. Toxicol.*, *9*(1), 39–52.

56. Hu, X., Li, S., Yang, G., Liu, H., Boden, G., & Li, L., (2014). Efficacy and safety of aldose reductase inhibitor for the treatment of diabetic cardiovascular autonomic neuropathy: Systematic review and meta-analysis. *PLoS One*, *9*(2), e87096.

57. Hwang, Y. C., Sato, S., & Tsai, J. Y., (2002). Aldose reductase activation is a key component of myocardial response to ischemia. *FASEB J.*, *16*(2), 243–245.

58. Iso, K., Tada, H., Kuboki, K., & Inokuchi, T., (2001). Long-term effect of epalrestat, an aldose reductase inhibitor, on the development of incipient diabetic nephropathy in Type 2 diabetic patients. *J. Diabetes Complications*, *15*(5), 241–244.

59. Jialal, I., Devaraj, S., & Venugopal, S. K., (2002). Oxidative stress, inflammation, and diabetic vasculopathies: The role of alpha-tocopherol therapy. *Free Radic. Res.*, *36*(12), 1331–1336.

60. Joshi, M. B., Lad, A., Bharath, P. A. S., Balakrishnan, A., Ramachandra, L., & Satyamoorthy, K., (2013). High glucose modulates IL-6 mediated immune homeostasis through impeding neutrophil extracellular trap formation. *FEBS Lett.*, *587*(14), 2241–2246.

61. Kaku, K., (2010). Pathophysiology of type 2 diabetes and its treatment policy. *J. Japan Med. Assoc.*, *60*(5 & 6), 361–368.

62. Kamalakkannan, N., & Prince, P. S. M., (2006). Antihyperglycaemic and antioxidant effect of rutin, a polyphenolic flavonoid, in streptozotocin-induced diabetic Wistar rats. *Basic Clin. Pharmacol. Toxicol.*, *98*(1), 97–103.

63. Kaneshige, H., Endoh, M., Tomino, Y., Nomoto, Y., Sakai, H., & Arimori, S., (1982). Impaired granulocyte function in patients with diabetes mellitus. *Tokai. J. Exp. Clin. Med.*, *7*(1), 77–80.

64. Karima, M., Kantarci, A., Ohira, T., Hasturk, H., Jones, V. L., Nam, B. H., Malabanan, A., et al., (2005). Enhanced superoxide release and elevated protein kinase C activity in neutrophils from diabetic patients: Association with periodontitis. *J. Leukoc. Biol.*, *78*(4), 862–870.

65. Kashima, K., Sato N., Sato K. E. N., Shimizu, H., & Mori, M., (1998). Effect of epalrestat, an aldose reductase inhibitor, on the generation of oxygen-derived free radicals in neutrophils from streptozotocin-induced diabetic rats. *Endocrinology*, *139*(8), 3404–3408.

66. Kassab, J. P., Guillot, R., & Andre, J., (1994). Renal and microvascular effects of an aldose reductase inhibitor in experimental diabetes. Biochemical, functional and ultrastructural studies. *Biochem. Pharmacol.*, *48*(5), 1003–1008.

67. Kaul, K., Hodgkinson, A., Tarr, J. M., Kohner, E. M., & Chibber, R., (2010). Is inflammation a common retinal-renal-nerve pathogenic link in diabetes? *Curr. Diabetes Rev.*, *6*(5), 294–303.

68. Kawai, T., Takei, I., Tokui, M., et al., (2010). Effects of epalrestat, an aldose reductase inhibitor, on diabetic peripheral neuropathy in patients with type 2 diabetes, in relation to suppression of Nε-carboxymethyl lysine. *J. Diabetes Complications*, *24*(6), 424–432.

69. Kelly, M. K., Brown, J. M., & Thong, Y. H., (1985). Neutrophil and monocyte adherence in diabetes mellitus, alcoholic cirrhosis, uraemia and elderly patients. *Int. Arch. Allergy Immunol.*, *78*(2), 132–138.

70. Kitabchi, A. E., Umpierrez, G. E., Miles, J. M., & Fisher, J. N., (2009). Hyperglycemic crises in adult patients with diabetes. *Diabetes Care*, *32*(7), 1335–1343.

71. Kitahara, M., Eyre, H. J., Lynch, R. E., Rallison, M. L., & Hill, H. R., (1980). Metabolic activity of diabetic monocytes. *Diabetes*, *29*(4), 251–256.

72. Koukoulitsa, C., Bailly, F., Pegklidou, K., Demopoulos, V. J., & Cotelle, P., (2010). Evaluation of aldose reductase inhibition and docking studies of 6'-nitro and 6,'6" -dinitrorosmarinic acids. *Eur. J. Med. Chem.*, *45*, 1663–1666.

73. Kumar, A., & Sharma, S. S., (2010). NF-κB inhibitory action of resveratrol: A probable mechanism of neuroprotection in experimental diabetic neuropathy. *Biochem. Biophys. Res Commun.*, *394*(2), 360–365.

74. Kumar, M., Laloo, D., Prasad, S. K., & Hemalatha, S., (2014). Aldose reductase inhibitory potential of different fractions of *Houttuynia cordata* Thunb. *J. Acute Dis.*, *3*(1), 64–68.

75. Kumar, S., Kumar, V., Rana, M., & Kumar, D., (2012). Review article enzymes inhibitors from plants : An alternate approach to treat diabetes. *Pharmacogn. Commun.*, *2*(2), 18–33.

76. Laugesen, E., Ostergaard, J. A., & Leslie, R. D. G., (2015). Latent autoimmune diabetes of the adult: Current knowledge and uncertainty. *Diabet. Med.*, *32*(7), 843–852.

77. Lecube, A., Pachón, G., Petriz, J., Hernández, C., & Simó, R., (2011). Phagocytic activity is impaired in type 2 diabetes mellitus and increases after metabolic improvement. *PLoS One*, *6*(8), 6–11.

78. Li, H. M., Hwang, S. H., Kang, B. G., Hong, J. S., & Lim, S. S., (2014). Inhibitory effects of *Colocasia esculenta* (L.) Schott constituents on aldose reductase. *Molecules*, *19*(9), 13212–13224.

79. Lim, A. K., (2014). Diabetic nephropathy-complications and treatment. *Int. J. Nephrol. Renovasc. Dis.*, *7*, 361–381.

80. Liu, W., Liu, P., & Tao, S., (2008). Berberine inhibits aldose reductase and oxidative stress in rat mesangial cells cultured under high glucose. *Arch. Biochem. Biophys.*, *475*(2), 128–134.

81. Lorenzi, M., (2007). The polyol pathway as a mechanism for diabetic retinopathy: Attractive, elusive, and resilient. *Exp. Diabesity. Res.*, *2007*, E-article ID–61038.

82. Luo, X., Wu, J., Jing, S., & Yan, L. J., (2016). Hyperglycemic stress and carbon stress in diabetic glucotoxicity. *Aging. Dis.*, *7*(1), 90–110.

83. Lyssenko, V., Jonsson, A., & Almgren, P., (2008). Clinical risk factors, DNA variants, and the development of type 2 diabetes. *N. Engl. J. Med.*, *359*(21), 2220–2232.

84. MacCari, R., Corso, A. D., Giglio, M., Moschini, R., Mura, U., & Ottan, R., (2011). *In vitro* evaluation of 5-arylidene–2-thioxo–4-thiazolidinones active as aldose reductase inhibitors. *Bioorganic Med. Chem. Lett.*, *21*(1), 200–203.

85. Mathebula, S., (2015). Polyol pathway: A possible mechanism of diabetes complications in the eye. *Afr. Vis. Eye Heal.*, *74*(1), 1–5.

86. McManus, L. M., Bloodworth, R. C., Prihoda, T. J., Blodgett, J. L., & Pinckard, R. N., (2001). Agonist-dependent failure of neutrophil function in diabetes correlates with extent of hyperglycemia. *J. Leukoc. Biol.*, *70*(3), 395–404.

87. Meng, S., Cao, J., Feng, Q., Peng, J., & Hu, Y., (2013). Roles of chlorogenic acid on regulating glucose and lipids metabolism: A review. *Evidence-Based Complement Alter Med.*, ID–801457.

88. Min, D., Brooks, B., Wong, J. R., Salomon, R., Bao, W., Harrisberg, B., Twigg, S. M., Yue, D. K., & McLennan, S. V., (2012). Alterations in monocyte CD16 in association with diabetes complications. *Mediators Inflamm.*, *2012*, ID–649083.

89. Miyoshi, A., Yamada, M., Shida, H., et al., (2016). Circulating neutrophil extracellular trap levels in well-controlled type 2 diabetes and pathway involved in their formation induced by high-dose glucose. *Pathobiology*, *83*(5), 243–251.

90. Mogensen, C. E., (1999). Microalbuminuria, blood pressure and diabetic renal disease: Origin and development of ideas. *Diabetologia*, *42*(3), 263–285.

91. Mukhopadhyay, P., & Prajapati, A. K., (2015). Quercetin in anti-diabetic research and strategies for improved quercetin bioavailability using polymer-based carriers - a review. *RSC Adv.*, *5*(118), 97547–97562.

92. Mulvihill, E. E., Allister, E. M., Sutherland, B. G., et al., (2009). Naringenin prevents dyslipidemia, apolipoprotein B overproduction, and hyperinsulinemia in LDL receptor-null mice with diet-induced insulin resistance. *Diabetes*, *58*(10), 2198–2210.

93. Muthenna, P., Suryanarayana, P., Gunda, S. K., Petrash, J. M., & Reddy, G. B., (2009). Inhibition of aldose reductase by dietary antioxidant curcumin: Mechanism of inhibition, specificity and significance. *FEBS Lett.*, *583*(22), 3637–3642.

94. Mysliwska, J., Zorena, K., Bakowska, A., Skuratowicz, K. A., & Mysliwski, A., (1998). Significance of tumor necrosis factor alpha in patients with long-standing type-I diabetes mellitus. *Horm. Metab. Res.*, *30*(0018–5043), 158–161.

95. Niture, N. T., Ansari, A. A., & Naik, S. R., (2014). Anti-hyperglycemic activity of Rutin in streptozotocin-induced diabetic rats: An effect mediated through cytokines, antioxidants and lipid biomarkers. *Indian J. Exp. Biol.*, *52*(7), 720–727.

96. O'Brien, M., (2014). Endocrine emergencies. In: *Handbook of Canine and Feline Emergency Protocols.* (2nd edn., pp. 35–44). John Wiley & Sons.

97. Oates, P. J., (2010). Aldose reductase inhibitors and diabetic kidney disease. *Curr. Opin. Investig. Drugs*, *11*(4), 402–416.

98. Oates, P. J., & Mara, L., (2009). The polyol pathway and diabetic retinopathy. *Contemporary Diabetes: Diabetic Retinopathy*, 3–25.

99. Obrosova, I. G., & Kador, P. F., (2011). Aldose reductase/polyol inhibitors for diabetic retinopathy. *Curr. Pharm. Biotechnol.*, *12*(3), 373–385.

100. Obrosova, I. G., Minchenko, A. G., Vasupuram, R., et al., (2003). Oxidative stress and vascular endothelial growth factor overexpression in streptozotocin-diabetic rats. *Diabetes*, *52*, 864–871.

101. Oishi, N., Kubo, E., Takamura, Y., Maekawa, K., Tanimoto, T., & Akagi, Y., (2002). Correlation between erythrocyte aldose reductase level and human diabetic retinopathy. *Br. J. Ophthalmol.*, *86*(12), 1363–1366.

102. Omori, K., Ohira, T., Uchida, Y. et al., (2008). Priming of neutrophil oxidative burst in diabetes requires preassembly of the NADPH oxidase. *J. Leukoc. Biol.*, *84*(1), 292–301.

103. Ong, K. W., Hsu, A., & Tan, B. K. H., (2013). Anti-diabetic and anti-lipidemic effects of chlorogenic acid are mediated by ampk activation. *Biochem. Pharmacol.*, *85*(9), 1341–1351.

104. Ortiz, A. R. R., Sánchez, S. J. C., Navarrete V. G., et al., (2008). Antidiabetic and toxicological evaluations of naringenin in normoglycaemic and NIDDM rat models and its implications on extra-pancreatic glucose regulation. *Diabetes, Obes. Metab.*, *10*(11), 1097–1104.

105. Ozcan, T., & Delikanli, B., (2014). Phenolics in human health. *International Journal of Chemical Engineering and Applications*, *5*(5), 393–396.

106. Ozougwu, J., Obimba, K., Belonwu, C., & Unakalamba, C., (2013). The pathogenesis and pathophysiology of type 1 and type 2 diabetes mellitus. *J. Physiol. Pathophysiol.*, *4*(4), 46–57.

107. Passariello, N., Sepe, J., Marrazzo, G., et al., (1993). Effect of aldose reductase inhibitor (tolrestat) on urinary albumin excretion rate and glomerular filtration rate in IDDM subjects with nephropathy. *Diabetes Care*, *16*(5), 789–795.

108. Patel, D. K., Kumar, R., Kumar, M., Sairam, K., & Hemalatha, S., (2012). Evaluation of *in vitro* aldose reductase inhibitory potential of different fraction of *Hybanthus enneaspermus* Linn F. Muell. *Asian Pac. J. Trop. Biomed.*, *2*(2), 134–139.

109. Pickup, J. C., & Crook, M. A., (1998). Is type II diabetes mellitus a disease of the innate immune system? *Diabetologia*, *41*(10), 1241–1248.

110. Puppala, M., Ponder, J., Suryanarayana, P., Reddy, G. B., Petrash, M., & Labarbera, D. V., (2012). The isolation and characterization of b-glucogallin as a novel aldose reductase inhibitor from *Emblica officinalis*. *PLoS One*, *7*(4), 1–9.

111. Ramana, K. V., Chandra, D., Srivastava, S., Bhatnagar, A., Aggarwal, B. B., & Srivastava, S. K., (2002). Aldose reductase mediates mitogenic signaling in vascular smooth muscle cells. *J. Biol. Chem.*, *277*(35), 32063–32070.

112. Ramana, K. V., & Srivastava, S. K., (2010). Aldose reductase: A novel therapeutic target for inflammatory pathologies. *Int. J. Biochem. Cell Biol.*, *42*(1), 17–20.

113. Rastelli, G., Antolini, L., Benvenuti, S., & Costantino, L., (2000). Structural bases for the inhibition of aldose reductase by phenolic compounds. *Bioorganic. Med. Chem.*, *8*, 1151–1158.

114. Reddy, G. B., Muthenna, P., Akileshwari, C., Saraswat, M., & Petrash, J. M., (2011). Inhibition of aldose reductase and sorbitol accumulation by dietary rutin. *Curr. Sci.*, *101*(9), 1191–1197.

115. Ristimäki, A., Sivula, A., & Lundin, J., (2002). Prognostic significance of elevated cyclooxygenase–2 expression in breast cancer. *Cancer Res.*, *62*(3), 632–635.

116. Robinson, W. G., Tillis, T. N., & Laver, N., (1990). Diabetes-related histopathologies of the rat retina prevented with an aldose reductase inhibitor. *Exp. Eye Res.*, *50*(4), 355–366.

117. Rondeau, J. M., Tête, F. F., & Podjarny, A., (1992). Novel NADPH-binding domain revealed by the crystal structure of aldose reductase. *Nature*, *355*(6359), 469–472.

118. Roth, F. R. J., & Czech, M. P., (2015). NETs and traps delay wound healing in diabetes. *Trends Endocrinol. Metab.*, *26*(9), 451–452.

119. Sangshetti, J. N., Chouthe, R., Sakle, N. S., Gonjari, I., & Shinde, D. B., (2014). Aldose reductase : A multi-disease target. *Curr. Enzym. Inhib.*, *10*, 2–12.

120. Saraswat, M., Muthenna, P., Suryanarayana, P., Petrash, J. M., & Reddy, G. B., (2008). Dietary sources of aldose reductase inhibitors: Prospects for alleviating diabetic complications. *Asia Pacific Journal of Clinical Nutrition*, *17*, 558–565.

121. Sattur, A. P., (2006). Sclerotiorin, from *Penicillium frequentans*, a potent inhibitor of aldose reductase. *Biotechnol. Lett.*, *28*, 1633–1636.

122. Seshiah, V., Sahay, B. K., Das, A. K., Balaji, V., Shah, S., Banerjee, S., et al., (2009). Diagnosis and management of gestational diabetes mellitus. *Am. Fam. Physician*, *80*(1), 57–62.

123. Seshiah, V., Sahay, B. K., Das, A. K., Shah, S., Banerjee, S., Rao, P. V., et al., (2009). Gestational diabetes mellitus-Indian guidelines. *J. Indian Med. Assoc.*, *107*(11), 799–806.

124. Shanmugam, N., Reddy, M. A., Guha, M., & Natarajan, R., (2003). High glucose-induced expression of proinflammatory cytokine and chemokine genes in monocytic cells. *Diabetes*, *52*, 1256–1264.

125. Sharma, V., & Sharma, P. L., (2013). Role of different molecular pathways in the development of diabetes-induced nephropathy. *J. Diabetes Metab.*, *59*, 1–7.

126. Shaw, J. E., Sicree, R. A., & Zimmet, P. Z., (2010). Global estimates of the prevalence of diabetes for 2010 and 2030. *Diabetes Res. Clin. Pract.*, *87*(1), 4–14.

127. Shimoshige, Y., Ikuma, K., Yamamoto, T. Takakura, S., Kawamura, I., Seki, J., & Mutoh, S., (2000). The effects of zenarestat, an aldose reductase inhibitor, on peripheral neuropathy in Zucker diabetic fatty rats. *Metabolism*, *49*(11), 1395–1399.

128. Singh, V. P., Bali, A., Singh, N., & Jaggi, A. S., (2014). Advanced glycation end products and diabetic complications. *Korean J. Physiol. Pharmacol.*, *18*, 1–14.

129. Solini, A., Dalla, V. M., Saller, A., Nosadini, R., Crepaldi, G., & Fioretto, P., (2002). The angiotensin-converting enzyme DD genotype is associated with glomerulopathy lesions in type 2 diabetes. *Diabetes, 51*(1), 251–255.

130. Soumya, D., & Srilatha, B., (2011). Late-stage complications of diabetes and insulin resistance. *Diabetes Metab. J., 2*(9), 1–7.

131. Sridevi, D., & Jialal, I., (2000). Alpha-tocopherol supplementation decreases serum c-reactive protein and monocyte interleukin–6 levels in normal volunteers and type 2 diabetic patients. *Science, 29*(8), 790–792.

132. Stefano, G. B., Challenger, S., & Kream, R. M., (2016). Hyperglycemia-associated alterations in cellular signaling and dysregulated mitochondrial bioenergetics in human metabolic disorders. *Eur. J. Nutr., 55*(8), 2339–2345.

133. Sugiyama, T., (2011). Management of gestational diabetes mellitus. *J. Japan Med. Assoc., 54139*(545), 2089–2094.

134. Sun, W., Oates, P. J., Coutcher, J. B., Gerhardinger, C., & Lorenzi, M., (2006). Selective aldose reductase inhibitor of a new structural class prevents or reverses early retinal abnormalities in experimental diabetic retinopathy. *Diabetes, 55*(10), 2757–2762.

135. Tammali, R., Ramana, K. V., Singhal, S. S., Awasthi, S., & Srivastava, S. K., (2006). Aldose reductase regulates growth factor-induced cyclooxygenase–2 expression and prostaglandin E2 production in human colon cancer cells. *Cancer Res., 66*(19), 9705–9713.

136. Tan, J. S., Anderson, J. L., Watanakunakorn, C., & Phair, J. P., (1975). Neutrophil dysfunction in diabetes mellitus. *J. Lab. Clin. Med., 85*(1), 26–33.

137. Tang, W. H., Martin, K. A., & Hwa, J., (2012). Aldose reductase, oxidative stress, and diabetic mellitus. *Front Pharmacol., 3*, 1–8.

138. Veeresham, C., Rao, R. A., & Asres, K., (2014). Aldose reductase inhibitors of plant origin. *Phytother. Res., 28*(3), 317–333.

139. Vinayagam, R., & Xu, B., (2015). Antidiabetic properties of dietary flavonoids: A cellular mechanism review. *Nutr. Metab. (Lond), 12*(1), 1–20.

140. Wild, S., Roglic, G., Green, A., Sicree, R., & King, H., (2004). Estimates for the year 2000 and projections for 2030. *World Health, 27*(5), 1047–1053.

141. Williams, M. D., & Nadler, J. L., (2007). Inflammatory mechanisms of diabetic complications. *Curr. Diab. Rep., 7*(3), 242–248.

142. Wilson, D. K., Bohren, K. M., Gabbay, K. H., & Quiocho, F. A., (1992). An unlikely sugar substrate site in the 1.65 A structure of the human aldose reductase holoenzyme implicated in diabetic complications. *Science, 257*(5066), 81–84.

143. Yabe, N. C., (1998). Aldose reductase in glucose toxicity : A potential. *Pharmacol. Rev., 50*(1), 21–33.

144. Yagihashi, S., Mizukami, H., & Sugimoto, K., (2011). Mechanism of diabetic neuropathy: Where are we now and where to go? *J Diabetes Investig., 2*(1), 18–32.

145. Yan, L. J., (2014). Pathogenesis of chronic hyperglycemia: From reductive stress to oxidative stress. *J. Diabetes Res.*, e-article 2014 (June 16):137919; doi: 10.1155/2014/137919.

146. Yong, D., Alpdogan, K., Hatice, H., Philip, C. T., Alan, M., & Dyke, T. E. V., (2007). Activation of RAGE induces elevated O2- generation by mononuclear phagocytes in diabetes. *J. Leukoc. Biol., 81*(2), 520–527.

147. Yong, S. B., Jung, M. P., Dong, H. B., Shigehiro, T., & Ju, H. Y., (1998). Assay of aldose reductase activity: Assay of antimicrobial activities. *J. Antibiot. (Tokyo)*, *51*(10), 902–907.

148. Zhang, L., Zhang, Z., Fu, Y., & Xu, Y., (2015). Research progress of trans-cinnamaldehyde pharmacological effects. *China Journal of Chinese Materia. Medica*, *40*(23), 4568–4572.

149. Zhou, Y. X., Zhang, H., & Peng, C., (2014). Puerarin: A review of pharmacological effects. *Phyther Res.*, *28*(7), 961–975.

150. Zhu, C., (2013). Aldose reductase inhibitors as potential therapeutic drugs of diabetic complications. *Diabetes Mellitus- Insights and Perspective InTech*, 17–46.

151. Ziegler, D., (2008). Treatment of diabetic neuropathy and neuropathic pain: How far have we come? *Diabetes Care*, *31*(2), 1–2.

152. Zozulinska, D., Majchrzak, A., Sobieska, M., Wiktorowicz, K., & Wierusz, W. B., (1999). Serum interleukin–8 level is increased in diabetic patients. *Diabetologia*, *42*(1), 117–118.

153. https://upload.wikimedia.org/wikipedia/commons/thumb/c/c6/Alpha-D-Glucopyranose.svg/180px-Alpha-D-Glucopyranose.svg.png (accessed on January, 5, 2017, 3.20 pm.).

154. https://upload.wikimedia.org/wikipedia/commons/thumb/e/e6/Sorbitol.png (accessed on January 5, 2017, 3.23 pm.).

155. https://upload.wikimedia.org/wikipedia/commons/thumb/6/67/Beta-D-Fructofuranose. svg/300px-Beta-D-Fructofuranose.svg.png (accessed on January, 5, 2017, 3.29 pm.).

156. https://upload.wikimedia.org/wikipedia/commons/thumb/3/39/Fidarestat_structure. svg/333px-Fidarestat_structure.svg.png (accessed on January, 6, 2017, 9.20 am.).

157. https://en.wikipedia.org/wiki/Epalrestat#/media/File:Epalrestat.svg (accessed on January, 6, 2017, 9.23 am.).

158. https://upload.wikimedia.org/wikipedia/commons/thumb/d/d8/Sorbinil.svg/225px-Sorbinil.svg.png (accessed on January, 6, 2017, 9.30 am.).

159. https://upload.wikimedia.org/wikipedia/commons/thumb/b/b7/Zopolrestat. svg/768px-Zopolrestat.svg.png (accessed on January, 6, 2017, 9.32 am.).

160. https://upload.wikimedia.org/wikipedia/commons/thumb/4/4c/Zenarestat_structure. svg/333px-Zenarestat_structure.svg.png (accessed on January, 6, 2017, 9.35 am.).

161. https://en.wikipedia.org/wiki/Ranirestat#/media/File:Ranirestat.svg (accessed on January, 6, 2017, 9.50 am.).

162. https://en.wikipedia.org/wiki/File:Tolrestat_structure.svg.png (accessed on January, 6, 2017, 9.56 am.).

163. http://www.chemspider.com/Chemical-Structure.165724.html (accessed on January, 6, 2017, 10.04 am.).

164. https://upload.wikimedia.org/wikipedia/commons/thumb/e/e6/Apigenin.svg/300px-Apigenin.svg.png (accessed on January, 6, 2017, 10.07 am.).

165. https://upload.wikimedia.org/wikipedia/commons/thumb/3/3a/Rosmarinic_acid. png/300px-Rosmarinic_acid.png (accessed on January, 6, 2017, 10.11 am.).

166. https://upload.wikimedia.org/wikipedia/commons/thumb/f/fe/Zimtaldehyd_-_cinnamaldehyde.svg/300px-Zimtaldehyd_-_cinnamaldehyde.svg.png (accessed on January, 6, 2017, 1.12 pm.).

167. https://upload.wikimedia.org/wikipedia/commons/thumb/a/ab/Baicalin.svg/330px-Baicalin.svg.png (accessed on January, 6, 2017, 1.20 pm.).

168. https://upload.wikimedia.org/wikipedia/commons/thumb/0/0f/Berberin.svg/300px-Berberin.svg.png (accessed on January, 6, 2017, 1.24 pm.).

169. https://upload.wikimedia.org/wikipedia/commons/thumb/2/26/Naringenin.svg/300px-Naringenin.svg.png (accessed on January, 6, 2017, 1.32 pm.).

170. https://upload.wikimedia.org/wikipedia/commons/thumb/0/09/Rutin_structure.svg/375px-Rutin_structure.svg.png (accessed on January, 6, 2017, 2.10 pm.).

171. https://upload.wikimedia.org/wikipedia/commons/thumb/9/90/Puerarin.svg/300px-Puerarin.svg.png (accessed on January, 6, 2017, 2.14 pm.).

172. https://upload.wikimedia.org/wikipedia/commons/thumb/1/17/Resveratrol.svg/375px-Resveratrol.svg.png (accessed on January, 6, 2017, 2.19 pm.).

173. https://upload.wikimedia.org/wikipedia/commons/thumb/d/da/Ellagic_acid.svg/330px-Ellagic_acid.svg.png (accessed on January, 6, 2017, 2.24 pm.).

174. https://upload.wikimedia.org/wikipedia/commons/thumb/7/7c/Ferulic_acid_acsv.svg/300px-Ferulic_acid_acsv.svg.png (accessed on January, 6, 2017, 2.31 pm.).

175. https://upload.wikimedia.org/wikipedia/commons/thumb/8/81/Quercetin.svg/375px-Quercetin.svg.png (accessed on January, 6, 2017, 2.37 pm.).

176. https://upload.wikimedia.org/wikipedia/commons/thumb/4/40/Curcumin.svg/360px-Curcumin.svg.png (accessed on January, 6, 2017, 2.43 pm.).

177. https://upload.wikimedia.org/wikipedia/commons/thumb/9/94/Eugenol2DCSD.svg/360px-Eugenol2DCSD.svg.png (accessed on January, 6, 2017, 2.52 pm.).

178. https://upload.wikimedia.org/wikipedia/commons/thumb/2/2f/Epigallocatechin_gallate_structure.svg/375px-Epigallocatechin_gallate_structure.svg.png (accessed on January, 6, 2017, 3.10 pm.).

179. https://upload.wikimedia.org/wikipedia/commons/thumb/d/da/Kaempferol.svg/375px-Kaempferol.svg.png (accessed on January, 6, 2017, 3.18 pm.).

180. https://upload.wikimedia.org/wikipedia/commons/thumb/1/1b/Chlorogenic-acid-from-CAS–2D-skeletal.png/300px-Chlorogenic-acid-from-CAS–2D-skeletal.png (accessed on January, 6, 2017, 3.27 pm.).

181. https://en.wikipedia.org/wiki/AKR1B1#/media/File:Aldose_reductase_1us0.png (accessed on January, 6, 2017, 3.33 pm.).

INDEX